Eukaryotic Transcription Factors

David S. Latchman

Director, Medical Molecular Biology Unit and Reader in Molecular Biology, University College and Middlesex School of Medicine, London

ACADEMIC PRESS
Harcourt Brace Jovanovich, Publishers
London San Diego New York
Boston Sydney Tokyo Toronto

This book is printed on acid-free paper

ACADEMIC PRESS LIMITED
24–28 Oval Road
London NW1 7DX

United States Edition published by
ACADEMIC PRESS INC.
San Diego, CA 92101

A catalogue record of this book is available from the British Library

ISBN 0–12–437170–1
ISBN 0–12–437171–X (pbk)

Typeset by J&L Composition Ltd
Printed in Great Britain by
the University Press, Cambridge

Contents

Preface ix
Acknowledgements xi

CHAPTER ONE
DNA Sequences and Transcription Factors 1
1.1 THE IMPORTANCE OF TRANSCRIPTION 1
1.2 DNA SEQUENCE ELEMENTS 2
1.3 INTERACTION BETWEEN FACTORS BOUND AT VARIOUS LEVELS 10
1.4 CONCLUSIONS 12
REFERENCES 12

CHAPTER TWO
Methods For Studying Transcription Factors 16
2.1 INTRODUCTION 16
2.2 METHODS FOR STUDYING DNA-PROTEIN INTERACTIONS 16
2.3 METHODS FOR STUDYING THE TRANSCRIPTION FACTOR ITSELF 27
2.4 CONCLUSIONS 38
REFERENCES 38

CHAPTER THREE
Transcription Factors and Constitutive Transcription 41
3.1 RNA POLYMERASES 41
3.2 THE STABLE TRANSCRIPTIONAL COMPLEX 43

3.3 UPSTREAM PROMOTER ELEMENTS 51
3.4 CONCLUSIONS 57
REFERENCES 57

CHAPTER FOUR
Transcription Factors and Inducible Gene Expression 62
4.1 INDUCIBLE GENE EXPRESSION 62
4.2 HEAT-INDUCIBLE TRANSCRIPTION 63
4.3 STEROID-INDUCIBLE TRANSCRIPTION 69
4.4 CONCLUSIONS 82
REFERENCES 83

CHAPTER FIVE
Transcription Factors and Cell Type-Specific Transcription 87
5.1 CELL TYPE-SPECIFIC GENE EXPRESSION 87
5.2 TRANSCRIPTION FACTORS AND IMMUNOGLOBULIN GENE EXPRESSION 89
5.3 MyoD AND THE CONTROL OF MUSCLE-SPECIFIC GENE EXPRESSION 94
5.4 REGULATION OF YEAST MATING TYPE 105
5.5 CONCLUSIONS 118
REFERENCES 118

CHAPTER SIX
Transcription Factors and Developmentally Regulated Gene Expression 124
6.1 DEVELOPMENTALLY REGULATED GENE EXPRESSION 124
6.2 THE HOMEOBOX-CONTAINING GENES OF *DROSOPHILA* 126
6.3 HOMEOBOX-LIKE GENES IN OTHER ORGANISMS 140
6.4 CONCLUSIONS 149
REFERENCES 149

CHAPTER SEVEN
Transcription Factors and Cancer 153
7.1 CELLULAR ONCOGENES AND CANCER 153
7.2 FOS, JUN AND AP1 156
7.3 V-*ERBA* AND THE THYROID HORMONE RECEPTOR 162
7.4 THE *MYB* ONCOGENE 165
7.5 CONCLUSIONS 172
REFERENCES 173

CHAPTER EIGHT
DNA Binding By Transcription Factors 177
8.1 INTRODUCTION 177

8.2 THE HELIX–TURN–HELIX MOTIF IN THE HOMEOBOX 178
8.3 THE ZINC FINGER MOTIF 182
8.4 THE LEUCINE ZIPPER AND THE BASIC DNA BINDING DOMAIN 194
8.5 OTHER DNA-BINDING MOTIFS 201
8.6 CONCLUSIONS 201
REFERENCES 205

CHAPTER NINE
Activation and Repression of Gene Expression by Transcription Factors 209
9.1 INTRODUCTION 209
9.2 ACTIVATION OF TRANSCRIPTION 209
9.3 REPRESSION OF TRANSCRIPTION 227
9.4 CONCLUSIONS 233
REFERENCES 234

CHAPTER TEN
What Activates The Activators? 238
10.1 INTRODUCTION 238
10.2 REGULATION OF SYNTHESIS 239
10.3 REGULATION OF ACTIVITY 248
10.4 CONCLUSIONS 260
REFERENCES 262

CHAPTER ELEVEN
Conclusions and Future Prospects 265
INDEX 267

To Maurice and Vivienne
In admiration

Preface

In my previous book, *Gene Regulation: A Eukaryotic Perspective* (Unwin-Hyman Ltd, 1990), I described the mechanisms by which the expression of eukaryotic genes is regulated during processes as diverse as steroid treatment and embryonic development. Although some of this regulation occurs at the post-transcriptional level, it is clear that the process of gene transcription itself is the major point at which gene expression is regulated. In turn this has focused attention on the protein factors, known as transcription factors, which control both the basal processes of transcription and its regulation in response to specific stimuli or developmental processes. The characterization of many of these factors and in particular the cloning of the genes encoding them has resulted in the availability of a bewildering array of information on these factors, their mechanism of action and their relationship to each other. Despite its evident interest and importance, however, this information could be discussed only relatively briefly in *Gene Regulation*, whose primary purpose was to provide an overview of the process of gene regulation and the various mechanisms by which this is achieved.

It is the purpose of this book, therefore, to discuss in detail the available information on transcription factors, emphasizing common themes and mechanisms to which new information can be related as it becomes available. As such it is hoped the work will appeal to final-year undergraduates and postgraduate students entering the field as well as to those moving into the area from other scientific or clinical fields who wish to know how transcription factors may regulate the gene in which they are interested.

In order to provide a basis for the discussion of transcription factors, the first two chapters focus respectively on the DNA sequences with which the factors interact and on the experimental methods which are used to study these factors and obtain the information about them provided in subsequent chapters. The remainder of the work is divided into two distinct portions. Thus Chapters 3 to 7 focus on the role of transcription factors in particular processes. These include constitutive and inducible gene expression, cell type-specific and developmentally regulated gene expression and the role of transcription factors in cancer. Chapters 8 to 10 adopt a more mechanistic approach and consider the features of transcription factors which allow them to fulfil their function. These include the ability to bind to DNA and modulate transcription either positively or negatively as well as the ability to respond to specific stimuli and thereby activate gene expression in a regulated manner.

Although this dual approach to transcription factors from both a process-oriented and mechanistic point of view may lead to some duplication, it is the most efficient means of providing the necessary overview both of the nature of transcription factors and the manner in which they achieve their role of modulating gene expression in many diverse situations.

Finally I would like to thank Mrs Rose Lang for typing the text and coping with the continual additions necessary in this fast-moving field and Mrs Jane Templeman for her outstanding skill in preparing the illustrations.

David S. Latchman

Acknowledgements

I would like to thank all the colleagues, listed below, who have given permission for material from their papers to be reproduced in this book and have provided prints suitable for reproduction.

Figure 4.3, photograph kindly provided by Dr C. Wu from Zimarino and Wu, *Nature* **327**, 727 (1987), by permission of Macmillan Magazines Ltd; Figure 4.9, photograph kindly provided by Professor M. Beato from Willmann and Beato, *Nature* **324**, 688 (1986) by permission of Macmillan Magazines Ltd. Figures 5.9 and 5.11, photographs kindly provided by Dr R. L. Davis from Davis *et al.*, *Cell* **51**, 987 (1987) by permission of Cell Press. Figures 6.1 and 10.2, photographs kindly provided by Professor W. J. Gehring from Gehring, *Science* **236**, 1245 (1987) by permission of the American Association for the Advancement of Science. Figure 6.15, photograph kindly provided by Dr P. Holland from Holland and Hogan, *Nature* **321**, 251 (1986) by permission of Macmillan Magazines Ltd. Figures 6.16 and 6.17, photographs kindly provided by Dr R. Krumlauf from Graham *et al.*, *Cell* **57**, 367 (1989) by permission of Cell Press. Figure 8.6 redrawn from Redemann *et al.*, *Nature* **332**, 90 (1988) by kind permission of Dr H. Jackle and Macmillan Magazines Ltd. Figures 8.11 and 8.15 redrawn from Schwabe *et al.*, *Nature* **348**, 458 (1990) by kind permission of Dr D. Rhodes and Macmillan Magazines Ltd; Figure 8.18 redrawn from Abel and Maniatis, *Nature* **341**, 24 (1989) by kind permission of Professor T. Maniatis and Macmillan Magazines Ltd. Figure 9.9 redrawn from Lin *et al.*, *Nature* **345**, 359 (1990) and Carey *et al.*, *Nature* **345**, 361 (1990) by kind permission of Professor M. Ptashne and Macmillan Magazines Ltd.

CHAPTER ONE

DNA sequences and transcription factors

1.1. THE IMPORTANCE OF TRANSCRIPTION

The fundamental dogma of molecular biology is that DNA produces RNA which in turn produces protein. Hence if the genetic information which each individual inherits as DNA (the genotype) is to be converted into the proteins which produce the corresponding characteristics of the individual (the phenotype), it must first be converted into an RNA product. The process of transcription whereby an RNA product is produced from the DNA is therefore an essential element in gene expression. The failure of this process to occur will obviously render redundant all the other steps which follow the production of the initial RNA transcript in eukaryotes, such as RNA splicing, transport to the cytoplasm or translation into protein (for a review of these stages see Nevins (1983)).

The central role of transcription in the process of gene expression also renders it an attractive control point for regulating the expression of genes in particular cell types or in response to a particular signal. Indeed, it is now clear that in the vast majority of cases where a particular protein is produced only in a particular tissue or in response to a particular signal this is achieved by control processes which ensure that its corresponding gene is transcribed only in that tissue or in response to such a signal (reviews: see Darnell, 1982; Latchman, 1990). For example, the genes encoding the immunoglobulin heavy and light chains of the antibody molecule are transcribed at high level only in the antibody-producing B cells (Gillis *et al.*, 1983), whilst the increase in somatostatin production in response to treatment of cells with cyclic AMP is mediated by increased transcription of the corresponding gene (Montminy *et al.*, 1986). Therefore, while post-transcriptional regulation affecting, for example, RNA splicing or stability plays some role in the regulation of gene expression (reviews: Brawerman, 1987; Breitbart *et al.*, 1987), the major control point lies at the level of transcription.

1.2 DNA SEQUENCE ELEMENTS

1.2.1 The gene promoter

The central role of transcription both in the basic process of gene expression in and its regulation in particular tissues has led to considerable study of this process. Initially such studies focused on the nature of the DNA sequences within individual genes which were essential for either basal or regulated gene expression. In prokaryotes, such sequences are found immediately upstream of the start site of transcription and form part of the promoter directing expression of the genes. Sequences found at this position include both elements found in all genes which are involved in the basic process of transcription itself and those found in a more limited number of genes which mediate their response to a particular signal (Schmitz and Galas, 1979; Miller and Reznikoff, 1980; Ptashne, 1986).

Early studies of cloned eukaryotic genes therefore concentrated on the region immediately upstream of the transcribed region where, by analogy, sequences involved in transcription and its regulation should be located. Putative regulatory sequences were identified by comparison between different genes, and the conclusions reached in this way confirmed either by destroying these sequences by deletion or mutation or by transferring them to another gene in an attempt to alter its pattern of regulation.

This work, carried out on a number of different genes encoding specific proteins, identified many short-sequence elements involved in transcriptional control (reviews: Davidson *et al.*, 1983; Jones *et al.*, 1988). The elements of this type present in two typical examples, the human gene encoding the 70-kDa heat-inducible (heat-shock) protein (Williams *et al.*, 1989) and the human metallothionein IIA gene (Lee *et al.*, 1987a), are illustrated in Figure 1.1.

Comparisons of these and many other genes revealed that, as in bacteria, their upstream regions contain two types of elements: firstly, sequences found in very many genes exhibiting distinct patterns of regulation which are likely to be involved in the basic

a)

SP1 CCAAT AP2 HSE CCAAT SP1 TATA AP2

-158 -105 -74 -28 0

b)

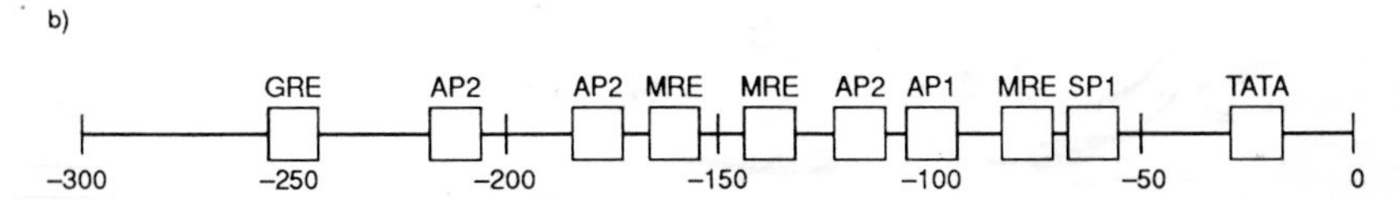

Figure 1.1 Transcriptional control elements upstream of the transcriptional start site in the human genes encoding hsp70 (a) and methallothionein IIA (b). The TATA, Sp1 and CCAAT boxes bind factors which are involved in constitutive transcription (see also Chapter 3) whilst the glucocorticoid response element (GRE), metal response element (MRE), heat shock element (HSE) and the AP1 and AP2 sites bind factors involved in the induction of gene expression in response to specific stimuli (see also Chapter 4 and Section 7.2).

process of transcription itself; and secondly, those found only in genes transcribed in a particular tissue or in response to a specific signal which are likely to produce this specific pattern of expression. These will be discussed in turn.

1.2.2 Sequences involved in the basic process of transcription.

Although they are regulated very differently, the *hsp70* and metallothionein genes both contain a TATA box. This is an AT-rich sequence (consensus TATAA/TAA/T) which is found about 30 base pairs upstream of the transcriptional start site in very many but not all genes. Mutagenesis or relocation of this sequence has shown that it plays an essential role in accurately positioning the start site of transcription (Breathnach and Chambon, 1981). The region of the gene bracketed by the TATA box and the site of transcriptional initiation (the Cap site) has been operationally defined as the gene promoter (Goodwin *et al.*, 1990). It is likely that this region binds several proteins essential for transcription, as well as RNA polymerase II itself, which is the enzyme responsible for transcribing protein-coding genes (Lewis and Burgess, 1982).

Although the TATA box is found in most eukaryotic genes, it is absent in some genes, notably housekeeping genes expressed in all tissues and in some tissue-specific genes (reviews: Sehgal *et al.*,

1988; Smale and Baltimore, 1989). In these promoters, the actual sequence over the start site of transcription itself appears to play a critical role in determining the initiation point and acts as a minimal promoter capable of producing basal levels of transcription (Smale and Baltimore, 1989).

In promoters which contain a TATA box and in those which lack it, the very low activity of the promoter itself is dramatically increased by other elements located upstream of the promoter. These elements are found in a very wide variety of genes with different patterns of expression, indicating that they play a role in stimulating the constitutive activity of promoters. Thus inspection of the *hsp70* and metallothionein IIA genes reveals that both contain one or more copies of a GC-rich sequence known as the Sp1 box which is found upstream of the promoter in many genes both with and without TATA boxes (review: Dynan and Tjian, 1985).

In addition the *hsp70* promoter but not the metallothionein promoter contains another sequence, the CCAAT box, which is also found in very many genes with disparate patterns of regulation. Both the CCAAT box and the Sp1 box are typically found upstream of the TATA box, as in the metallothionein and *hsp70* genes. Some genes, as in the case of hsp70, may have both of these elements, whereas others, such as the metallothionein gene, have single or multiple copies of one or the other (review: McKnight and Tjian, 1986). In every case, however, these elements are essential for transcription of the genes, and their elimination by deletion or mutation abolishes transcription (McKnight and Kingsbury, 1982; McKnight *et al.*, 1984). Hence these sequences play an essential role in efficient transcription of the gene and have been termed upstream promoter elements (UPE) (Goodwin *et al.*, 1990). The role of the promoter and UPE sequences and the protein factors which bind to them are discussed further in Chapter 3.

1.2.3. Sequences involved in regulated transcription

Inspection of the *hsp70* promoter (Figure 1.1) reveals several other sequence elements which are only shared with a much more limited number of other genes and which are interdigitated with the upstream promoter elements discussed above. Indeed, one of these, which is located approximately 90 bases upstream of the transcriptional start site, is shared only with other heat-shock genes whose transcription is increased in response to elevated temperature. This suggests that this heat-shock consensus element may be essential for the regulated transcription of the *hsp70* gene in response to heat.

To directly prove this, however, it is necessary to transfer this sequence to a non-heat-inducible gene and show that this transfer renders the recipient gene heat-inducible. Pelham (1982) successfully achieved this by linking the heat-shock consensus element to the non-heat-inducible thymidine kinase gene of the eukaryotic virus herpes simplex. This hybrid gene could be activated following its introduction into mammalian cells by raising the temperature (Figure 1.2). Hence the heat-shock consensus element can confer heat inducibility on another gene, directly proving that its presence in the *hsp* gene promoters is responsible for their heat inducibility.

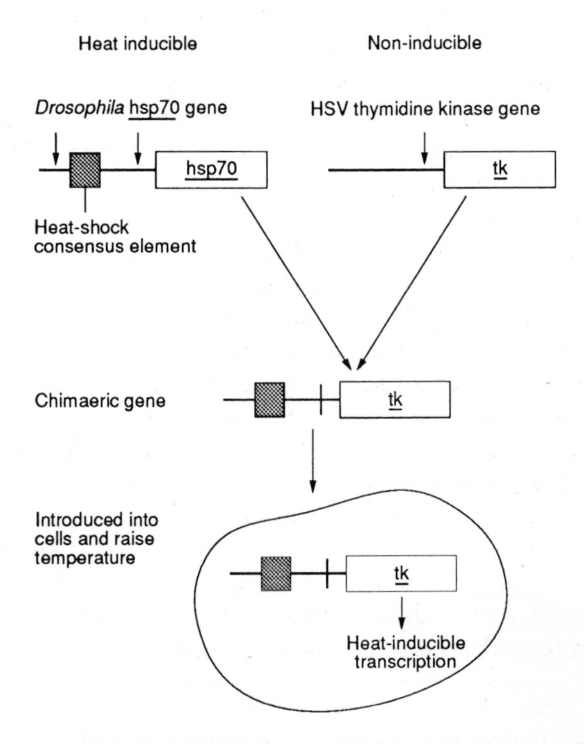

Figure 1.2 Demonstration that the heat-shock consensus element mediates heat inducibility. Transfer of this sequence to a gene (thymidine kinase) which is not normally inducible renders this gene heat-inducible.

Moreover, although these experiments used a heat-shock consensus element taken from the *hsp70* gene of the fruit fly *Drosophila melanogaster*, the hybrid gene was introduced into mammalian cells. Not only does the successful functioning of the fly element in mammalian cells indicate that this process is evolutionarily conserved but it permits a further conclusion about the way in which the effect operates. Thus in the cold-blooded *Drosophila*, 37°C represents a thermally stressful temperature and the heat-shock response would normally be active at this temperature. The hybrid gene was inactive at 37°C in the mammalian cells, however, and was only induced at 42°C, the heat-shock temperature characteristic of the cell into which it was introduced. Hence this sequence does not act as a thermostat, set to go off at a particular temperature, since this would occur at the *Drosophila* heat-shock temperature (Figure 1.3a). Rather, this sequence must act by being recognized by a cellular protein which is activated only at an elevated temperature characteristic of the mammalian cell heat-shock response (Figure 1.3b).

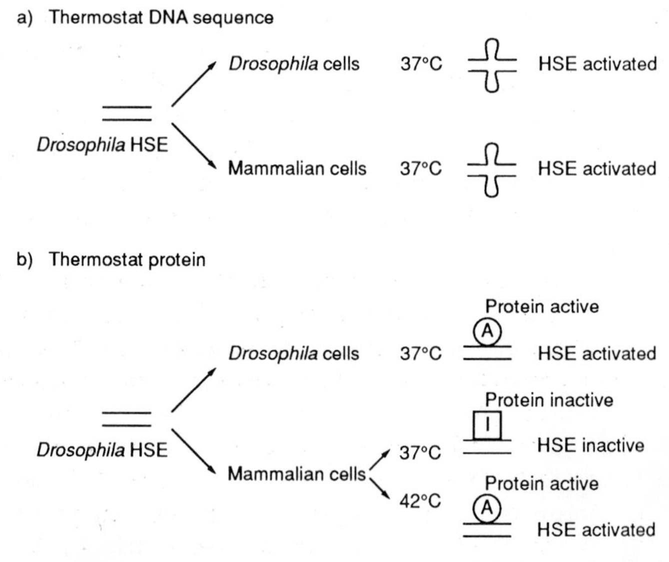

Figure 1.3 Predicted effects of placing the *Drosophila* heat-shock element in a mammalian cell if the element acts as a thermostat detecting elevated temperature directly (a) or if it acts by binding a protein which is activated by elevated temperature (b). Note that only possibility (b) can account for the observation that the *Drosophila* heat-shock element only activates transcription in mammalian cells at the mammalian heat-shock temperature of 42°C and not at the *Drosophila* heat-shock temperature of 37°C.

This experiment therefore not only directly proves the importance of the heat-shock response element in producing the heat inducibility of the *hsp70* gene but also shows that this sequence acts by binding a cellular protein which is activated in response to elevated temperature. The binding of this transcription factor then activates transcription of the *hsp70* gene. The manner in which this factor activates transcription of the *hsp70* gene and the other heat-shock genes is discussed further in Section 4.2.

The presence of specific DNA sequences which can bind particular proteins will therefore confer on a specific gene the ability to respond to particular stimuli. Thus the lack of a heat-shock consensus element in the metallothionein IIA gene (Figure 1.1) means that this gene is not heat-inducible. In contrast, however, this gene, unlike the *hsp70* gene, contains a glucocorticoid response element (GRE). Hence it can bind the complex of the glucocorticoid receptor and the hormone itself which forms following treatment of cells with glucocorticoid (review: Beato, 1989). Its transcription is therefore activated in response to glucocorticoid whereas that of the *hsp70* gene is not (see Section 4.3). Similarly, only the metallothionein gene contains metal response elements (MRE), allowing it to be activated in response to treatment with heavy metals such as zinc and cadmium (Karin *et al.*, 1984). In contrast, both genes contain binding sites for the transcription factor AP2 which mediates gene activation in response to cyclic AMP and phorbol esters (Imagawa *et al.*, 1987). The manner in which the binding of specific transcription factors to different regulatory sequences modulates gene expression in response to specific inducing factors is discussed further in Chapter 4.

Similar DNA sequence elements in the promoters of tissue-specific genes play a critical role in producing their tissue-specific pattern of expression by binding transcription factors which are present in an active form only in a particular tissue where the gene will be activated. For example, the promoters of the immunoglobulin heavy- and light-chain genes contain a sequence known as the octamer motif, ATGCAAAT (Parslow *et al.*, 1984; Mason *et al.*, 1985) which can confer B cell-specific expression on an unrelated promoter (Wirth *et al.*, 1987). This sequence acts by binding a transcription factor known as Oct-2 which is absent in most cell types but is expressed at high levels in immunoglobulin-producing B cells (Staudt *et al.*, 1986; Landolfi *et al.*, 1986) (see Section 5.2.2 for further details). Similarly, the related sequence ATGAATAA/T is found in two genes expressed specifically in the anterior pituitary gland, the prolactin gene and the growth hormone gene (Nelson *et al.*, 1988), and binds a transcription factor known as Pit-1, which is expressed only in the anterior pituitary (Ingraham *et al.*, 1988). If this short sequence is inserted upstream of a promoter, the gene is expressed only in pituitary cells. In contrast, the octamer motif, which differs by only two bases, will direct expression only in B cells when inserted upstream of the same promoter (Elsholtz *et al.*, 1990; Figure 1.4). Hence small differences in control-element sequences can produce radically different patterns of gene expression.

The role of transcription factors in producing tissue specific gene expression is discussed further in Chapter 5.

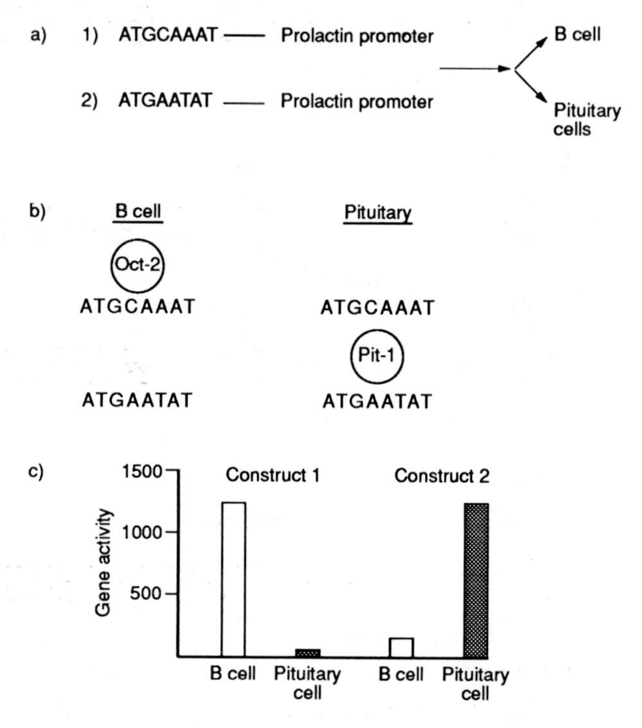

Figure 1.4 Linkage of the octamer-binding motif ATGCAAAT (1) and the related Pit-1-binding motif ATGAATAT (2) to the prolactin promoter and introduction into B cells and pituitary cells (a). In B cells, the octamer motif binds the B-cell-specific octamer binding protein Oct-2, whereas the Pit-1 motif fails to bind any protein. In contrast, in pituitary cells the Pit-1 motif binds the Pit-1 protein whereas the octamer motif fails to bind any protein (b). This results in only the octamer containing construct 1 directing a high level of activity in B cells, whereas only construct 2 containing the Pit-1-binding site directs a high level of gene activity in pituitary cells (c). Data from Elsholtz *et al.*, (1990).

1.2.4 Enhancers

One of the characteristic features of eukaryotic gene expression is the existence of sequence elements located at great distances from the start site of transcription which can influence the level of gene expression. These elements can be located upstream, downstream or within a transcription unit and function in either orientation relative to the start site of transcription (Figure 1.5). They act by

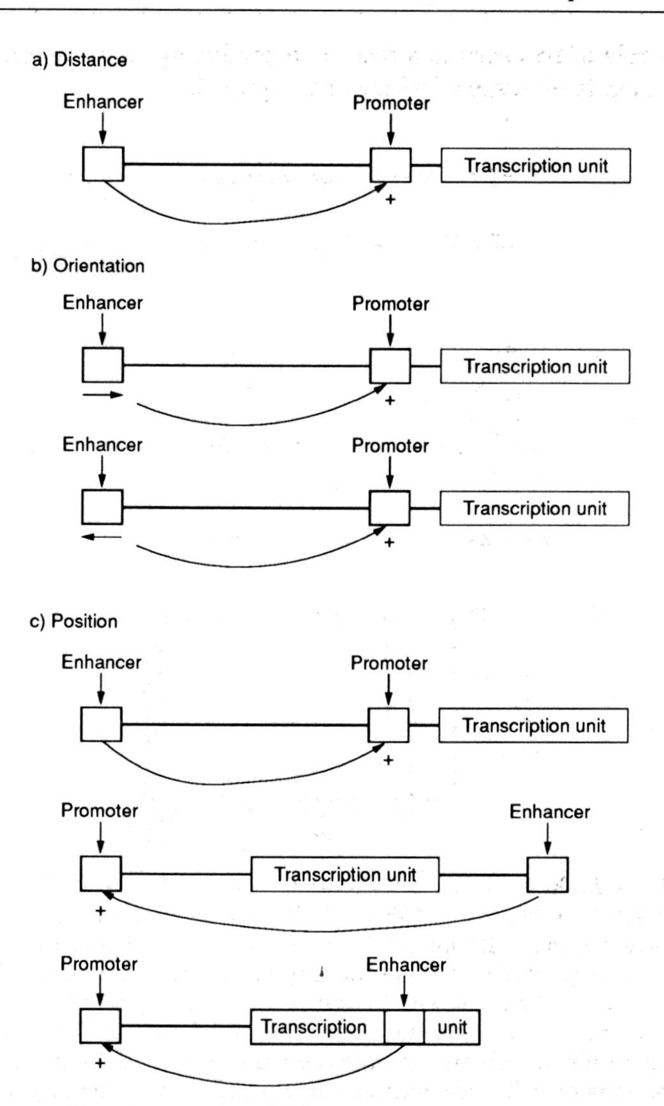

Figure 1.5 Characteristics of an enhancer element which can activate a promoter at a distance (a); in either orientation relative to the promoter (b); and when positioned upstream, downstream, or within a transcription unit (c).

increasing the activity of a promoter, although they lack promoter activity themselves and are hence referred to as enhancers (reviews: Serfling *et al.*, 1985; Hatzopoulos *et al.*, 1988). Some enhancers are active in all tissues and increase the activity of a promoter in all cell types whilst others function as tissue-specific enhancers which activate a particular promoter only in a specific cell type. Thus the enhancer located in the intervening region of the immunoglobulin genes is active only in B cells (Gillis *et al.*, 1983) and the B cell-specific expression of the immunoglobulin genes is produced by the interaction of this enhancer and the immunoglobulin promoter which, as we have previously seen, is also B cell-specific (Garcia *et al.*, 1986) (see Section 5.2 for further discussion).

As with promoter elements, enhancers contain multiple binding sites for transcription factors which interact together (review: Dynan, 1989). In many cases these elements are identical to those contained immediately upstream of gene promoters. Thus the immunoglobulin-heavy chain enhancer contains a copy of the octamer sequence (Sen and Baltimore, 1986) which is also found in the immunoglobulin promoters (Section 1.2.3). Similarly, multiple copies of the heat-shock consensus element are located far upstream of the start site in the *Xenopus hsp70* gene and function as a heat-inducible enhancer when transferred to another gene (Bienz and Pelham, 1986).

Enhancers therefore consist of sequence elements which are also present in similarly regulated promoters and may be found within the enhancer associated with other control elements or in multiple copies.

1.3 INTERACTION BETWEEN FACTORS BOUND AT VARIOUS SITES

The typical eukaryotic gene will therefore consist of up to four distinct transcriptional control elements (Figure 1.6). These are:

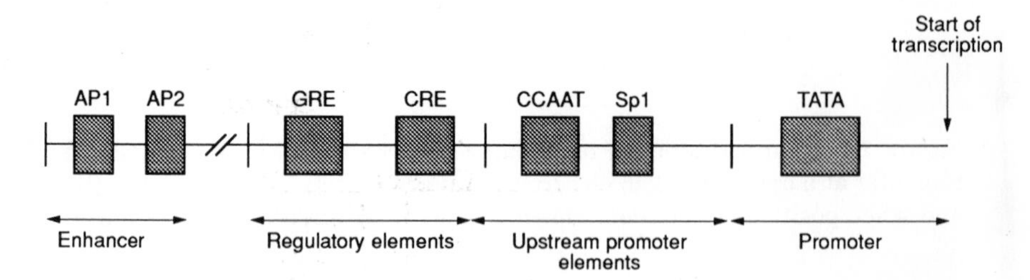

Figure 1.6 Structure of a typical gene with a TATA box-containing promoter, upstream promoter elements such as the CCAAT and Sp1 boxes, regulatory elements inducing expression in response to treatment with substances such as glucocorticoid (GRE) and cyclic AMP (CRE) and other elements within more distant enhancers. Note that, as discussed in the text and illustrated in Figure 1.1, the upstream promoter elements are often interdigitated with the regulatory elements whilst the same regulatory elements can be found upstream of the promoter and in enhancers.

firstly, the promoter itself; secondly, upstream promoter elements (UPE) located close to it which are required for efficient transcription in any cell type; thirdly, other elements adjacent to the promoter which are interdigitated with the UPEs and which activate the gene in particular tissues or in response to particular stimuli; and lastly, more distant enhancer elements which increase gene activity either in all tissues or in a regulated manner.

The binding of specific transcription factors to these sequences is necessary for transcription to occur. In some cases binding of the RNA polymerase and associated factors to the promoter and of other factors to the UPEs will be sufficient for transcription to occur and the gene will be expressed constitutively. In other cases, however, such interactions will be insufficient and transcription of the gene will occur only in response to the binding, to another DNA sequence, of a factor which is activated in response to a particular stimulus or is present only in a particular tissue. These regulatory factors will then interact with the constitutive factors, allowing transcription to occur. Hence their binding will result in the observed tissue-specific or inducible pattern of gene expression.

Such interaction is well illustrated by the metallothionein IIA gene. As illustrated in Figure 1.1, this gene contains a binding site for the transcription factor AP1 which produces induction of gene expression in response to phorbol ester treatment (Angel *et al.*, 1987; Lee *et al.*, 1987b). The action of AP1 on the expression of the metallothionein gene is abolished, however, both by mutations in its binding site and by mutations in the adjacent Sp1 motif which prevent this motif from binding its corresponding transcription factor Sp1 (Lee *et al.*, 1987a). Although these mutations in the Sp1 motif do not abolish AP1 binding, they do prevent its action,

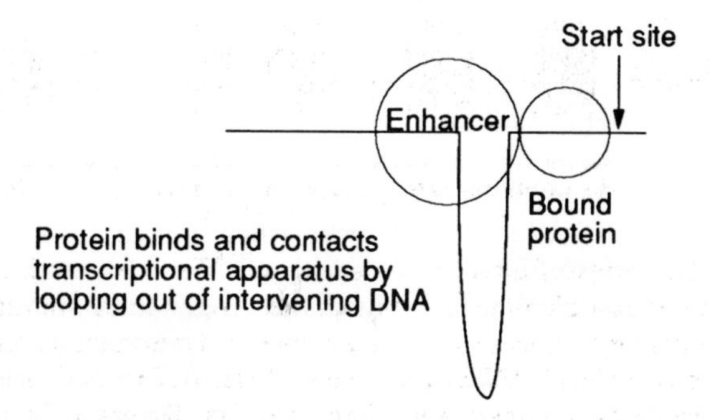

Figure 1.7 Contact between proteins bound at the promoter and those bound at a distant enhancer can be achieved by looping out of the intervening DNA.

indicating that the inducible AP1 factor interacts with the constitutive Sp1 factor to activate transcription.

Clearly such interactions between bound transcription factors need not be confined to factors bound to regions adjacent to the promoter but can also involve the similar factors bound to more distant enhancers. It is likely that this is achieved by a looping out of the intervening DNA, allowing contact between factors bound at the promoter and those bound at the enhancer (Figure 1.7); for further discussion see Latchman (1990).

1.4 CONCLUSIONS

It is clear that both the process of transcription itself and its regulation in particular tissues or in response to particular signals are controlled by short DNA sequence elements located adjacent to the promoter or in enhancers. In turn such sequences act by binding proteins which are either active constitutively or are present in an active form only in a specific tissue or following a specific inducing signal. Such DNA-bound transcription factors then interact with each other and the RNA polymerase itself in order to produce constitutive or regulated transcription. The nature of these factors, the manner in which they function and their role in different biological processes forms the subject of this book.

REFERENCES

Angel, P., Imagawa, M., Chiu, R., Stein, B., Imbra, R. J., Rahmsdorf, H. J., Jonat, C., Herrlich, P. and Karin, M. (1987). Phorbol ester-inducible genes contain a common cis element recognized by a TPA-modulated transacting factor. *Cell* **49**, 729–739.

Beato, M. (1989). Gene regulation by steroid hormones. *Cell* **56**, 335–344.

Bienz, M. and Pelham, H. R. B. (1986). Heat shock regulatory elements function as an inducible enhancer when linked to a heterologous promoter. *Cell* **45**, 753–760.

Brawerman, G. (1987). Determinants of messenger RNA stability. *Cell* **48**, 5–6.

Breathnach, R. and Chambon, P. (1981). Organization and expression of eukaryotic split genes coding for proteins. *Annual Review of Biochemistry* **50**, 349–383.

Breitbart, R. E., Andreadis, A. and Nadal-Ginard, B. (1987). Alternative splicing: a ubiquitous mechanism for the generation of multiple protein isoforms from different genes. *Annual Review of Biochemistry* **56**, 467–495.

Darnell, J. E. (1982). Variety in the level of gene control in eukaryotic cells. *Nature* **297**, 365–371.

Davidson, E. H., Jacobs, H. T. and Britten, R. J. (1983). Very short repeats and coordinate induction of genes. *Nature* **301**, 468–470.

Dynan, W. S. (1989). Modularity in promoters and enhancers. *Cell* **58**, 1–4.

Dynan, W. S. and Tjian, R. (1985). Control of eukaryotic messenger RNA synthesis by sequence-specific DNA-binding proteins. *Nature* **316**, 774–778.

Elsholtz, H. P., Albert, V. R., Treacy, M. N. and Rosenfeld, M. G. (1990). A two-base change in a POU factor binding site switches pituitary-specific to lymphoid-specific gene expression. *Genes and Development* **4**, 43–51.

Garcia, J. V., Bich-Thuy, L., Stafford, J. and Queen, C. (1986). Synergism between immunoglobulin enhancers and promoters. *Nature* **322**, 383–385.

Gillis, S. D., Morrison, S. L., Oi, U. T. and Tonegawa, S. (1983). A tissue-specific transcription enhancer element is located in the major intron of a rearranged immunoglobulin heavy chain gene. *Cell* **33**, 717–728.

Goodwin, G. H., Partington, G. A. and Perkins, N. D. (1990). Sequence specific DNA binding proteins involved in gene transcription. In: *Chromosomes: Eukaryotic, Prokaryotic and Viral* (Adolph, K. W., ed.), vol. 1, pp. 31–85. Boca Raton, Florida: CRC Press.

Hatzopoulous, A. K., Schlokat, U. and Gruss, P. (1988). Enhancers and other cis-acting sequences. In: *Transcription and Splicing* (Hames, B. D. and Glover, D. M., eds), pp. 43–96. Oxford: IRL Press.

Imagawa, M., Chiu, R. and Karin, M. (1987). Transcription factor AP-2 mediates induction by two different signal-transduction pathways: Protein kinase C and cAMP. *Cell* **51**, 251–260.

Ingraham, J. A., Chan, R., Mangalam, H. J., Elsholtz, H. P., Flynn, S. E., Linn, C. R., Simmons, D. M., Swanson, L. and Rosenfeld, M. G. (1988). A tissue specific factor containing a homeo domain specifies a pituitary phenotype. *Cell* **55**, 519–529.

Jones, N. C., Rigby, P. W. J. and Ziff, E. B. (1988). Trans-acting protein factors and the regulation of eukaryotic transcription. *Genes and Development* **2**, 267–281.

Karin, M., Haslinger, A., Holtgreve, J., Richards, R. I., Krauter, P., Westphal, H. M. and Beato, M. (1984). Characterization of DNA sequences through which cadmium and glucocorticoid hormones induce metallothionein IIA gene. *Nature* **308**, 513–519.

Landolfi, N. F., Capra, D. J. and Tucker, P. W. (1986). Interaction of cell type-specific nuclear proteins with immunoglobulin VH promoter region sequences. *Nature* **323**, 548–551.

Latchman, D. S. (1990). *Gene Regulation: A Eukaryotic Perspective*. London: Unwin Hyman.

Lee, W., Haslinger, A., Karin, M. and Tjian, R. (1987a). Activation of transcription by two factors that bind promoter and enhancer sequences of the human metallothionein gene and SV40. *Nature* **325**, 369–372.

Lee, W., Mitchell, P. and Tjian, R. (1987b). Purified transcription factor AP-1 interacts with TPA-inducible enhancer elements. *Cell* **49**, 741–752.

Lewis, M. K. and Burgess, R. R. (1982). Eukaryotic RNA polymerases In: *The enzymes*, vol. 15 (Boyer, P., ed.), pp. 109–153. New York: Academic Press.

Maniatis, T., Goodboun, S. and Fischer, J. A. (1987). Regulation of inducible and tissue specific gene expression. *Science* **236**, 1237–1245.

Mason, J. O., Williams, G. T. and Neuberger, M. S. (1985). Transcription cell type specificity is conferred by an immunoglobulin VH gene promoter that includes a functional consensus sequence. *Cell* **41**, 479–487.

McKnight, S. L. and Kingsbury, R. (1982). Transcriptional control signals of a eukaryotic protein coding gene. *Science* **217**, 316–324.

McKnight, S. and Tjian, R. (1986). Transcriptional selectivity of viral genes in mammalian cells. *Cell* **46**, 795–805.

McKnight, S. L., Kingsbury, R. C., Spence, A. and Smith, M. (1984). The distal transcription signals of the herpes virus tK gene share a common hexanucleotide control sequence. *Cell* **37**, 253–262.

Miller, J. and Reznikoff, W. K. (eds) (1980). *The Operon*. New York: Cold Spring Harbor Laboratory.

Montminy, M. R., Sevarno, K. A., Wagner, J. A., Mondel, G. and Goodman, R. H. (1986). Identification of a cyclic AMP responsive element within the rat somatostatin gene. *Proceedings of the National Academy of Sciences, USA* **86**, 6682–6686.

Nelson, C. R., Albert, V. R., Elsholtz, H. P., Lu, L. E. W. and Rosenfeld, M. G. (1988). Activation of cell specific expression of rat growth hormone and prolactin genes by a common transcription factor. *Science* **239**, 1400–1505.

Nevins, J. R. (1983). The pathway of eukaryotic mRNA transcription. *Annual Review of Biochemistry* **52**, 441–446.

Parslow, T. G., Blair, D. L., Murphy, W. J. and Granner, D. K. (1984). Structure of the 5′ ends of immunoglobulin genes: a novel conserved sequence. *Proceedings of the National Academy of Sciences, USA* **81**, 2650–2654.

Pelham, H. R. B. (1982). A regulatory upstream promoter element in the Drosophila hsp70 heat-shock gene. *Cell* **30**, 517–528.

Ptashne, M. (1986). *A Genetic Switch*. Cambridge and Palo Alto, California, USA: Cell Press and Blackwell Scientific Publications.

Schmitz, A. and Galas, D. (1979). The interaction of RNA polymerase and lac repressor with the lac control region. *Nucleic Acids Research* **6**, 111–137.

Sehgal, A., Patil, N. and Chau, M. (1988). A constitutive promoter directs expression of the nerve growth factor receptor. *Molecular and Cellular Biology* **8**, 3160–3167.

Sen, R. and Baltimore, D. (1986). Multiple nuclear factors interact with the immunoglobulin enhancer sequences. *Cell* **46**, 705–716.

Serfling, E., Jason, M. and Schaffner, W. (1985). Enhancers and eukaryotic gene transcription. *Trends in Genetics* **1**, 224–230.

Smale, S. T. and Baltimore, D. (1989). The initiator as a transcription control element. *Cell* **57**, 103–113.

Staudt, L. H., Singh, H., Sen, R., Wirth, T., Sharp, P. A. and Baltimore, D. (1986). A lymphoid-specific protein binding to the octamer motif of immunoglobulin genes. *Nature* **323**, 640–643.

Williams, G. T., McClanahan, T. K. and Morimoto, R. I. (1989). E1a transactivation of the human hsp70 promoter is mediated through the basal transcriptional complex. *Molecular and Cellular Biology* **9**, 2574–2587.

Wirth, T., Staudt, L. and Baltimore, D. (1987). An octamer oligonucleotide upstream of a TATA motif is sufficient for lymphoid specific promoter activity. *Nature* **329**, 174–178.

CHAPTER TWO

Methods for studying transcription factors

2.1 INTRODUCTION

The explosion in the available information on transcription factors which has occurred in the last few years has arisen primarily because of the availability of new or improved methods for studying these factors. Initially such studies normally focus on identifying a factor which interacts with a particular DNA sequence and characterizing this interaction. Subsequently the protein identified in this way is further characterized and purified and its corresponding gene isolated for further study. The methods involved in these two types of study will be considered in turn.

2.2 METHODS FOR STUDYING DNA–PROTEIN INTERACTIONS

2.2.1 DNA mobility shift assay

As discussed in Section 1.2, the initial stimulus to identify a transcription factor normally comes from the identification of a particular DNA sequence that confers a specific pattern of expression on a gene which carries it. The next step, therefore, following the identification of such a sequence, will be to define the protein factors which bind to it. This can be readily achieved by the DNA mobility shift or gel retardation assay (Fried and Crothers, 1981; Garner and Revzin, 1981).

This method relies on the obvious principle that a fragment of DNA to which a protein has bound will move more slowly in gel electrophoresis than the same DNA fragment without bound protein. The DNA mobility shift assay is carried out, therefore, by first radioactively labelling the specific DNA sequence whose protein binding properties are being investigated. The labelled DNA is then incubated with a nuclear (Dignam *et al.*, 1983) or whole cell (Manley *et al.*, 1980) extract of cells prepared in such a way as to contain the DNA binding proteins. In this way DNA–protein complexes are allowed to form. The complexes are then electrophoresed on a non-denaturing polyacrylamide gel and the position of the radioactive DNA visualized by autoradiography. If no protein has bound to the DNA, all the radioactive label will be at the bottom of the gel, whereas if a protein–DNA complex has formed, radioactive DNA to which the protein has bound will migrate more slowly and hence will be visualized near the top of the gel (Figure 2.1).

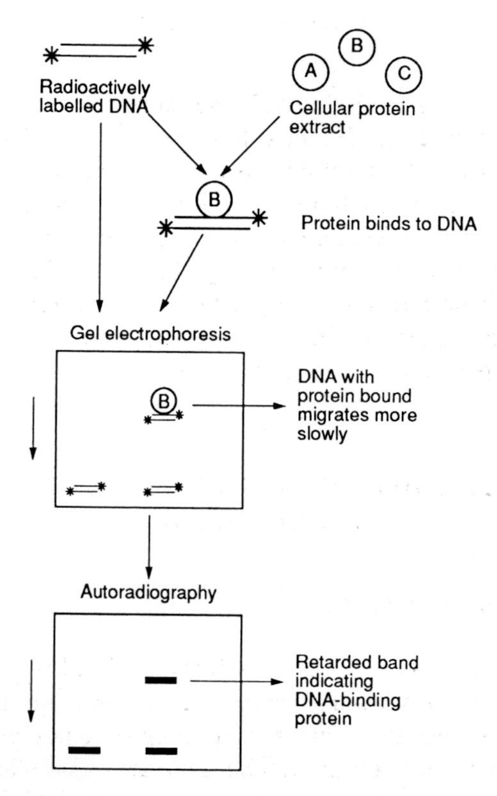

Figure 2.1 DNA mobility shift or gel retardation assay. Binding of a cellular protein (B) to the radioactively labelled DNA causes it to move more slowly upon gel electrophoresis and hence results in the appearance of a retarded band upon autoradiography to detect the radioactive label.

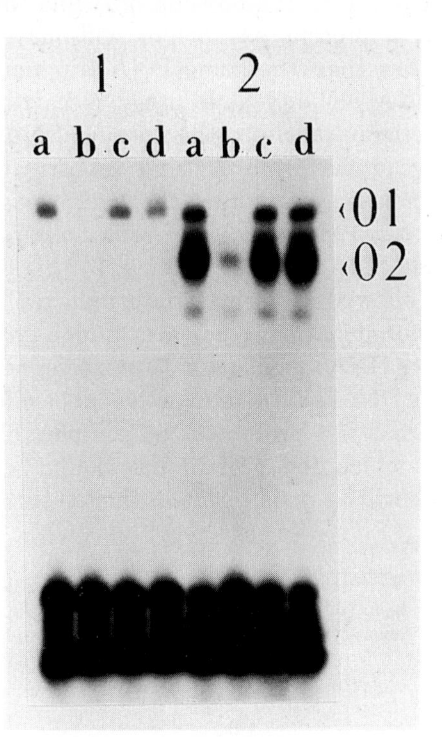

Figure 2.2 DNA mobility shift assay using a radioactively labelled probe containing the binding site for octamer-binding proteins (ATGCAAAT) and extracts prepared from fibroblast cells (1) or B cells (2). Note that fibroblast cells contain only one protein, Oct-1 (01), capable of producing a retarded band, whereas B cells contain both Oct-1 and an additional tissue-specific protein Oct-2 (02). The complexes formed by Oct-1 and Oct-2 on the labelled oligonucleotide in the absence of unlabelled oligonucleotide (track a) are readily removed by a 100-fold excess of unlabelled octamer oligonucleotide (track b). They are not removed, however, by a similar excess of a mutant octamer oligonucleotide (ATAATAAT) which is known not to bind octamer-binding proteins (track c) (Lenardo *et al.*, 1987) or of the binding site for the unrelated transcription factor Sp1 (track d) (Dynan and Tjian, 1983). This indicates that the retarded bands are produced by sequence-specific DNA-binding proteins which bind specifically to the octamer motif and not to mutant or unrelated motifs.

This technique can be used, therefore, to identify proteins which can bind to a particular DNA sequence in extracts prepared from specific cell types. Thus, for example, in the case of the octamer sequence discussed in Section 1.2.3, a single retarded band is detected when this sequence is mixed, for example, with a fibroblast extract. In contrast, when an extract from immunoglobulin-producing B cells is used, two distinct retarded bands are seen

(Figure 2.2). Since each band is produced by a distinct protein binding to the DNA, this indicates that in addition to the ubiquitous octamer-binding protein Oct-1 which is present in most cell types, B cells also contain an additional octamer-binding protein, Oct-2, which is absent in many other cells (Staudt *et al.*, 1986; Landolfi *et al.*, 1986). The role of the Oct-2 protein in immunoglobulin gene expression is discussed further in Section 5.2.2.

As well as defining the proteins binding to a particular sequence, the DNA mobility shift assay can also be used to investigate the

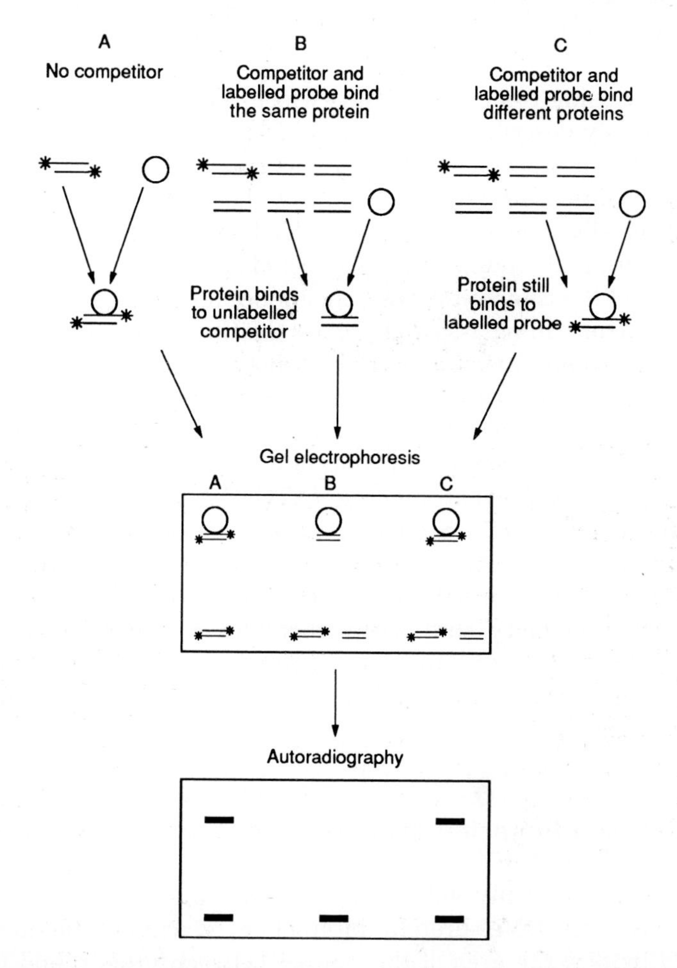

Figure 2.3 Use of unlabelled competitor DNAs in the DNA mobility shift assay. If an unlabelled DNA sequence is capable of binding the same protein as is bound by the labelled probe, it will do so (B) and the radioactive retarded band will not be observed, whereas if it cannot bind the same protein (C), the radioactive retarded band will form exactly as in the absence of competitor (A).

precise sequence specificity of this binding. This can be done by including in the binding reaction a large excess of a second DNA sequence which has not been labelled. If this DNA sequence can also bind the protein bound by the labelled DNA, it will do so. Moreover, binding to the unlabelled DNA will predominate since it is present in large excess. Hence the retarded band will not appear in the presence of the unlabelled competitor since only protein–DNA complexes containing labelled DNA are visualized on autoradiography (Figure 2.3b). In contrast, if the competitor cannot bind the same sequence as the labelled DNA, the complex with the labelled DNA will form and the labelled band will be visualized as before (Figure 2.3c).

Thus by using competitor DNAs which contain the binding sites for previously described transcription factors, it can be established whether the protein detected in a particular mobility shift experiment is identical or related to any of these factors. Similarly, if competitor DNAs are used which differ in only one or a few bases from the original binding site the effect of such base changes on the efficiency of the competitor DNA and hence on binding of the transcription factor can be assessed. Figure 2.2 illustrates an example of this type of competition approach showing that the octamer-binding proteins Oct-1 and Oct-2 are efficiently competed away from the labelled octamer probe by an excess of identical unlabelled competitor but not by a competitor containing three base changes in this sequence which prevent binding (ATGCAAAT to ATAATAAT) (Lenardo *et al.*, 1987). Similarly, no competition is observed, as expected, when the binding site of an unrelated transcription factor Sp1 (Dynan and Tjian, 1983) is used as the competitor DNA.

The DNA mobility shift assay therefore provides an excellent means of initially identifying a particular factor binding to a specific sequence and characterizing both its tissue distribution and its sequence specificity.

2.2.2 DNase I footprinting assay

Although the mobility shift assay provides a means of obtaining information on DNA–protein interaction, it cannot be used to directly localize the area of the contact between protein and DNA. For this purpose, the DNase I footprint assay is used (Galas and Schmitz, 1978; Dynan and Tjian, 1983).

In this assay, DNA and protein are mixed as before, the DNA being labelled however, only at the end of one strand of the double-stranded molecule. Following binding, the DNA is treated with a

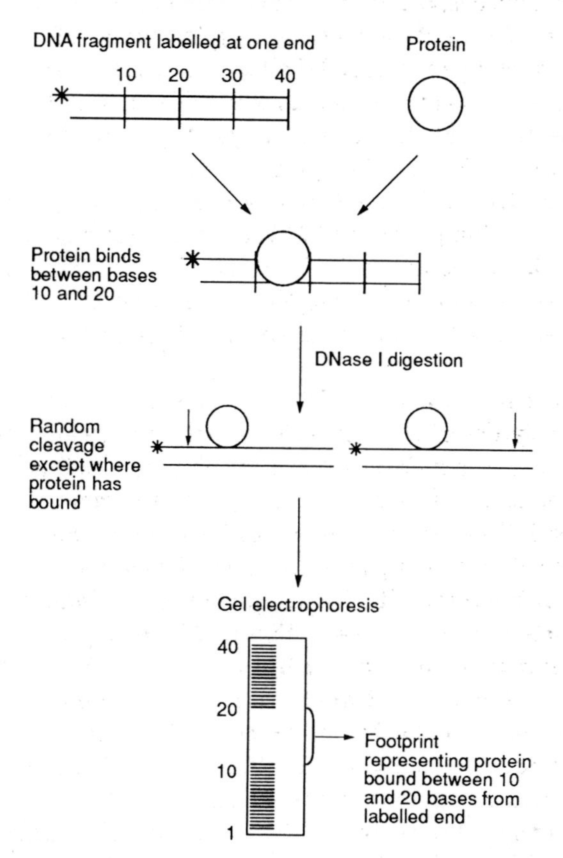

Figure 2.4 DNase I footprinting assay. If a protein binds at a specific site within a DNA fragment labelled at one end, the region of DNA at which the protein binds will be protected from digestion with DNase I. Hence this region will appear as a footprint in the ladder of bands produced by the DNA being cut at all other points by DNase I.

small amount of the enzyme deoxyribonuclease I (DNase I), which will digest DNA. The digestion conditions are chosen, however, so that each molecule of DNA will be cut once or a very few times by the enzyme. Following digestion the bound protein is removed and the DNA fragments separated by electrophoresis on a polyacrylamide gel capable of resolving DNA fragments differing in size by only one base. This produces a ladder of bands representing the products of DNase I cutting either one or two or three or four etc. bases from the labelled end. Where a particular piece of the DNA has bound a protein, however, it will be protected from digestion and hence the bands corresponding to cleavage at these points will be absent. This will be visualized on electrophoresis as a blank area on the gel

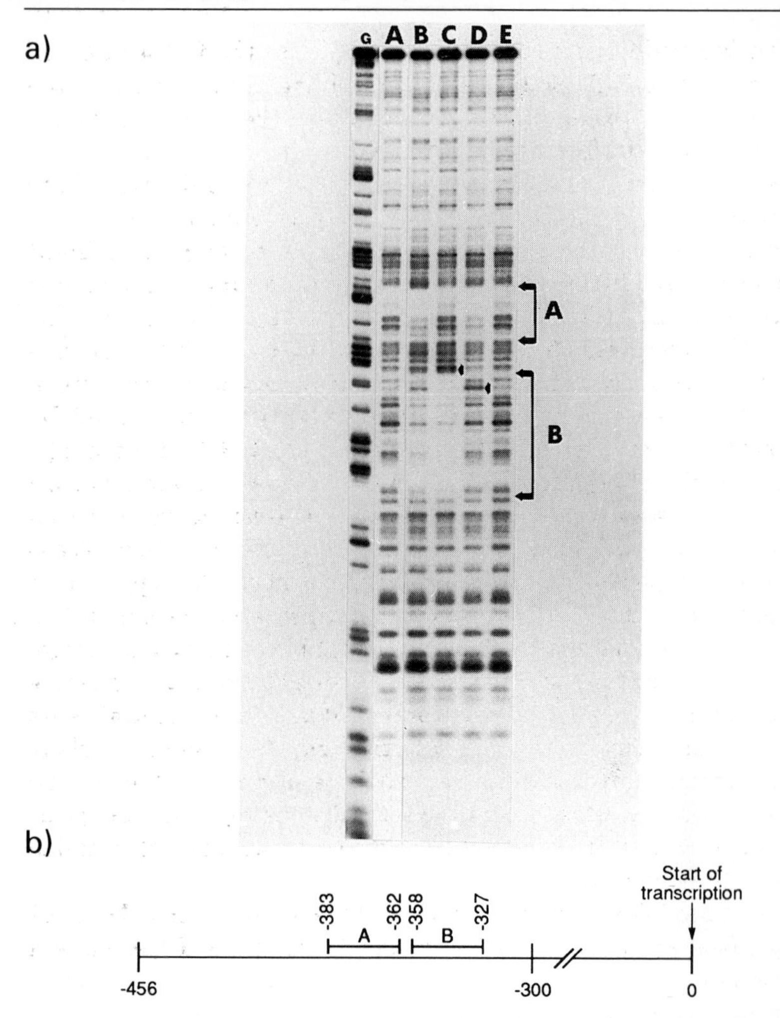

Figure 2.5 (a) DNase I footprinting assay carried out on a region of the human immunodeficiency virus (HIV) control element. The two footprints (A and B) are not observed when no cell extract is added to the reaction (track A) but are observed when cellular extract is added in the absence of competitor (track B). Addition of unlabelled oligonucleotide competitor containing the DNA sequence of site A removes the site A footprint without affecting site B (track C), whilst an unlabelled oligonucleotide containing the site B DNA sequence has the opposite effect (track D). Both footprints are removed by a mixture of unlabelled site A and B oligonucleotides (track E). Arrows indicate the position of sites at which cleavage with DNase I is enhanced in the presence of protein bound to an adjacent site, indicating the existence of conformational changes induced by protein binding. The track labelled G represents a marker track consisting of the same DNA fragment chemically cleaved at every guanine residue. (b) Position of sites A and B within the HIV control element. The arrow indicates the start site of transcription.

lacking labelled fragments and is referred to as the footprint of the protein (Figure 2.4). Similar labelling of the other strand of the DNA molecule will allow the interaction of the protein with the other strand of the DNA to be assessed.

The footprinting technique therefore allows a visualization of the interaction of a particular factor with a specific piece of DNA. By using a sufficiently large piece of DNA, the binding of different proteins to different DNA sequences within the same fragment can be assessed. An analysis of this type is shown in Figure 2.5. This shows the footprints (A and B) produced by two cellular proteins binding to two distinct sequences within a region of the human immunodeficiency virus (HIV) control element which has an inhibiting effect on promoter activity (Orchard *et al.*, 1990). Interestingly, some insights into the topology of the DNA–protein interaction are also obtained in this experiment since bands adjacent to the protected region appear more intense in the presence of the protein. These regions of hypersensitivity to cutting are likely to represent a change in the structure of the DNA in this region when the protein has bound and rendered the DNA more susceptible to enzyme cleavage.

As with the mobility shift assay, unlabelled competitor sequences can be used to remove a particular footprint and determine its sequence specificity. In the HIV case illustrated in Figure 2.5, short DNA competitors containing the sequence of one or other of the footprinted areas were used to specifically remove each footprint without affecting the other, indicating that two distinct proteins produce the two footprints.

The DNase I footprint thus offers an advance on the mobility shift assay, allowing a more precise visualization of the DNA–protein interaction.

2.2.3 Methylation interference assay

Having identified the area of contact between a specific DNA sequence and a particular protein, the pattern of DNA–protein interaction can be studied in more detail using the methylation interference assay (Siebenlist and Gilbert, 1980). This method is based on assessing whether the methylation of specific G residues in the target DNA affects protein binding. If such methylation does affect binding of the protein, this identifies the particular G residue as being an important component of the binding site. Hence this method offers an advance over the footprinting technique in that it allows assessment of the interaction of the DNA-binding protein with specific residues within its binding site.

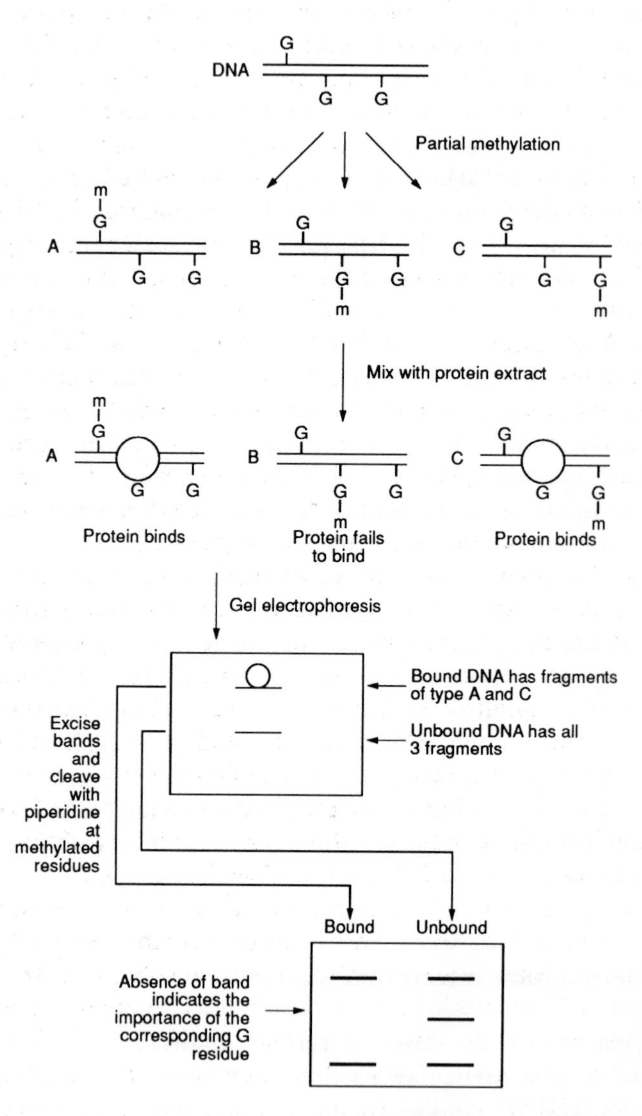

Figure 2.6 Methylation interference assay. Partially methylated DNA is used in a DNA mobility shift assay and both the DNA which has failed to bind protein and that which has bound protein and formed a retarded band are subsequently cleaved at methylated G residues with piperidine. If methylation at a specific G residue has no effect on protein binding (types A and C) the bound and unbound DNA will contain equal amounts of methylated G at this position. In contrast, if methylation at a particular G prevents binding of the protein (type B), only the unbound DNA will contain methylated G at this position.

To do this, the target DNA is partially methylated using dimethylsulphate so that on average only one G residue per DNA molecule is methylated, methylation taking place at position 7 on the purine ring (Maxam and Gilbert, 1980). Each individual DNA molecule will therefore contain some methylated G residues, with the particular residues which are methylated being different in each molecule. These partially methylated DNAs are then used in a DNA mobility shift experiment with an appropriate cell extract containing the DNA-binding protein. Following electrophoresis, the band produced by the DNA which has bound protein and that produced by the DNA which has not are excised from the gel and treated with piperidine, which cleaves the DNA only at the methylated G residues and not at unmethylated Gs (Maxam and Gilbert, 1980). Clearly, if methylation of a particular G prevents protein binding then cleavage at this particular methylated G will be observed only in the DNA which failed to bind the protein. Conversely, if a particular G residue plays no role in binding, then cleavage at this G residue will be observed equally in both the DNA which bound the protein and that which failed to do so (Figure 2.6).

Figure 2.7 shows this type of analysis applied to the protein binding to site B within the negatively acting element in the human immunodeficiency virus promoter (for the footprint produced by the binding of this protein see Figure 2.5). In this case the footprinted sequence was palindromic (Figure 2.7), suggesting that the DNA–protein interaction may involve similar binding to the two halves of the palindrome. The methylation interference analysis of site B confirms this by showing that methylation of equivalent G residues in each half of the palindrome interferes with binding of the protein, indicating that these residues are critical for binding.

The insights obtained using the methylation interference assay can be confirmed by using the reciprocal method of methylation protection. In this case the DNA is first incubated with protein extract and then methylated with dimethylsulphate. G residues which are in contact with the protein should be protected from methylation and hence from subsequent piperidine cleavage; see Lassar *et al.* (1989) for an example of the use of this approach.

Similarly, by modifying the technique, interference analysis can be used to study the interaction of DNA-binding proteins with A residues in the binding site. This can be done either by methylating all purines to allow study of interference at A and G residues simultaneously (e.g. Ares *et al.*, 1987) or by using diethylpyrocarbonate to specifically modify A residues (probably by carboxyethylation) rendering them susceptible to piperidine cleavage (e.g. Sturm *et al.*, 1988). These techniques are of particular value when studying

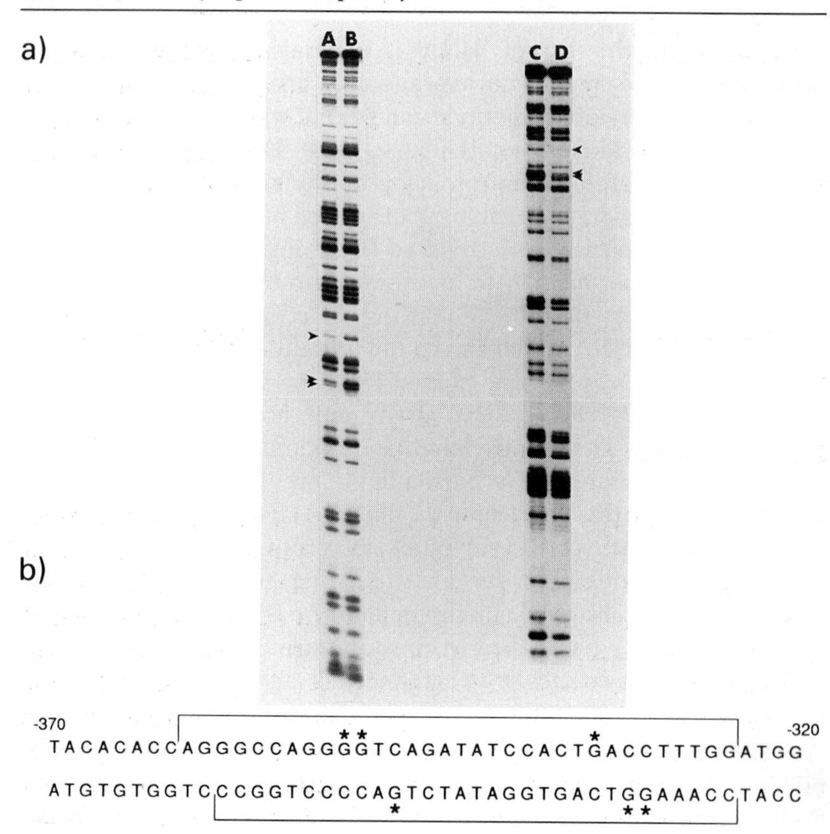

Figure 2.7 (a) Methylation interference assay applied to the DNA of site B in the HIV control element as defined in the footprinting experiment shown in Figure 2.5. Both the upper (tracks A and B) and lower (tracks C and D) strands of the double-stranded DNA sequence were analysed. Tracks B and C show the methylation pattern of the unbound DNA which failed to bind protein, whereas tracks A and B show the methylation pattern of DNA which has bound protein. The arrows show G residues whose methylation is considerably lower in the bound compared to the unbound DNA and which are therefore critical for binding the specific cellular protein which interacts with this DNA sequence. (b) DNA sequence of site B. The extent of the footprint region is indicated by the square brackets and the critical G residues defined by the methylation interference assay in (a) are asterisked. Note the symmetrical pattern of critical G residues within the palindromic DNA sequence.

sequences such as the octamer motif in which there are relatively few G residues, hence limiting the information which can be obtained by studying interference at G residues alone (Sturm *et al.*, 1987; Baumruker *et al.*, 1988).

Chemical interference techniques can therefore be used to supplement DNase I footprinting by identifying the precise DNA–protein interactions within the footprinted region.

Taken together, therefore, these three methods of DNA mobility shift, DNase I footprinting and methylation interference can provide considerable information on the nature of the interaction between a particular DNA sequence and a transcription factor. They serve as an essential prelude to a detailed study of the transcription factor itself.

2.3 METHODS FOR STUDYING THE TRANSCRIPTION FACTOR ITSELF

2.3.1 Protein purification

As discussed above, once a particular DNA sequence has been shown to be involved in transcriptional regulation, a number of techniques are available for characterizing the binding of transcription factors to this sequence. Although such studies can be carried out on crude cellular extracts containing the protein, ultimately they need to be supplemented by studies on the protein itself. This can be achieved by purifying the transcription factor from extracts of cells containing it. Unfortunately, however, conventional protein purification techniques such as conventional chromatography and high-pressure liquid chromatography (HPLC) result in the isolation of transcription factors at only 1–2% purity (Kadonga and Tjian, 1986).

To overcome this problem and purify the transcription factor Sp1, Kadonga and Tjian (1986) devised a method involving DNA affinity chromatography. In this method (Figure 2.8), a DNA sequence containing a high-affinity binding site for the transcription factor is synthesized and the individual molecules joined to form a multimeric molecule. This very high-affinity binding site is then coupled to an activated sepharose support on a column and total cellular protein passed down the column. The Sp1 protein binds specifically to its corresponding DNA sequence, whilst all other cellular proteins do not bind. The bound Sp1 can be eluted simply by raising the salt concentration. Two successive affinity chromatography steps of this type successfully resulted in the isolation of Sp1 at 90% purity, 30% of the Sp1 in the original extract being recovered, representing a 500–1000-fold purification (Kadonga and Tjian, 1986).

Although this simple one-step method was successful in this case, it relies critically on the addition of exactly the right amount of non-specific DNA carrier to the cell extract. Thus this added carrier acts

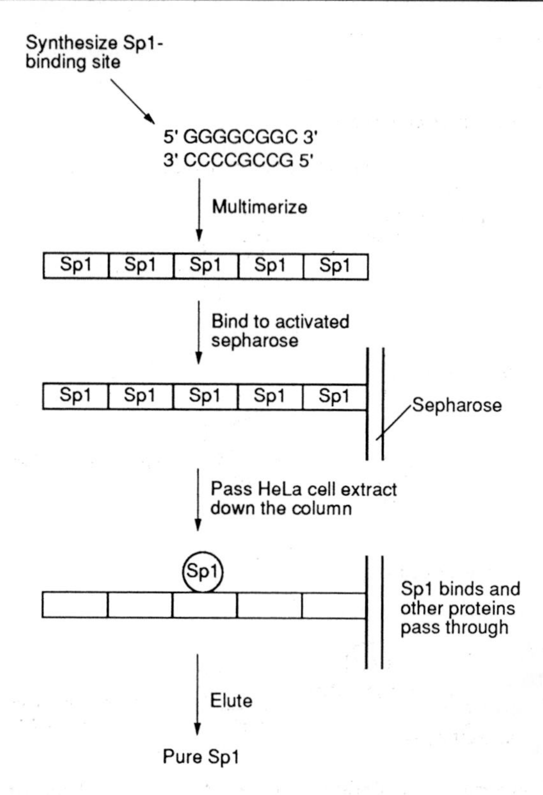

Figure 2.8 Purification of transcription factor Sp1 on an affinity column in which multiple copies of the DNA sequence binding Sp1 have been coupled to a sepharose support (Kadonga and Tjian, 1986).

to remove proteins which bind to DNA in a non-sequence-specific manner and which would hence bind non-specifically to the Sp1 affinity column and contaminate the resulting Sp1 preparation. This contamination will occur if too little carrier is added. If too much carrier is added, however, it will bind out the Sp1 since, like all sequence-specific proteins, Sp1 can bind with low affinity to any DNA sequence. Hence in this case no Sp1 will bind to the column itself (Figure 2.9).

To overcome this problem Rosenfeld and Kelley (1986) devised a method in which proteins capable of binding to DNA with high affinity in a non-sequence-specific manner are removed prior to the affinity column. To do this the bulk of cellular protein was removed on a Biorex 70 high-capacity ion-exchange column and proteins which can bind to any DNA with high affinity were then removed on a cellulose column to which total bacterial DNA had been bound. Subsequently the remaining proteins which had bound to non-sequence-specific DNA only with low affinity were applied to a

A Correct amount of carrier

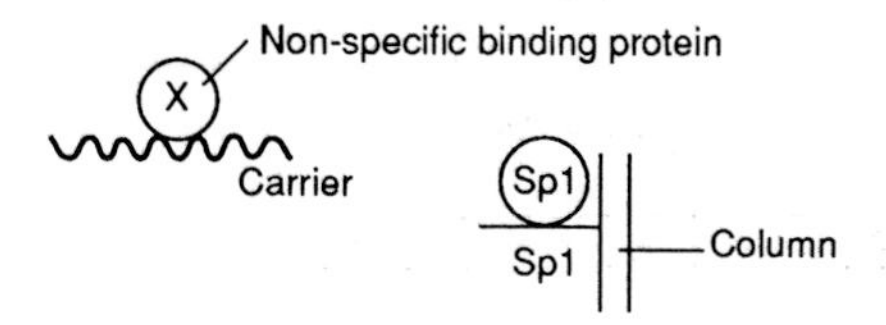

B Too little carrier

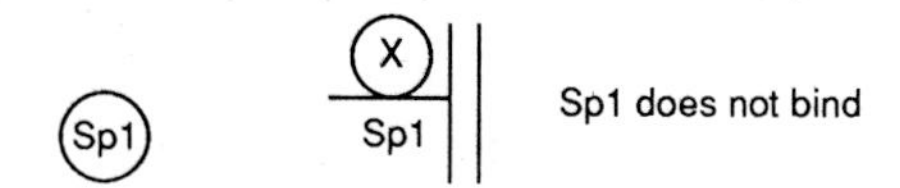

C Too much carrier

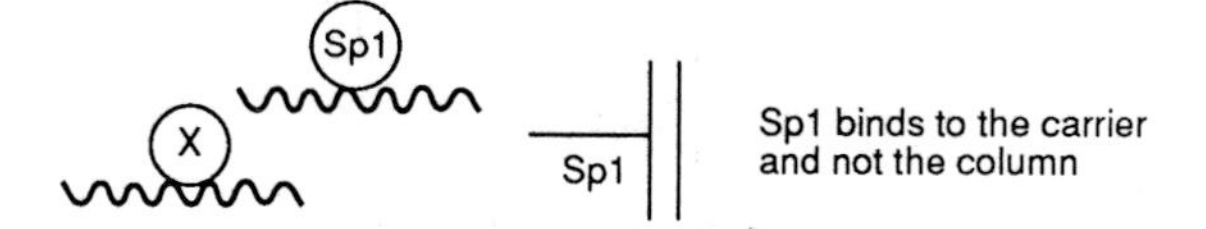

Figure 2.9 Consequences of adding different amounts of non-specific carrier DNA to the protein passing through the Sp1 affinity column. If the correct amount of non-specific carrier is added it will bind proteins which interact with DNA in a non-sequence-specific manner, allowing Sp1 to bind to the column (A). However, addition of too little carrier will result in non-sequence-specific proteins binding to the column and thereby preventing the binding of Sp1 (B), whereas in the presence of too much carrier both the non-specific proteins and Sp1 will bind to the carrier (C).

Table 2.1 Purification of transcription factor NF1 from HeLa cells

	Total protein (mg)	Specific binding of ^{32}P DNA (fmol/mg protein) $\times 10^{-3}$	Purification (fold)	Yield (%)
HeLa cell extract*	4590	3.1	1.0	100
Biorex 70 column	550	27.1	8.7	104
E. coli DNA-cellulose	65.2	181	58.4	83
NF1 Affinity matrix 1st passage	2.1	4510	1455	67
NF1 Affinity matrix 2nd passage	1.1	7517	2425	57

* Prepared from 6×10^{10} cells or 120 grams of cells.

column containing a high-affinity binding site for transcription factor NF1 (Figure 2.10). NF1 bound to this site with high affinity and could be eluted in essentially pure form by raising the salt concentration (Table 2.1). It should be noted that in this and other purification procedures the fractions containing the transcription factor can readily be identified by carrying out a DNA mobility shift or footprinting assay with each fraction using the specific DNA-binding site of the transcription factor.

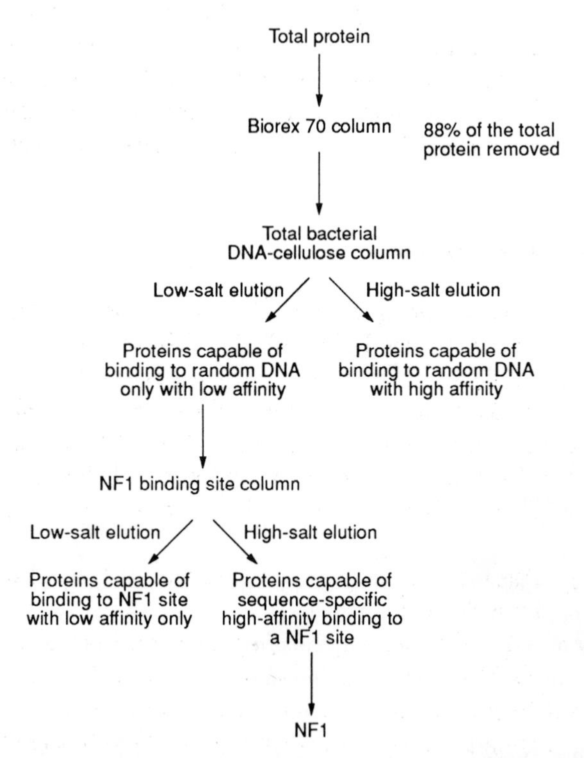

Figure 2.10 Purification of transcription factor NF1 (Rosenfeld and Kelley, 1986). Following removal of most cellular proteins on a Biorex 70 ion-exchange column, proteins which bind to all DNA sequences with high affinity were removed on a bacterial DNA-cellulose column. Subsequent application of the remaining proteins to a column containing the NF1 binding site results in the purification of NF1, since it is the only protein which binds with low affinity to random DNA but with high affinity to an NF1 site.

The purified protein obtained in this way can obviously be used to characterize the protein, for example by determining its molecular weight or by raising an antibody to it to characterize its expression pattern in different cell types. Similarly, the activity of the protein can be assessed by adding it to cellular extracts and assessing its effect on their ability to transcribe an exogenously added DNA in an *in vitro* transcription assay.

Unfortunately, however, because of the very low abundance of transcription factors in the cell, these purification procedures yield very small amounts of protein. For example, Treisman (1987) succeeded in purifying only 1.6 μg of the serum response factor starting with 2×10^{10} cells or 40 g of cells. Such difficulties clearly limit the experiments which can be done with purified material. Indeed, the primary use of purified factor in most cases has simply been to provide material to isolate the gene encoding the protein. This gene can then be expressed either *in vitro* or in bacteria to provide a far more abundant source of the corresponding protein than could be obtained from cells which naturally express it.

2.3.2. Gene cloning

In order to isolate the gene encoding a particular transcription factor, a complementary DNA (cDNA) library is prepared from mRNA isolated from a cell type expressing the factor (for details of the methods used in preparing these libraries see Sambrook *et al.* (1989)). Some means is then required to identify a clone derived from the mRNA encoding the factor amongst all the other clones in the library. Two methods are normally used for this purpose.

(a) Use of oligonucleotide probes predicted from the protein sequence of the factor

If a particular transcription factor has been purified it is possible to obtain portions of its amino acid sequence. In turn such sequences can be used to predict oligonucleotides containing DNA sequences capable of encoding these protein fragments. Due to the redundancy of the genetic code, whereby several different DNA codons can encode a particular amino acid, there will be multiple different oligonucleotides capable of encoding a particular amino acid sequence. All these possible oligonucleotides are synthesized chemically, made radioactive and used to screen the cDNA library. The oligonucleotide in the mixture which corresponds to the transcription factor amino acid sequence will hybridize to the corresponding sequence in a cDNA clone derived from mRNA encoding the factor. Hence such a clone can be readily identified in the cDNA library (Figure 2.11).

In cases where purified protein is available, as in those discussed in the previous section, this approach represents a relatively simple method for isolating cDNA clones. It has therefore been widely used to isolate cDNA clones corresponding to purified factors such as Sp1 (Kadonga *et al.*, 1987; Figure 2.11), NF1 (Santoro *et al.*, 1988) and the serum response factor (Norman *et al.*, 1988).

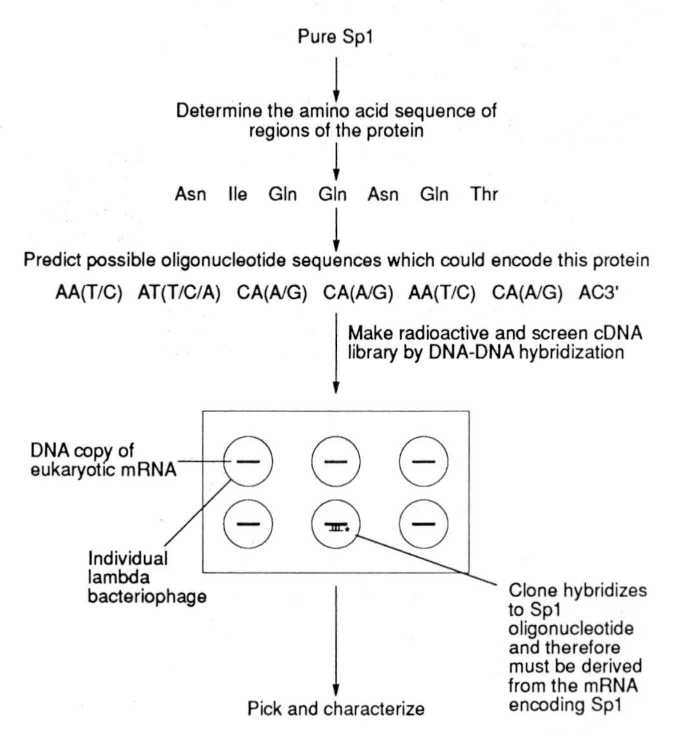

Figure 2.11 Isolation of cDNA clones for the Sp1 transcription factor by screening with short oligonucleotides predicted from the protein sequence of Sp1. Because several different triplets of bases can code for any given amino acid, multiple oligonucleotides that contain every possible coding sequence are made. Positions at which these oligonucleotides differ from one another are indicated by the brackets containing more than one base.

(b) Use of oligonucleotide probes derived from the DNA-binding site of the factor

Although relatively simple, the use of oligonucleotides derived from protein sequence does require purified protein. As we have seen, purification of a transcription factor requires a vast quantity of cells and is technically difficult. Moreover, eventual determination of the partial amino acid sequence of the protein requires access to expensive protein-sequencing apparatus.

To bypass these problems Singh *et al.* (1988) devised a procedure which is based on the fact that information is usually available about the specific DNA sequence to which a particular transcription

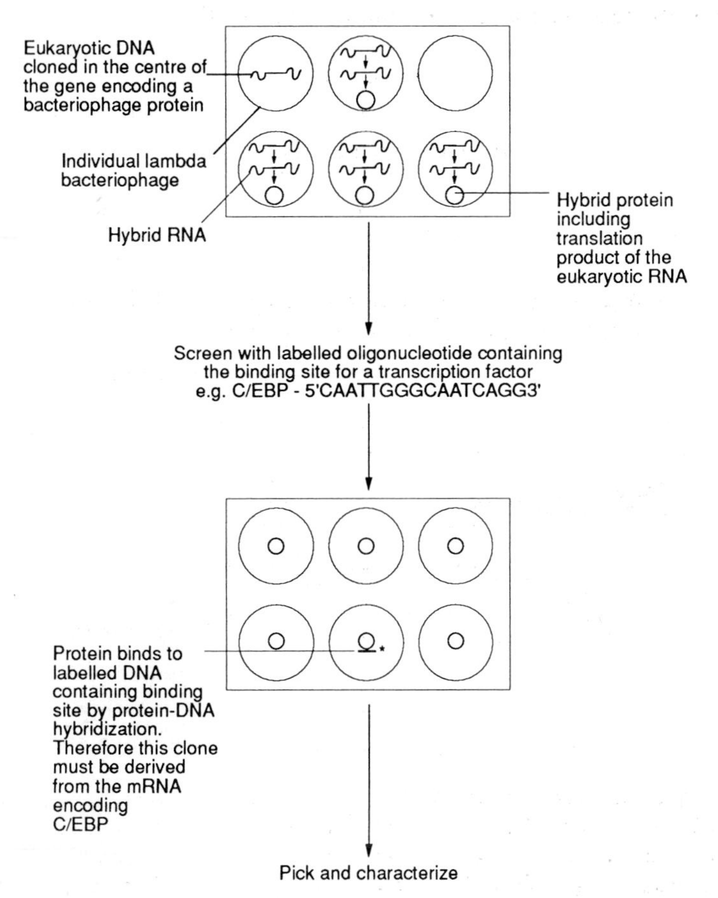

Figure 2.12 Isolation of cDNA clones for the C/EBP transcription factor by screening an expression library with a DNA probe containing the binding site for the factor.

factor binds. Hence a cDNA clone expressing the factor can be identified in a library by its ability to bind the appropriate DNA sequence. This method relies, therefore, on DNA–protein binding rather than DNA–DNA binding. Hence the library must be prepared in such a way that the cloned cDNA inserts are translated by the bacteria into their corresponding proteins. This is normally achieved by inserting the cDNA into the coding region of the bacteriophage lambda β-galactosidase gene, resulting in its translation as part of the bacteriophage protein (see Young and Davis (1983) for a further description of this bacteriophage vector and its other applications). The resulting fusion protein binds DNA with the same specificity as the original factor. Hence a cDNA clone encoding a particular factor can be identified in the library by screening with a radioactive oligonucleotide containing the binding site (Figure 2.12). This technique has been used to isolate cDNA clones encoding several transcription factors such as the CCAAT box-binding factor C/EBP (Vinson *et al.*, 1988) and the octamer-binding proteins Oct-1 (Sturm *et al.*, 1988) and Oct-2 (Staudt *et al.*, 1988).

The development of these two methods of screening with oligonucleotides derived from the protein sequence or oligonucleotides derived from the binding site has therefore resulted in the isolation of cDNA clones corresponding to very many transcription factors.

2.3.3. Use of cloned genes

This isolation of cDNA clones has in turn resulted in an explosion of information on these factors. Thus once a clone has been isolated, its DNA sequence can be obtained, allowing prediction of the corresponding protein sequence and comparison with other factors. Similarly, the clone can be used to identify the mRNA encoding the protein and examine its expression in various tissues by Northern blotting, to study the structure of the gene itself within genomic DNA by Southern blotting and as a probe to search for related genes expressed in other tissues or other organisms.

Most importantly, however, the isolation of cDNA clones provides a means of obtaining large amounts of the corresponding protein for functional study. This can be achieved either by coupled *in vitro* transcription and translation (Figure 2.13A) (e.g. Sturm *et al.*, 1988) or by expressing the gene in bacteria either in the original expression vector used in the screening procedure (see Section 2.3.2b) or more commonly by sub-cloning the cDNA into a plasmid expression vector (Figure 2.13B) (e.g. Kadonga *et al.*, 1987).

The protein produced in this way has similar activity to the natural protein, being capable of binding to DNA in footprinting or mobility shift assays (e.g. Kadonga *et al.*, 1987) and of stimulating the transcription of appropriate DNAs containing its binding site when added to a cell-free transcription system (e.g. Mueller *et al.*, 1990).

Moreover, once a particular activity has been identified in a protein produced in this way, it is possible to analyse the features of the protein which produce this activity in a way that would not be

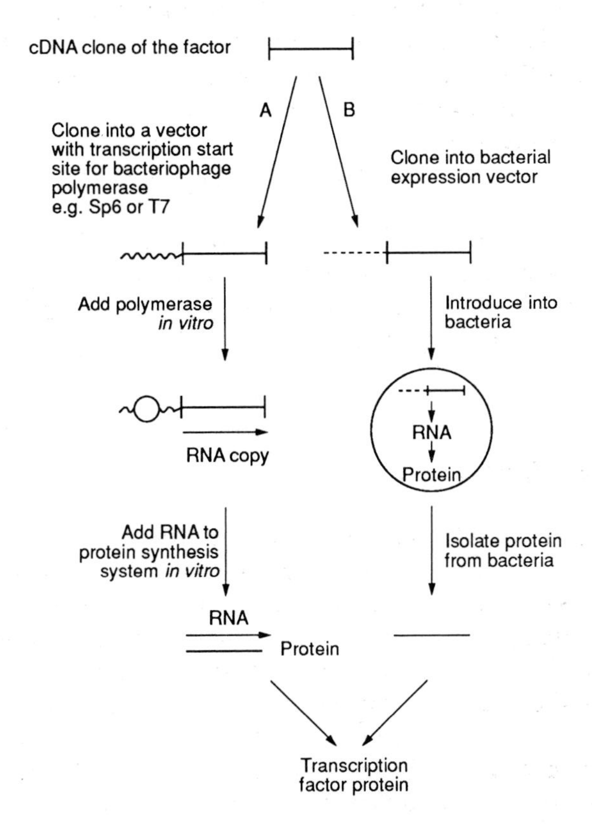

Figure 2.13 Methods of producing transcription factor protein from a cloned transcription factor cDNA. In the coupled *in vitro* transcription and translation method (A) the cDNA is cloned downstream of a promoter recognized by a bacteriophage polymerase and transcribed *in vitro* by addition of the appropriate polymerase. The resulting RNA is translated in an *in vitro* protein synthesis system to produce transcription factor protein. Alternatively (B) the cDNA can be cloned downstream of a prokaryotic promoter in a bacterial expression vector. Following introduction of this vector into bacteria, the bacteria will transcribe the cDNA into RNA and translate the RNA into protein which can be isolated from the bacteria.

possible using the factor purified from cells which normally express it. Thus because the cDNA clone of the factor can be readily cut into fragments and each fragment expressed as a protein in isolation, particular features exhibited by the intact protein can readily be mapped to a particular region. Using the approach outlined in Figure 2.14, for example, it has proved possible to map the DNA-binding abilities of specific transcription factors such as the

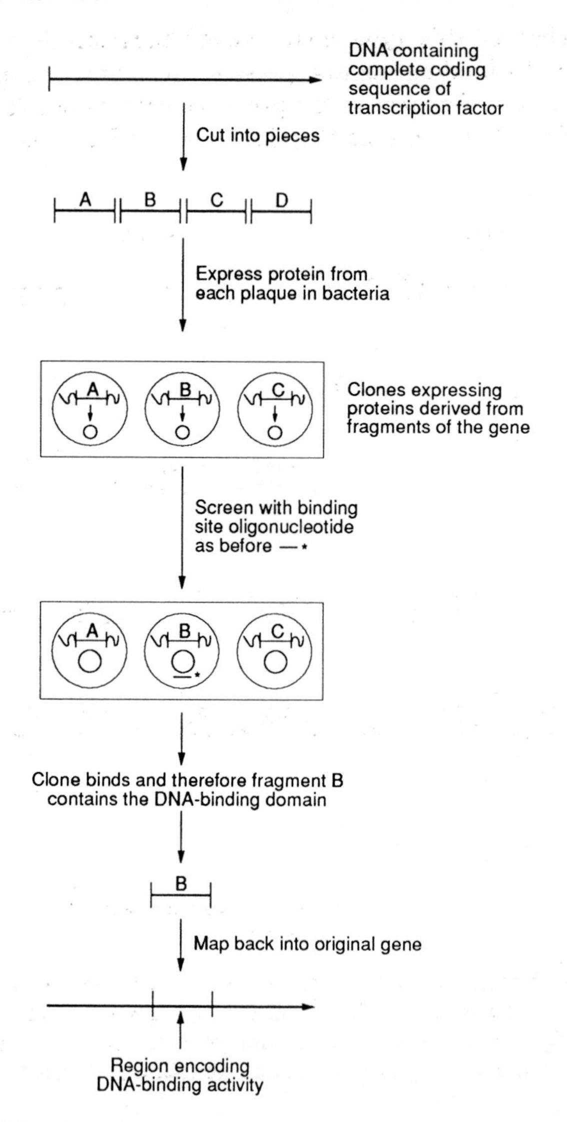

Figure 2.14 Mapping of the DNA-binding region of a transcription factor by testing the ability of different regions to bind to the appropriate DNA sequence when expressed in bacteria.

octamer-binding proteins Oct-1 (Sturm *et al.*, 1988) and Oct-2 (Clerc *et al.*, 1988) to a specific short region of the protein. Once this has been done, particular bases in the DNA encoding the DNA-binding domain of the factor can then be mutated so as to alter its amino acid sequence, and the effect of these mutations on DNA binding can be assessed as before by expressing the mutant protein and measuring its ability to bind to DNA.

Approaches of this type have proved particularly valuable in defining DNA-binding motifs present in many factors and in analysing how differences in the protein sequence of related factors define which DNA sequence they bind. This is discussed in Chapter 8.

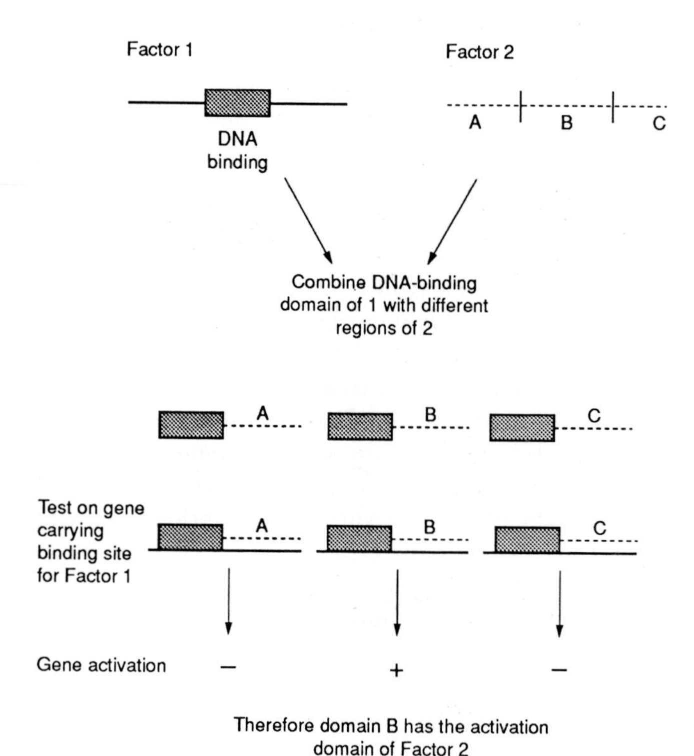

Figure 2.15 Domain-swapping experiment in which the activation domain of factor 2 is mapped by combining different regions of factor 2 with the DNA-binding domain of factor 1 and assaying the hybrid proteins for the ability to activate transcription of a gene containing the DNA-binding site of factor 1.

One other piece of information to emerge from these studies is that the binding to DNA of a small fragment of the factor does not normally result in the activation of transcription. Thus a 60-amino-acid region of the yeast transcription factor GCN4 can bind to DNA in a sequence-specific manner but does not activate transcription of genes bearing its binding site (Hope and Struhl, 1986). Although DNA binding is necessary for transcription, therefore, it is not sufficient. This indicates that transcription factors have a modular structure in which the DNA-binding domain is distinct from another domain of the protein which mediates transcriptional activation.

The identification of the activation domain in a particular factor is complicated by the fact that DNA binding is necessary prior to activation. Hence the activation domain cannot be identified simply by expressing fragments of the protein and monitoring their activity. Rather the various regions of the cDNA encoding the factor must each be linked to the region encoding the DNA-binding domain of another factor and the hybrid proteins produced. The ability of the hybrid factor to activate a target gene bearing the DNA binding site of the factor supplying the DNA-binding domain is then assessed (Figure 2.15). In these so-called 'domain-swap' experiments, binding of the factor to the appropriate DNA binding site will be followed by gene activation only if the hybrid factor contains the region encoding the activation domain of the factor under test, allowing the activation domain to be identified.

Thus if another 60-amino-acid region of GCN4 distinct from the DNA-binding domain is linked to the DNA-binding domain of the bacterial Lex A protein, it can activate transcription in yeast from a gene containing a binding site for Lex A. This cannot be achieved by the Lex A DNA-binding domain or this region of GCN4 alone, indicating that this region of GCN4 contains the activation domain of the protein which can activate transcription following DNA binding and is distinct from the GCN4 protein DNA-binding domain (Hope and Struhl, 1986).

As with DNA-binding domains, the identification of activation domains and comparisons between the domains in different factors has provided considerable information on the nature of activation domains and the manner in which they function. This is discussed in Chapter 9.

2.4 CONCLUSIONS

This chapter has described a number of methods which allow the investigation of the interaction of a transcription factor with DNA, its purification, its gene cloning and dissection of its functional domains. The information obtained by the application of these procedures to particular factors is discussed in subsequent chapters.

REFERENCES

Ares, M. Jr, Chung, J.-S., Giglio, L. and Weiner, A. M. (1987). Distinct factors with Sp1 and NF-A specificities bind to adjacent functional elements of the human U2 snRNA gene enhancer. *Genes and Development* **1**, 808–817.

Baumruker, T., Sturm, R. and Herr, W. (1988). OBP 100 binds remarkably degenerate octamer motifs through specific interactions with flanking sequences. *Genes and Development* **2**, 1400–1413.

Clerc, R. G., Corcoran, L. M., LeBowitz, J. H., Baltimore, D. and Sharp, P. A. (1988). The B-cell specific Oct-2 protein contains POU box and homeo box type domains. *Genes and Development* **2**, 1570–1581.

Dignam, J. D., Lebovitz, R. M. and Roeder, R. G. (1983). Accurate transcription initiation by RNA polymerase II in a soluble extract from isolated mammalian nuclei. *Nucleic Acids Research* **11**, 1575–1589.

Dynan, W. S. and Tjian, R. (1983). The promoter specific transcription factor Sp1 binds to upstream sequences in the SV40 promoter. *Cell* **35**, 79–87.

Fletcher, C., Heintz, N. and Roeder, R. G. (1987). Purification and characterization of OTF-1, a transcription factor regulating cell cycle expression of human H2B gene. *Cell* **51**, 773–781.

Fried, M. and Crothers, D. M. (1981). Equilibria and kinetics of lac repressor–operator interactions by polyacrylamide gel electrophoresis. *Nucleic Acids Research* **9**, 6505–6525.

Galas, D. and Schmitz, A. (1978). DNAse footprinting: A simple method for the detection of protein–DNA binding specificity. *Nucleic Acids Research* **5**, 3157–3170.

Garner, M. M. and Revzin, A. (1981). A gel electrophoresis method for quantifying the binding of proteins to specific DNA regions: application to components of the *Escherichia coli* lactose operon regulatory system. *Nucleic Acids Research* **9**, 3047–3060.

Hope, I. A. and Struhl, K. (1986). Functional dissection of a eukaryotic transcriptional activator GCN4 of yeast. *Cell* **46**, 885–894.

Kadonga, J. T. and Tjian, R. (1986). Affinity purification of sequence-specific DNA binding proteins. *Proceedings of the National Academy of Sciences, USA* **83**, 5889–5893.

Kadonga, J. T., Carner, K. R., Masiarz, F. R. and Tjian, R. (1987). Isolation of cDNA encoding the transcription factor Sp1 and functional analysis of the DNA binding domain. *Cell* **51**, 1079–1090.

Landolfi, N. F., Capra, D. J. and Tucker, P. W. (1987). Interaction of cell type-specific nuclear proteins with immunoglobulin VH promoter region sequences. *Nature* **323**, 548–551.

Lassar, A. B., Buskin, J. N., Lockshun, D., Davis, R. L., Apone, S., Hanaschka, S. D. and Weintraub, H. (1989). Myo D is a sequence-specific DNA binding protein requiring a region of myc homology to bind to the muscle creatine kinase enhancer. *Cell* **58**, 823–831.

Lenardo, M., Pierce, J. W. and Baltimore, D. (1987). Protein binding sites in Ig gene enhancers determine transcriptional activity and inducibility. *Science* **236**, 1573–1577.

Manley, J. L., Fire, A., Cano, A., Sharp, P. A. and Gefter, M. L. (1980). DNA-dependent transcription of adenovirus genes in a soluble whole-cell extract. *Proceedings of the National Academy of Sciences, USA* **77**, 3855–3859.

Maxam, A. M. and Gilbert, W. (1980). Sequencing end labelled DNA with base-specific chemical cleavages. *Methods in Enzymology* **65**, 499–560.

Mueller, C. R., Macre, P. and Schibler, U. (1990). DBP a liver-enriched transcriptional activator is expressed late in ontogeny and its tissue specificity is determined post-transcriptionally. *Cell* **61**, 279–291.

Norman, C., Runswick, M., Pollock, R. and Treisman, R. (1988). Isolation and properties of cDNA clones encoding SRF, a transcription factor that binds to the c-*fos* serum response element. *Cell* **55**, 989–1003.

Orchard, K., Perkins, N. D., Chapman, C., Harris, J., Emery, V., Goodwin, G., Latchman, D. S. and Collins, M. K. L. (1990). A novel T cell protein recognizes a palindromic element in the negative regulatory element of the HIV-1 LTR. *Journal of Virology* **64**, 3234–3239.

Rosenfeld, P. J. and Kelley, T. J. (1986). Purification of nuclear factor 1 by DNA recognition site affinity chromatography. *Journal of Biological Chemistry* **261**, 1398–1408.

Sambrook, J., Fritsch, E. F. and Maniatis, T. (1989). *Molecular Cloning: A Laboratory Manual*. New York: Cold Spring Harbor Laboratory Press.

Santoro, C., Mermod, N., Andrews, P. C. and Tjian, R. (1988). A family of human CCAAT box binding proteins active in transcription and DNA replication: cloning and expression of multiple cDNAs. *Nature* **334**, 218–224.

Siebenlist, U. and Gilbert, W. (1980). Contacts between the RNA polymerase and an early promoter of phage T7. *Proceedings of the National Academy of Sciences, USA* **77**, 122–126.

Singh, H., Le Bowitz, J. H., Baldwin, A. S. and Sharp, P. A. (1988). Molecular cloning of an enhancer binding protein: isolation by screening of an expression library with a recognition site DNA. *Cell* **52**, 415–429.

Staudt, L. H., Singh, H., Sen, R., Wirth, T., Sharp, P. A. and Baltimore, D. (1986). A lymphoid-specific protein binding to the octamer motif of immunoglobulin genes. *Nature* **323**, 640–643.

Staudt, L. M., Clerc, R. G., Singh, H., Le Bowitz, J. H., Sharp, P. A. and Baltimore, D. (1988). Cloning of a lymphoid-specific cDNA encoding a protein binding the regulatory octamer DNA motif. *Science* **241**, 577–580.

Sturm, R. A., Das, G. and Herr, W. (1988). The ubiquitous octamer-binding protein Oct-1 contains a POU domain with a homeobox subdomain. *Genes and Development* **2**, 1582–1599.

Sturm, R., Baumruker, T., Franza, R. Jr and Herr, W. (1987). A 100–kD HeLa cell octamer binding protein (OBP 100) interacts differently with two separate octamer-related sequences within the SV40 enhancer. *Genes and Development* **1**, 1147–1160.

Treisman, R. (1987). Identification and purification of a polypeptide that binds to the c-fos serum response element. *EMBO Journal* **6**, 2711–2717.

Vinson, C. R., La Marco, K. L., Johnson, P. F., Landschulz, W. H. and McKnight, S. Z. (1988). In situ detection of sequence-specific DNA binding activity specified by a recombinant bacteriophage. *Genes and Development* **2**, 801–806.

Young, R. A. and Davis, R. W. (1983). Yeast polymerase II genes: isolation with antibody probes. *Science* **222**, 778–782.

CHAPTER THREE

Transcription factors and constitutive transcription

3.1 RNA POLYMERASES

Transcription involves the polymerization of ribonucleotide precursors into an RNA molecule using a DNA template. The enzymes which carry out this reaction are known as RNA polymerases. In eukaryotes three different enzymes of this type exist which are active on different sets of genes and can be distinguished on the basis of their different sensitivities to the fungal toxin α-amanitin (Table 3.1) (reviews: Lewis and Burgess, 1982; Sentenac, 1985). All

Table 3.1 Eukaryotic RNA polymerases

	Genes transcribed	Sensitivity to α-amanitin
I	Ribosomal RNA (45*S* precursor of 28*S*, 18*S* and 5.8*S* rRNA)	Insensitive
II	All protein-coding genes, small nuclear RNAs U1, U2, U3, etc	Very sensitive (inhibited 1 μg/ml)
III	Transfer RNA, 5*S* ribosomal RNA, small nuclear RNA U6, repeated DNA sequences: Alu, B1, B2, etc., 7SK, 7SL RNA.	Moderately sensitive (inhibited 10 μg/ml)

the genes which code for proteins as well as those encoding some of the small nuclear RNAs involved in splicing are transcribed by RNA polymerase II. Because of the very wide variety of regulatory processes which these genes exhibit, much of this book is concerned with the interaction of different transcription factors with RNA polymerase II. Information is also available, however, on the interaction of such factors with RNA polymerase I, which transcribes the genes encoding the 28*S*, 18*S* and 5.8*S* ribosomal RNAs (Sommerville, 1984), and with RNA polymerase III, which transcribes the transfer RNA and 5*S* ribosomal RNA genes (Cilberto *et al.*, 1983). These interactions are therefore discussed where appropriate.

All three RNA polymerases are large multi-subunit enzymes, RNA polymerase II, for example, having 8–10 subunits with sizes ranging from 240 to 10 kDa in size (Sentenac, 1985; Saltzman and Weinmann, 1989). Interestingly, the cloning of the genes encoding the largest subunits of each of the three polymerases has revealed that they show homology to one another (Memet *et al.*, 1988). Similarly, chemical labelling experiments have indicated that the second largest subunit of each polymerase contains the active site of the enzyme (Riva *et al.*, 1987), whilst recent studies have shown that at least three smaller, non-catalytic subunits are shared by the three yeast polymerases (Woychik *et al.*, 1990). Such relationships evidently indicate a basic functional similarity between the three eukaryotic RNA polymerases and may also be indicative of a common evolutionary origin.

In addition to the conservation of function between the three eukaryotic enzymes, each individual enzyme exhibits a strong conservation between different organisms. Thus the largest subunit of the mammalian RNA polymerase II enzyme is 75% homologous to that of the fruit fly *Drosophila* (Saltzman and Weinmann, 1989) and also shows homology to the equivalent enzymes in yeast (Memet *et al.*, 1988) and even *E. coli* (Ahearn *et al.*, 1987). Interestingly, all the eukaryotic RNA polymerase II enzymes contain a repeated region at the C-terminus of the largest subunit which contains multiple copies of the sequence Tyr-Ser-Pro-Thr-Ser-Pro-Ser. This sequence is unique to the largest subunit of RNA polymerase II and is present in multiple copies, being repeated 52 times in the mouse protein (Ahearn *et al.*, 1987) and 26 times in the yeast protein (Allison *et al.*, 1985). As expected from its evolutionary conservation, the repeated region is essential for the proper functioning of the enzyme and hence for cell viability, although its size can be reduced to some extent without affecting the activity of the enzyme (Allison *et al.*, 1988; Nonet *et al.*, 1987). The functional role of this domain is unclear, however, although, as will be discussed in Section 9.3.2, it has been suggested that it serves as a target for the activation domains of specific transcription factors which interact with it in order to increase the activity of the polymerase (Sigler, 1988).

Whether this is the case or not, it is clear that whilst the RNA polymerases possess the enzymatic activity necessary for transcription, they cannot function independently. Rather, transcription involves numerous transcription factors which must interact with the polymerase and with each other if transcription is to occur. The role of these factors is to organize a stable transcriptional complex containing the RNA polymerase and which is capable of repeated rounds of transcription.

3.2 THE STABLE TRANSCRIPTIONAL COMPLEX

3.2.1 Characteristics of the stable transcriptional complex

For all three eukaryotic polymerases, the initiation of transcription requires a multi-component complex containing the RNA polymerase and transcription factors. This complex has several characteristics which have led to it being referred to as a stable transcriptional complex (Brown, 1984). These are as follows:

1. The assembled complex is stable to treatment with low concentrations of specific detergents or to the presence of a competing DNA template, both of which would prevent its assembly.
2. The complex contains factors which are necessary for its assembly but not for transcription itself. These factors can therefore be dissociated once the complex has formed without affecting transcription.
3. The complex of RNA polymerase and other factors necessary for transcription is stable through many rounds of transcription, resulting in the production of many RNA copies from the gene.

These characteristics are illustrated in Figure 3.1.

Much of the information on these complexes have been obtained by studying the relatively simple systems of RNA polymerases I and III and applying the information obtained to the RNA polymerase II situation. The stable complex formed by each of these enzymes will therefore be discussed in turn.

3.2.2 RNA polymerase I

The simplest complex known is found for the transcription of the ribosomal RNA genes by RNA polymerase I in *Acanthamoeba* (review: Paule, 1990). In this organism only one transcription

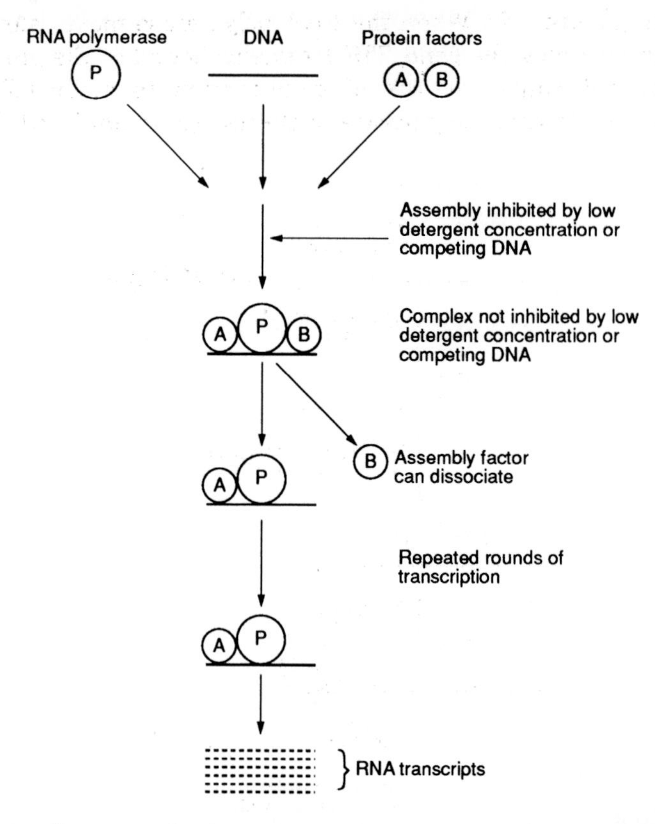

Figure 3.1 Stages in the formation of the stable transcriptional complex. The initial binding of the transcription factor (A) and the assembly factor (B) results in a metastable complex which can be dissociated by low levels of detergent or competing DNA. Following RNA polymerase binding, however, a stable complex is formed. This complex cannot be dissociated by low levels of detergent or competing DNA, is stable through multiple rounds of transcription and retains activity if the assembly factor (B) is removed.

factor, known as TIF-1, is required for transcription by the polymerase (Bateman *et al.*, 1985). This factor binds to the ribosomal RNA promoter, protecting a region from 12 to 70 bases upstream of the transcriptional start site from DNase I digestion. Subsequently the polymerase itself binds to the DNA just downstream of TIF-1, protecting a region between 18 and 52 bases upstream of the start site. Interestingly, binding of the polymerase is not dependent on the specific DNA sequence within this region since it can be replaced with a completely random sequence without affecting binding of the polymerase (Kownin *et al.*, 1990). Hence RNA polymerase is positioned on the promoter by protein–protein interaction with TIF-1 which has previously bound in a sequence-specific manner (Figure 3.2). When the RNA polymerase moves along the DNA transcribing the gene, TIF-1 remains bound at the promoter, allowing subsequent rounds of transcription to occur following binding of another polymerase molecule (Bateman and Paule, 1986).

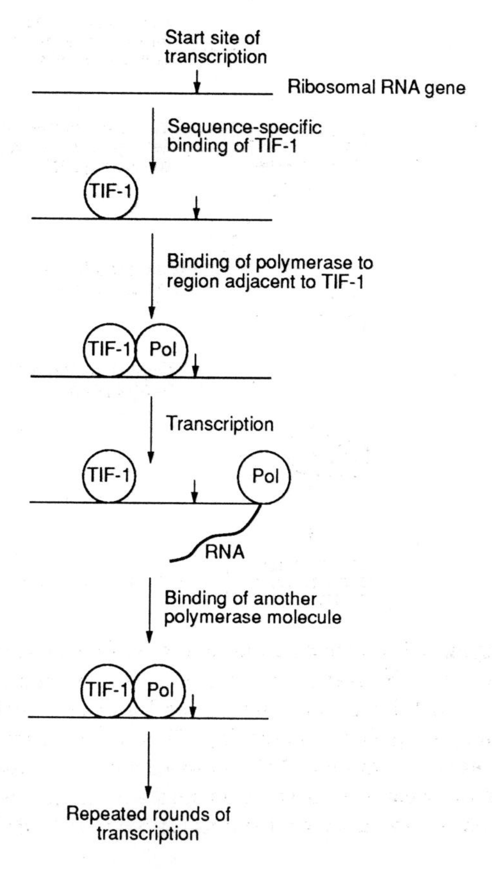

Figure 3.2 Transcription by *Acanthamoeba* RNA polymerase I involves the binding of transcription factor TIF-1 to a specific DNA sequence followed by binding of the polymerase in a non-sequence-specific manner to the DNA region adjacent to TIF-1. When the RNA polymerase moves away as it transcribes the gene, TIF-1 remains bound at the promoter, allowing another RNA polymerase molecule to bind and initiate a new round of transcription.

This system therefore represents a simple one in which one single factor is necessary for transcription and is active through multiple rounds of transcription. In vertebrate rRNA gene transcription, the situation is more complex, however, with an additional factor UBF (upstream binding factor) also being involved. UBF binds specifically to the promoter and upstream elements of the ribosomal RNA genes and stimulates transcription (Jantzen *et al.*, 1990). This is achieved, however, by interaction with the vertebrate TIF-1 homologue, known as SL1. Thus, although a low basal rate of transcription is observed in the absence of UBF (Jantzen *et al.*, 1990), no transcription is detectable unless SL1 is present (Learned *et al.*, 1985). Unlike TIF-1, SL1 does not exhibit sequence-specific binding to the ribosomal RNA promoter (Learned *et al.*, 1986). Hence UBF acts by binding to the DNA in a sequence-specific manner and facilitating the binding of SL1. Thus whilst both SL1 and its homologue TIF-1 act as transcription factors necessary for polymerase I binding, UBF is an additional assembly factor required for binding of SL1 in vertebrates but not of TIF-1 in *Acanthamoeba*. This example therefore illustrates the distinction between factors required only for assembly of the complex or for binding of the polymerase and transcription itself (Figure 3.3).

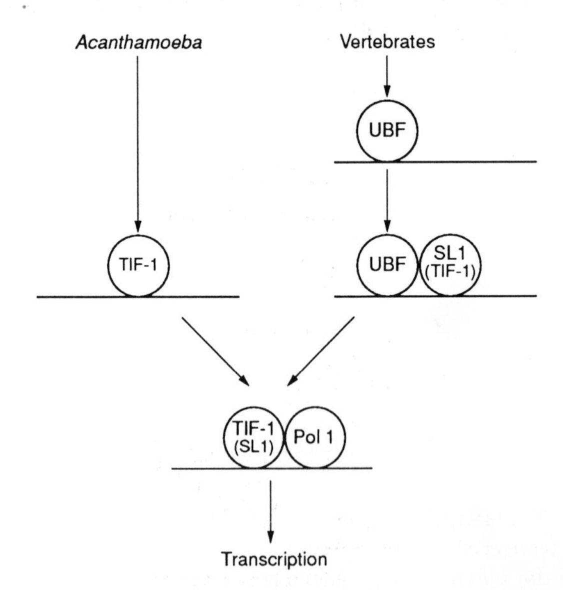

Figure 3.3 Comparison of ribosomal RNA gene transcription in *Acanthamoeba* and vertebrates; in vertebrates transcription requires both the TIF-1 homologue SL1 and an additional assembly factor UBF whose prior binding is necessary for subsequent binding of SL1.

3.2.3 RNA polymerase III

The different roles of transcription factors and assembly factors are also well illustrated by the RNA polymerase III system (reviews:

Cilberto *et al.*, 1983; Paule 1990). In the case of the 5*S* ribosomal RNA genes, for example, transcription by RNA polymerase III required the binding to the DNA of three additional factors, TFIIIA, TFIIIB and TFIIIC. Although both TFIIIA and TFIIIC exhibit the ability to bind to 5*S* DNA in a sequence-specific manner (Sakonju and Brown, 1982; Lassar *et al.*, 1983), TFIIIB, like SL1, cannot do so unless TFIIIC has already bound (Bieker *et al.*, 1980). Once the complex of all these factors has formed and the RNA polymerase has bound, TFIIIA and TFIIIC can be removed and transcription continued with only TFIIIB and the polymerase bound to the DNA (Kassavetis *et al.*, 1990). Hence, like UBF, TFIIIA and TFIIIC are assembly factors which are required for the binding of the transcription factor TFIIIB. In turn, bound TFIIIB is recognized by the polymerase itself and transcription begins (Figure 3.4). As with RNA polymerase I, RNA polymerase III binds to the region of DNA adjacent to that which has bound the transcription factor, binding of the polymerase being independent of the DNA sequence in this region (Sakonju *et al.*, 1980).

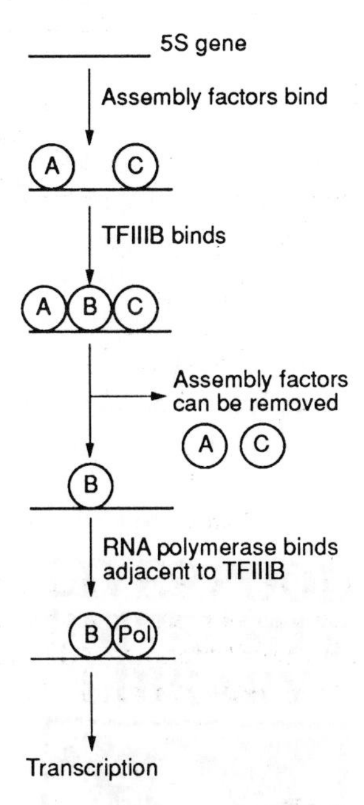

Figure 3.4 Binding of factors to the 5*S* RNA gene. Transcription requires the initial binding of the assembly factors TFIIIA and TFIIIC with subsequent binding of the transcription factor TFIIIB and of RNA polymerase III itself.

Although the transcription of other genes, such as those encoding the tRNAs by RNA polymerase III, is similar to that described for the 5*S* RNA genes, TFIIIA is not required. Rather, transcription is dependent only upon TFIIIB and TFIIIC, with binding of TFIIIC being sufficient for subsequent binding of TFIIIB and the polymerase (Segall *et al.*, 1980).

The process of transcription by RNA polymerases I and III therefore involves the binding of a single transcription factor to the promoter, allowing subsequent binding of the RNA polymerase to an adjacent region of DNA. The transcription factor remains bound at the promoter as the polymerase moves down the DNA, allowing repeated binding of polymerase molecules and hence repeated rounds of transcription. Binding of the polymerase to the promoter requires prior binding of the transcription factor, since the polymerase does not recognize a specific sequence in the promoter but rather makes protein–protein contact with the transcription factor and binds to the adjacent region of the DNA.

In different systems, however, different requirements exist for the binding of the transcription factor itself. Thus in the *Acanthamoeba* system, TIF-1 can bind to DNA in a sequence-specific manner and hence is the only factor required. In the other systems, this is not the case and the transcription factors do not bind to the DNA unless other assembly factors which exhibit sequence-specific DNA binding are present (Table 3.2). Once the transcription factor has bound, these assembly factors can be removed, for example by detergent treatment, without affecting subsequent transcription. It is unclear, however, whether these factors do actually dissociate from the complex under normal conditions *in vivo* once the transcription factor has bound (for discussion see Paule (1990)). Whatever the case, the transcription factor itself remains bound at the promoter even after the polymerase has moved down the gene, allowing repeated binding of polymerase molecules and hence repeated rounds of transcription.

Table 3.2 Transcription factors and assembly factors for RNA polymerase I and III

Gene	Polymerase	Transcription factor	Assembly factor
Acanthamoeba			
ribosomal RNA genes	I	TIF-1	None
Vertebrate			
ribosomal RNA genes	I	SL1	UBF
5*S* rRNA genes	III	TFIIIB	TFIIIA
			TFIIIC
tRNA genes	III	TFIIIB	TFIIIC

Although assembly factors play only an accessory role in transcription itself, they are essential if the complex is to assemble. Hence both assembly factors and transcription factors can be the target for processes which regulate the rate of transcription. Thus whilst the high rate of polymerase III transcription in embryonal carcinoma cells is dependent on a high level of transcription factor TFIIIB (White *et al.*, 1989), the increase in transcription by this polymerase following adenovirus infection is due to an increase in the activity of the assembly factor TFIIIC (Hoeffler *et al.*, 1988). Similarly, alterations in the level of TFIIIA during *Xenopus* development control the nature of the 5*S* rRNA genes which are transcribed at different developmental stages (Sakonju and Brown, 1982).

3.2.4 RNA polymerase II

As noted above, the very wide variety of regulatory events affecting the activity of genes transcribed by RNA polymerase II results in a bewildering array of transcription factors interacting with this enzyme and conferring particular patterns of regulation. These factors will be discussed in Chapters 4 to 7. Interestingly, however, even the basic transcriptional complex which is essential for any transcription by this enzyme contains far more components than is the case for the other RNA polymerases (reviews: Parker, 1989; Saltzman and Weinmann, 1989).

One component of this complex which has been intensively studied and plays an essential role in RNA polymerase II-mediated transcription is TFIID. In promoters containing a TATA box (see Section 1.2.2), TFIID binds to this element, protecting a region from 35 to 19 bases upstream of the start site of transcription in the human *hsp70* promoter, for example (Nakajima *et al.*, 1988). It is also apparently required, however, for the transcription of genes which do not contain a TATA box. Thus it is capable of binding to the promoters of these genes but with less affinity compared to binding to the TATA box (Carcamo *et al.*, 1989).

TFIID is highly conserved in evolution, the equivalent factor from yeast being able to substitute for the mammalian TFIID and allow transcription to occur in an *in vitro* extract from mammalian cells (Buratowski *et al.*, 1988; Cavallini *et al.*, 1988). Interestingly, when this is done, the transcription of the added DNA template begins 25–30 bases downstream of the TATA box, as is typical of mammalian systems, rather than the 50 bases downstream characteristic of the yeast system. Hence, although the TATA box plays a critical role in determining the initiation point of transcription (see Chapter 1), the precise distance between the TATA box and the initiation site must be determined by other factors present in the mammalian extract rather than by TFIID itself.

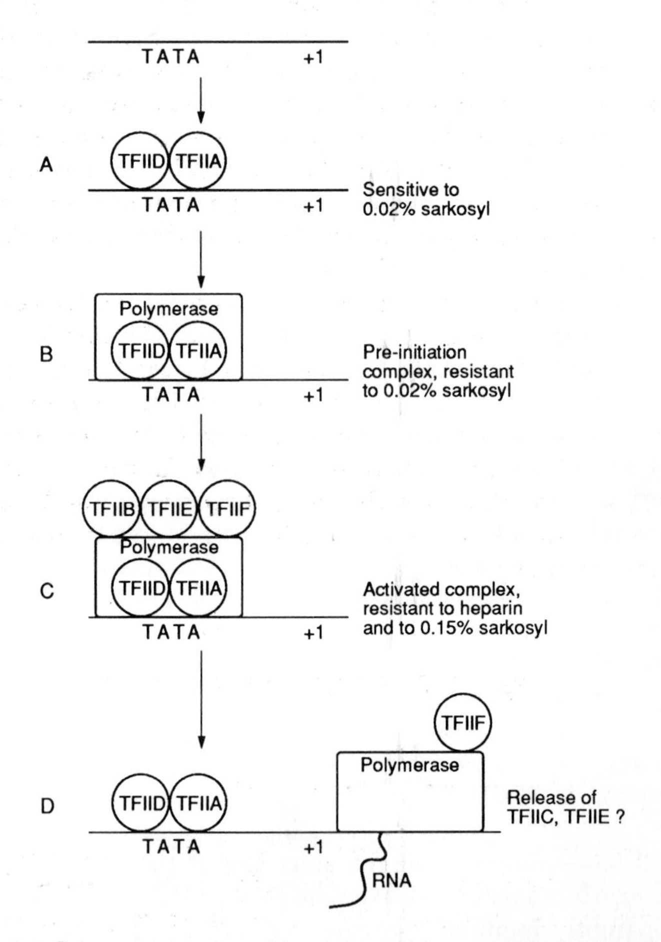

Figure 3.5 Stages in the assembly of the stable transcriptional complex for RNA polymerase II transcription. Note the progressive increase in the stability of the complex as more factors bind. As the polymerase moves away from the promoter to transcribe the gene, TFIIF remains associated with it whilst TFIIA and TFIID remain bound at the TATA box, allowing the formation of a new stable complex and further rounds of transcription.

The binding of TFIID to the TATA box or equivalent region is the earliest step in the formation of the stable transcriptional complex, such binding being facilitated by another factor, TFIIA, which is also essential for transcription (Figure 3.5A; Reinberg *et al.*, 1987). Subsequently RNA polymerase itself binds, forming a pre-initiation complex which is committed to transcribing the DNA template and can do so in the presence of low levels of the detergent sarkosyl (Figure 3.5B; Davison *et al.*, 1983). Following polymerase binding, three other essential transcription factors, TFIIB, TFIIE and TFIIF, rapidly associate with the pre-initiation complex to form an activated complex which is resistant to treatment with heparin or higher concentrations of sarkosyl (Figure 3.5C). This complex of the five factors and the polymerase is sufficient for transcription to occur. As the polymerase moves down the gene during this process, TFIIF remains associated with it whilst TFIIA and TFIID remain bound at the promoter and are capable of binding another molecule of polymerase, allowing repeated rounds of transcription as with the other polymerases (Figure 3.5D).

Hence the assembly of the basic stable transcriptional complex for RNA polymerase II requires at least five factors in addition to the polymerase itself (Reinberg *et al.*, 1987) and is therefore much more complex than that of RNA polymerase I or III. Because of this complexity it remains unclear which of these factors are accessory factors and which are actual transcription factors. It remains possible, therefore, that as with RNA polymerases I and III only a single transcription factor is required for polymerase II binding and subsequent transcription.

3.3 UPSTREAM PROMOTER ELEMENTS

3.3.1 Constitutive transcription factors

The complexity of the RNA polymerase II stable transcriptional complex is paralleled by the existence of numerous factors binding to elements upstream of the TATA box and increasing or decreasing the level of transcription (see Section 1.2). Many of these factors are active only in the presence of an inducing stimulus or in a specific tissue, thereby producing a specific pattern of inducible or tissue-specific gene expression. In addition, however, other factors which are constitutively active bind to specific upstream sequences. The binding of these factors and their interaction with the basal transcriptional complex results in increased levels of transcription in all tissues. This may be dependent on increased stability of the complex in the presence of these factors or to its increased activity. In the absence of such factors, therefore, the basal transcriptional complex can produce only a very low level of transcription and the binding of one or more of these factors is therefore necessary if significant levels of transcription are to occur. Two factors of this type will be discussed, namely those which bind to the Spl and CCAAT box elements found upstream of many promoters.

3.3.2 Spl

The Spl box, which has the consensus sequence GGGCGG, was originally defined in the promoter of the eukaryotic virus SV40, which contains six copies of this motif, each of which binds the transcription factor Spl (Dynan and Tjian, 1983; McKnight and Tjian, 1986). Subsequently this sequence has been found in a variety of other promoters, such as the herpes simplex virus thymidine kinase promoter, the mouse dihydrofolate reductase promoter and the human metallothionein IIA promoter (reviews: La Thangue and Rigby, 1988; Jones *et al.*, 1988). As expected, the Sp1 protein is present in all cell types, two closely related Sp1 polypeptides of 105 kDa and 95 kDa having been purified, for example, from HeLa cells (Kadonga and Tjian, 1986). The addition of these proteins to an *in vitro* transcription reaction specifically stimulates the transcription of genes containing Sp1-binding sites (Dynan and Tjian, 1983), and the single gene encoding them has been cloned (Kadonga *et al.*, 1987). Using the techniques described in Section 2.3.3, the DNA-binding activity of the protein encoded by the Sp1 gene has been shown to be dependent upon a region at the C-terminus which contains three so-called zinc finger motifs (Kadonga *et al.*, 1987) (see Section 8.3.1 for further details of this DNA-binding motif). Similarly, the ability of the protein to activate transcription has been mapped to two glutamine-rich regions of the protein (Courey and Tjian, 1988) (see Section 9.2.2 for further details of this activation motif). Hence the presence of the Sp1 protein in all cell types allows it to bind to its binding sites within specific gene promoters and results in their constitutive activation.

3.3.3 CCAAT box-binding proteins

Like the Sp1 motif, the CCAAT box has been found in a wide variety of promoters such as those of the thymidine kinase and *hsp70* genes and plays an essential role in their activity (reviews: McKnight and Tjian, 1986; Jones *et al.*, 1988). Unlike the Sp1 box, the CCAAT box binds a number of different proteins, some of which are expressed in all tissues whilst others are expressed in a tissue-specific manner (reviews: La Thangue and Rigby, 1988; Johnson and McKnight, 1989).

Initially, two distinct factors binding to this sequence were identified. One of these, CTF (CCAAT box transcription factor), consists of a family of polypeptides of 52–66 kDa in size (Rosenfeld and Kelley, 1986) which are encoded by multiple genes (Rupp *et al.*, 1990) with diversity being increased by alternative splicing of the primary transcripts (Santoro *et al.*, 1988). As well as being able to stimulate transcription of genes containing a CCAAT box, these proteins can also stimulate DNA replication, CTF having been shown to be identical to nuclear factor 1, a protein which can stimulate the replication of adenovirus DNA *in vitro* (Jones *et al.*, 1987). Hence, like SV40 large T antigen and the octamer-binding protein-Oct-1 (also referred to as nuclear factor III), this protein is capable of stimulating both transcription and DNA replication and is hence referred to as CTF/NF1.

The second CCAAT box-binding protein to be defined, C/EBP, differs from CTF/NF1 in a number of properties such as its size (42 kDa), heat stability and the production of a different DNase I footprint on the CCAAT box (Graves *et al.*, 1987; Johnson *et al.*, 1987). Most importantly, however, this protein differs from CTF/NF1 in its sequence specificity. Thus a C to G change in the first base of the CCAAT box impedes biding of CTF/NF1 but actually enhances the binding of C/EBP (Graves *et al.*, 1987). This enhanced binding of C/EBP reflects its ability to bind also to a core motif present in a wide variety of eukaryotic viral enhancers which has the consensus sequence TGTGGA/TA/TA/TG (Johnson *et al.*, 1987). The ability of C/EBP to bind with high affinity to these two distinct motifs led to its being termed CCAAT/enhancer binding protein (C/EBP) (for discussion see Johnson and McKnight (1989)).

The distinction between CTF/NF1 and C/EBP was confirmed when the genes encoding them were cloned and shown to be completely non-homologous to one another (Santoro *et al.*, 1988; Landschulz *et al.*, 1988). Indeed, although these two proteins can bind the identical DNA sequence in the CCAAT box, they do so using two completely different DNA-binding motifs. Thus C/EBP contains the leucine zipper motif with adjacent basic DNA-binding domain common to a number of transcription factors, whilst CTF/NF1 contains a distinct DNA-binding motif not so far described in any other factor (see Chapter 8 for discussion of these and other DNA-binding motifs). Hence sequence-specific binding to the same DNA sequence can be mediated by two distinct DNA-binding structures.

In addition to CTF/NF1 and C/EBP, numerous other CCAAT

box-binding proteins have been reported in various tissues and cell types (e.g. Chodosh *et al.*, 1988a; Rajmondjean *et al.*, 1988). Of particular interest was the report by Chodosh *et al.* (1988a) of two multi-component CCAAT-binding activities. One of these, CP1, was formed by the combined activity of two factors CP1A and CP1B, while the other, CP2, was formed by combining two other factors, CP2A and CP2B. Each of the four individual factors (CP1A, CP1B, CP2A and CP2B) or incorrect combinations (e.g. CP1A and CP2B) were unable to bind to the CCAAT box (Figure 3.6).

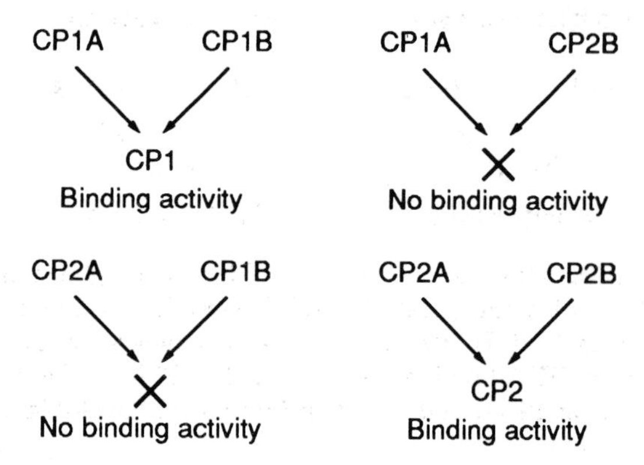

Figure 3.6 Generation of a functional CCAAT box-binding protein by mixing the appropriate subunits of CP1 or of CP2 but not by mixing one subunit of CP1 with a subunit of CP2.

Most interestingly, the yeast catabolite derepression gene *cycl* contains a sequence related to the CCAAT box (CCAAC) which is also bound only by a dimer of two proteins HAP2 and HAP3, resulting in gene activation (Guarente *et al.*, 1984). Mixing of HeLa cell CP1A and yeast HAP2 or of CP1B and HAP3 regenerates a functional CCAAT box-binding activity (Chodosh *et al.*, 1988b; Figure 3.7). Hence HAP2 and HAP3 are the yeast homologues of the human CP1B and CP1A proteins respectively. In agreement with this close relationship of the two pairs of proteins, mutation of the yeast *cycl* gene sequence from CCAAC to CCAAT creating a perfect CCAAT box, increases the binding affinity of the HAP2–HAP3 complex (Guarente *et al.*, 1984). This indicates, therefore, that a pair of proteins capable of interacting with one another and binding to the CCAAT box has been conserved from yeast to humans, such conservation having been sufficiently strong to allow correct interactions of yeast and human subunits.

The existence of multiple CCAAT box proteins clearly suggested that this sequence might play an important role in gene regulation

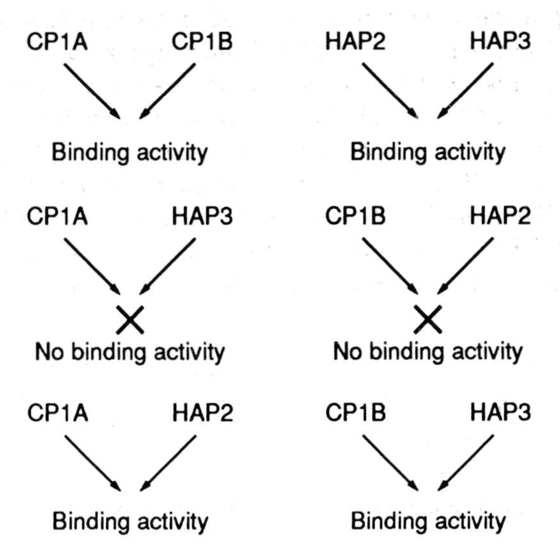

Figure 3.7 Generation of functional CCAAT box-binding protein by mixing subunits of CP1 with the yeast CCAAT box-binding subunits HAP2 and HAP3, indicating that HAP2 is the yeast homologue of CP1B and can therefore form a DNA-binding complex with CP1A whilst HAP3 is the yeast homologue of CP1A and can therefore form a DNA-binding complex with CP1B.

apart from acting simply as an activator of constitutive gene regulation in all tissues, and evidence is available that this is the case. Thus whilst the CTF/NF1 protein, like the Sp1 protein, is present in all tissues, C/EBP is expressed in the liver at much higher levels than in other tissues (Xanthopoulos *et al.*, 1989). Hence, while binding of CTF/NF1 to the CCAAT box is likely to play a role in constitutive expression of particular genes, C/EBP may activate the expression of genes expressed specifically in the liver. In agreement with this, the CCAAT box in several genes expressed specifically in the liver has been shown to play a role in their tissue-specific pattern of expression (e.g. Gorski *et al.*, 1986; Lichtsteiner *et al.*, 1987). Interestingly, the disruption of C/EBP binding by a mutation in its binding site within one such liver-specific gene (that encoding factor IX) has recently been shown to be the cause of the human disease haemophilia B (Crossley and Brownlee, 1990). Similarly, C/EBP is specifically expressed during the terminal phase of adipocyte differentiation and appears to play a critical role in this process (Umek *et al.*, 1991).

As well as functioning in a positive manner, the tissue-specific CCAAT box-binding factors can also be involved in tissue-specific gene regulation by acting negatively to prevent the binding of a constitutively expressed, positively acting factor. Thus in the sea urchin a factor known as the CCAAT displacement protein is present in embyronic tissues and binds to a sequence overlapping the CCAAT box in the sperm histone H2B gene. This binding prevents the binding of the positively acting CCAAT box-binding protein and thereby prevents gene transcription (Figure 3.8). Hence the gene is expressed only in sperm when the displacement protein is absent and not in other tissues even though the positively acting CCAAT box-binding protein is present (Barberis *et al.*, 1987). A negative effect of C/EBP itself on the expression of human hepatitis B virus has also been reported, suggesting that the negative effect of CCAAT box-binding proteins may be widespread (Pei and Shih, 1990).

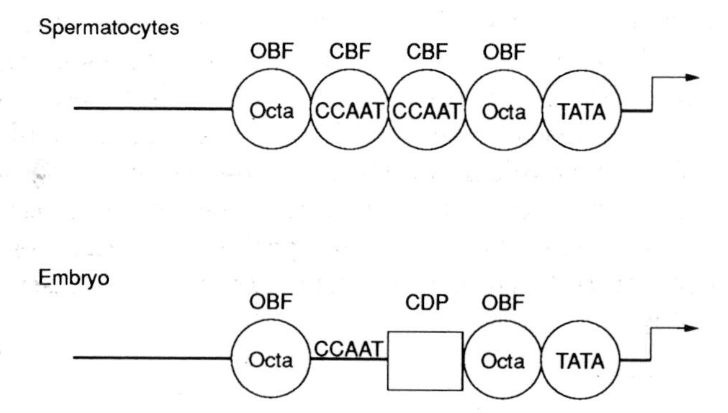

Figure 3.8 Binding of transcription factors to the sea urchin sperm histone H2B gene in spermatocytes where the gene is expressed and in embryos where it is not expressed. Note that the presence of a CCAAT displacement protein (CDP) in embryos prevents the binding of the positively acting CCAAT box-binding factor (CBF) to the two CCAAT boxes in the promoter. Expression of the H2B gene is obligately dependent on the binding of CBF and hence the gene is expressed only in sperm, where it is present even though the octamer-binding factor (OBF) is bound to both copies of the octamer motif and TFIID is bound to the TATA box in both cell types.

It is clear, therefore, that by binding a variety of constitutively expressed and tissue-specific proteins, the CCAAT box can contribute not only to constitutive expression of the genes which contain it but also to a variety of tissue-specific and developmental patterns of gene regulation.

3.4 CONCLUSIONS

The binding of each of the three eukaryotic RNA polymerases to appropriate gene promoters and subsequent transcription is dependent on the prior binding of a specific transcription factor to the promoter. Binding of the polymerase to the DNA adjacent to this factor occurs by recognition of the bound protein rather than by recognition of the specific DNA sequence in this region. In most cases, the binding of the transcription factor itself requires the prior binding of other factors to the DNA. These assembly factors therefore play a critical role in the formation of the stable transcriptional complex but can be dissociated once the complex has formed without affecting its activity. In the case of RNA polymerase II transcription, either the stability of the complex or its activity is greatly affected by the binding of other proteins to sequences upstream of the promoter. Some of these proteins, such as Sp1 and CTF/NF1, which have been discussed in this chapter are present in all tissues and therefore play a role in constitutive gene expression. Others are present or activated only in specific cell types or in response to a specific signal. The role of these proteins in the regulation of inducible or cell type-specific gene expression is discussed in Chapters 4 to 6.

REFERENCES

Ahearn, J. M. Jr, Bartolomei, M. S., West, M. L., Cisek, L. J. and Corden, J. L. (1987). Cloning and sequence analysis of the mouse genomic locus encoding the large subunit of RNA polymerase II. *Journal of Biological Chemistry* **262**, 10695–10705.

Allison, L. A., Moyle, M., Shales, M. and Inglis, C. J. (1985). Extensive homology among the largest subunits of eukaryotic and prokaryotic RNA polymerases. *Cell* **42**, 599–610.

Allison, L. A., Wong, J.K.-C., Fitzpatrick, U. D., Moyle, M. and Inglis, C. J. (1988). The C-terminal domain of the largest subunit of RNA polymerase II of *Saccharomyces cerevisiae, Drosophila melanogaster* and mammals: a conserved structure with an essential function. *Molecular and Cellular Biology* **8**, 321–329.

Barberis, A., Superti-Furga, G. and Busslinger, M. (1987). Mutually exclusive interaction of the CCAAT binding factor and of a displacement protein with overlapping sequences of a histone gene promoter. *Cell* **50**, 347–359.

Bateman, E. and Paule, M. R. (1986) Regulation of eukaryotic ribosomal RNA transcription by RNA polymerase modification. *Cell* **47**, 445–450.

Bateman, E., Lida, C. T., Kownin, P. and Paule, M. R. (1985). Footprinting of ribosomal RNA genes by transcription initiation factor and RNA polymerase I. *Proceedings of the National Academy of Sciences, USA* **82**, 8004–8008.

Bieker, J. J., Martin, P. L. and Roeder, R. G. (1985). Formation of a rate-limiting intermediate in 5*S* RNA gene transcription. *Cell* **40**, 119–127.

Brown, D. D. (1984). The role of stable complexes that repress and activate eukaryotic genes. *Cell* **37**, 359–365.

Buratowski, S., Hahn, S., Sharp, P. A. and Guarente, L. (1988). Function of a yeast TATA element-binding protein in a mammalian transcription system. *Nature* **334**, 37–42.

Carcamo, J., Lobos, S., Merino, A., Buckbinder, L., Weinmann, R., Natarajan, V. and Reinberg, D. (1989). Factors involved in specific transcription by mammalian RNA polymerase II. *Journal of Biological Chemistry* **264**, 7704–7714.

Cavallini, B., Huet, J., Plassat, J.-L., Sentenac, A., Egly, J.-M. and Chambon, P. (1988). A yeast activity can substitute for the HeLa cell TATA box factor. *Nature* **334**, 77–80.

Chodosh, L. A., Baldwin, A. S., Carthew, R. W. and Sharp, P. A. (1988a). Human CCAAT-binding proteins have heterologous subunits. *Cell* **53**, 11–24.

Chodosh, L. A., Olesen, J., Hahn, S., Baldwin, A. S., Guarente, L. and Sharp, P. A. (1988b). A yeast and a human CCAAT-binding protein have heterologous subunits that are functionally interchangeable. *Cell* **53**, 25–35.

Cilberto, G., Castagnoli, L. and Cortese, R. (1983). Transcription by RNA polymerase III. *Current Topics in Developmental Biology* **18**, 59–88.

Courey, A. J. and Tjian, R. (1988). Analysis of Sp1 *in vivo* reveals multiple transcriptional domains including a novel glutamine-rich activation motif. *Cell* **55**, 887–898.

Crossley, M. and Brownlee, G. G. (1990). Disruption of a C/EBP binding site in the factor IX promoter is associated wth haemophilia B. *Nature* **345**, 444–446.

Davison, B. L., Egly, J. M., Mulvhill, E. R. and Chambon, P. (1983). Formation of stable pre-initiation complexes between eukaryotic class B transcription factors and promotoer sequences. *Nature* **301**, 680–686.

Dynan, W. S. and Tjian, R. (1983). The promoter-specific transcription factor Sp1 binds to upstream sequences in the SV40 early promoter. *Cell* **35**, 79–87.

Gorski, K., Carniero, M. and Schibler, U. (1986). Tissue-specific *in vitro* transcription from the mouse albumin promoter. *Cell* **47**, 767–776.

Graves, B. J., Johnson, P. F. and McKnight, S. L. (1987). Homologous recognition of a promoter domain common to the MSV LTR and the HSV tk gene. *Cell* **44**, 565–576.

Guarente, L., Lalonde, B., Gifford, P. and Alani, E. (1984). Distinctly regulated tandem upstream activation sites mediate catabolite repression of the cycl gene of S. cerevisiae. *Cell* **36**, 503–511.

Hoeffler, W. K., Kovelman, R. and Roeder, R. G. (1988). Activation of transcription factor IIIC by the adenovirus E1A protein. *Cell* **53**, 907–920.

Jantzen, H.-M., Admon, A., Bell, S. P. and Tjian, R. (1990). Nucleolar transcription factor UBF contains a DNA binding motif with homology to HMG proteins. *Nature* **344**, 830–836.

Johnson, P. F. and McKnight, S. L. (1989). Eukaryotic transcriptional regulatory proteins. *Annual Review of Biochemistry* **58**, 799–839.

Johnson, P. F., Landschulz, W. H., Graves, B. J. and McKnight, S. L. (1987). Identification of a rat liver nuclear protein that binds to the enhancer core element of three animal viruses. *Genes and Development* **1**, 133–146.

Jones, K. A., Kadonga, J. T., Rosenfeld, P. J., Kelley, T. J. and Tjian, R. (1987). A cellular DNA binding protein that activates eukaryotic transcription and DNA replication. *Cell* **48**, 79–89.

Jones, N. C., Rigby, P. W. J. and Ziff, E. B. (1988). Trans-acting protein factors and the regulation of eukaryotic transcription. *Genes and Development* **2**, 267–281.

Kadonga, J. T. and Tjian, R. (1986). Affinity purification of sequence-specific DNA binding proteins. *Proceedings of the National Academy of Sciences, USA* **83**, 5889–5893.

Kadonga, J. T. Carner, K. R., Masiarz, F. R. and Tjian, R. (1987). Isolation of cDNA encoding the transcription factor Spl and functional analysis of the DNA binding domain. *Cell* **51**, 1079–1090.

Kassavetis, G. A., Braun, B. R., Nguyen, L. H. and Geiduschek, E. P. (1990). *S. cerevisiae* TFIIIB is the transcription initiation factor of RNA polymerase III while TFIIIA and TFIIC are assembly factors. *Cell* **60**, 235–245.

Kownin, P., Bateman, E. and Paule, M. R. (1990). Eukaryotic RNA polymerase 1 promoter binding is directed by protein contacts with transcription initiation factor and is DNA sequence independent. *Cell* **50**, 693–699.

Landschulz, W. H., Johnson, P. F., Adashi, E. Y., Graves, B. J. and McKnight, S. L. (1988). Isolation of a recombinant copy of the gene encoding C/EBP. *Genes and Development* **2**, 786–800.

Lassar, A. B., Martin, P. L. and Roeder, R. G. (1983). Transcription of class III genes: Formation of pre-initiation complexes. *Science* **222**, 740–748.

La Thangue, N. B. and Rigby, P.W.J. (1988). Trans-acting protein factors and the regulation of eukaryotic transcription. In: *Transcription and Splicing* (Hames, B. D. and Glover, D., eds.), pp. 3–42. Oxford: IRL Press.

Learned, R. M., Cordes, S. and Tjian, T. R. (1985). Purification and characterization of a transcription factor that confers promoter specificity. *Molecular and Cellular Biology* **5**, 1358–1369.

Learned, R. M., Learned, T. K., Haltimer, M. M. and Tjian, R. R. (1986). Human RNA transcription is modulated by the coordinate binding of two factors to an upstream control element. *Cell* **45**, 847–857.

Lewis, M. K. and Burgess, R. R. (1982). Eukaryotic RNA polymerases. In: *The Enzymes*, vol. 15 (Boyer, P., ed.), pp. 109–153. New York: Academic Press.

Lichtsteiner, S., Wuari, J. and Schibler, U. (1987). The interplay of DNA-binding proteins on the promotor of the mouse albumin gene. *Cell* **51**, 963–973.

McKnight, S. and Tjian, R. (1986). Transcriptional selectivity of viral genes in mammalian cells. *Cell* **46**, 795–805.

Memet, S., Saurn, W. and Sentenac, A. (1988). RNA polymerases B and C are more closely related to each other than to RNA polymerase A. *Journal of Biological Chemistry* **263**, 10048–10051.

Nakajima, N., Horikoshi, M. and Roeder, R. G. (1988). Factors involved in specific transcription by mammalian RNA polymerase II: purification, genetic specificity and TATA box-promoter interactions of TFIID. *Molecular and Cellular Biology* **8**, 4028–4040.

Nonet, M., Sweetser, D. and Yong, R. A. (1987). Functional redundancy and structural polymorphism in the largest subunit of RNA polymerase II. *Cell* **50**, 909–915.

Parker, C. S. (1989). Transcription factors. *Current Opinion in Cell Biology* **1**, 512–518.

Paule, M. R. (1990). In search of the single factor. *Nature* **344**, 819–820.

Pei, D. and Shih, C. (1990). Transcriptional activation and repression by cellular DNA binding protein C/EBP. *Journal of Virology* **64**, 1517–1522.

Rajmondjean, M., Cereghini, C. and Yaniv, M. (1988). Several distinct 'CCAAT' box binding proteins coexist in eukaryotic cells. *Proceedings of the National Academy of Sciences, USA* **85**, 757–761.

Reinberg, D., Horkushi, M. and Roeder, R. G. (1987). Factors involved in specific transcription in mammalian RNA polymerase II. Functional analysis of initiation factors TFIIA and TFIID and identification of a new factor operating at sequences downstream of the initiation site. *Journal of Biological Chemistry* **262**, 3322–3330.

Riva, M., Schaffner, A. R., Sentenac, A., Hartmann, G. R., Mustner, A. A., Zaychikov, F. and Grachev, M. A. (1987). Active site labelling of the RNA polymerases A, B and C from yeast. *Journal of Biological Chemistry* **262**, 14377–14380.

Rosenfeld, P. J. and Kelley, T. J. (1986). Purification of nuclear factor 1 by DNA recognition site affinity chromatography. *Journal of Biological Chemistry* **261**, 1398–1408.

Rupp, R. A. W., Kruse, U., Multhaup, G., Govel, U., Beyreuther, K. and Sippel, A. E. (1990). Chicken NF1/TGGCA proteins are encoded by at least three independent genes: NF1-A, NF1-B and NF1-C with homologues in mammalian genomes. *Nucleic Acids Research* **18**, 2607–2616.

Sakonju, S. and Brown, D. D. (1982). Contact points between a positive transcription factor and the *Xenopus* 5S RNA gene. *Cell* **31**, 395–405.

Sakonju, S., Bogenhagen, D. F. and Brown, D. D. (1980). A control region in the center of the 5S RNA gene directs specific initiation of transcription. *Cell* **19**, 13–25.

Saltzman, A. G. and Weinmann, R. (1989). Promoter specificity and modulation of RNA polymerase II transcription. *FASEB Journal* **3**, 1723–1733.

Santoro, C., Mermod, N., Andrews, P. C. and Tjian, R. (1988). A family of human CCAAT box binding proteins active in transcription and DNA replication: cloning and expression of multiple cDNAs. *Nature* **334**, 218–224.

Segall, J., Matsui, T. and Roeder, R. G. (1980). Multiple factors are required for the accurate transcription of purified genes by RNA polymerase III. *Journal of Biological Chemistry* **255**, 11986–11991.

Sentenac, A. (1985). Eukaryotic RNA polymerases. *CRC Critical Reviews in Biochemistry* **1**, 31–90.
Sigler, P. B. (1988). Acid blobs and negative noodles. *Nature* **333**, 210–212.
Somerville, J. (1984). RNA polymerase I promoters and cellular transcription factors. *Nature* **310**, 189–190.
Umik, R. M., Freidman, A. D. and McKnight, S. L. (1991). CCAAT-enhancer binding protein: a component of a differentiation switch. *Science* **251**, 288–292.
White, R. J., Stott, P. and Rigby, P. W. J. (1989). Regulation of RNA polymerase III transcription in response to F9 embryonal carcinoma stem cell differentiation. *Cell* **59**, 1081–1092.
Woychik, N. A., Liao, S.-M., Koldrieg, P. A. and Young, R. A. (1990). Subunits shared by eukaryotic RNA polymerases. *Genes and Development* **4**, 313–323.
Xanthopoulos, K. G., Mirkovitch, J., Decker, T., Kuo, C. F. and Darnell, J. E. Jr (1989). Cell-specific transcriptional control of the mouse DNA binding protein mC/EBP. *Proceedings of the National Academy of Sciences USA* **86**, 4117–4121.

CHAPTER FOUR

Transcription factors and inducible gene expression

4.1 INDUCIBLE GENE EXPRESSION

All cells, from bacteria to mammals, respond to various treatment by activating or repressing the expression of particular genes. As discussed in Chapter 1 (Section 1.2.3), genes which are activated in response to a specific treatment share a short DNA sequence in their promoters or enhancers whose transfer to another gene renders that gene inducible by the specific treatment. In turn, such sequences act by binding a specific transcription factor which becomes activated in response to the stimulus. Once activated, this factor interacts with the constitutive transcription factors discussed in Chapter 3, resulting in increased transcription of the gene.

A selection of DNA sequences which enable a gene to respond to a particular stimulus and the transcription factors which they bind is given in Table 4.1 (reviews: Davidson *et al.*, 1983; Jones *et al.*, 1988; Maniatis *et al.*, 1987). Rather than discuss each of these examples individually, we will focus on two cases of inducible gene expression. These two examples, the induction of gene activity by heat shock and by steroid hormones, have been chosen to illustrate differences in the manner in which a transcription factor can become active in response to a signal as well as in how the active factor induces increased transcription of the target gene.

Table 4.1 Sequences that confer response to a particular stimulus

Consensus sequences	Response to	Protein factor	Gene containing sequences
CTNGAATNTT CTAGA	Heat	Heat-shock transcription factor	*hsp70*, *hsp83*, *hsp27*, etc.
T/G T/A CGTCA	Cyclic AMP	CREB / ATF	Somatostatin, fibronectin, α–gonadotrophin, c–*fos*, *hsp70*
TGAGTCAG	Phorbol esters	AP1	Metallothionein IIA, α_1–antitrypsin, collagenase,
GATGTCCATATT AGGACATC	Growth factors in serum	Serum response factor	c–*fos*, *Xenopus* γ–actin
GGTACANNN TGTTCT	Glucocorticoid, progesterone	GR and PR receptors	Metallothionein IIA, tryptophan oxygenase, uteroglobin, lysozyme
AGGTCANNN TGACCT	Oestrogen	Oestrogen receptor	Ovalbumin, conalbumin, vitellogenin
TCAGGTCAT GACCTGA	Thyroid hormone, retinoic acid	TH and RA receptors	Growth hormone, myosin heavy chain
TGCGCCC GCC	Heavy metals	Not known	Metallothionein genes
AAGTGA	Viral infection	Not known	Interferon α and β, tumour necrosis factor

GR, glucorticoid receptor; PR, progesterone receptor; TH, thyroid hormone receptor; RA, retinoic acid receptor

4.2 HEAT-INDUCIBLE TRANSCRIPTION

4.2.1 The heat-shock factor

When cells from a variety of species, ranging from bacteria to mammals, are subjected to elevated temperatures (heat shock) or other stresses such as anoxia, they respond by inducing increased transcription of a small number of genes known as the heat shock-genes (reviews: Lindquist, 1986; Lindquist and Craig, 1988). As discussed in Section 1.2.3, these heat-inducible genes share a common DNA sequence which, when transferred to another gene, can render the second gene heat-inducible (review: Bienz and Pelham, 1987). This sequence is known as the heat shock-promoter element (HSE). The manner in which a *Drosophila* HSE, when introduced into mammalian cells, functioned at the mammalian rather than the *Drosophila* heat-shock temperature suggested that this sequence acted by binding a protein rather than by acting directly as a thermosensor (see Figure 1.3).

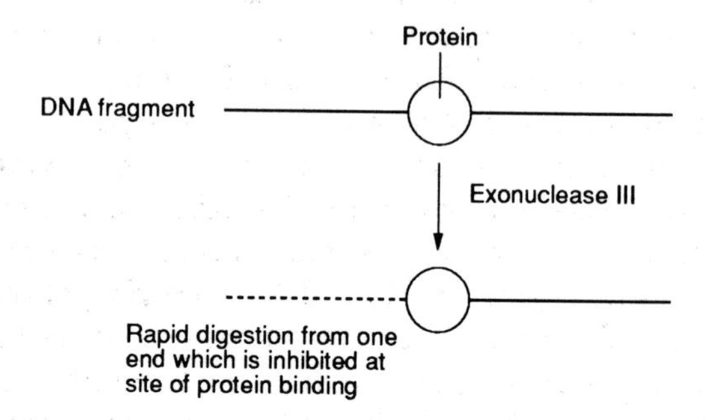

Figure 4.1 Detection of a protein binding to a DNA sequence by inhibition of DNA digestion with exonuclease III.

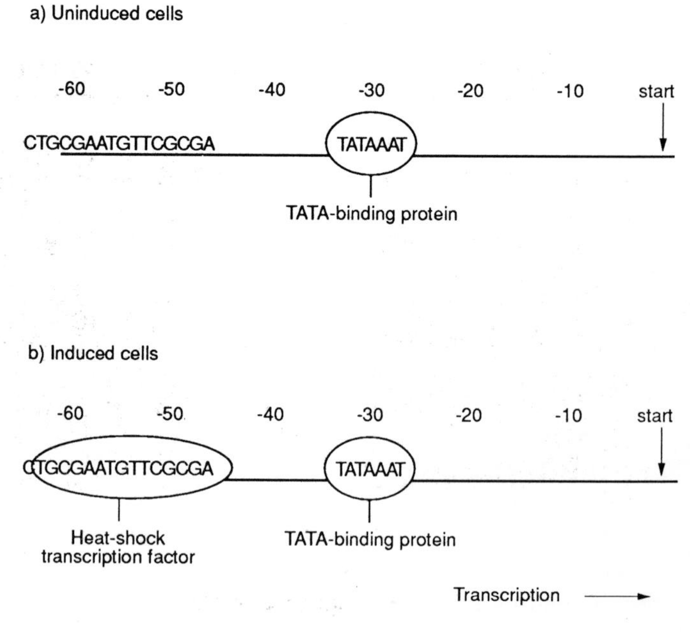

Figure 4.2 Proteins binding to the promoter of the *hsp70* gene before (a) and after (b) heat shock.

Direct evidence that this was the case was provided by Wu (1984, 1985), who used a technique involving digestion of the heat shock-promoter DNA with the enzyme exonuclease III (Figure 4.1). This enzyme will progressively digest DNA, beginning at one end. If a protein is bound to the DNA, however, it will impede the enzyme and hence protein-binding sites on the DNA can be mapped. When this was done using the heat shock-promoter in non-heat-shocked *Drosophila* cells, only a single protein was observed bound to the TATA box region of the DNA (Figure 4.2a).

Following heat shock, however, an additional factor is observed which is bound to the HSE (Figure 4.2b). The amount of this factor bound to the HSE increased with the time of exposure to elevated temperature and with the extent of temperature elevation. Moreover, increased protein binding to the HSE was also observed following exposure to other agents which also induce the transcription of the heat-shock genes such as 2,4–dinitrophenol (Figure 4.3). Thus activation of the heat-shock genes, mediated by the HSE, is accompanied by the binding of a specific transcription factor to this DNA sequence. This factor, which has been variously referred to as the heat-shock transcription factor or heat-shock activator protein, is now generally known as the heat-shock factor (HSF). The gene encoding the yeast factor has been cloned and shown to encode an

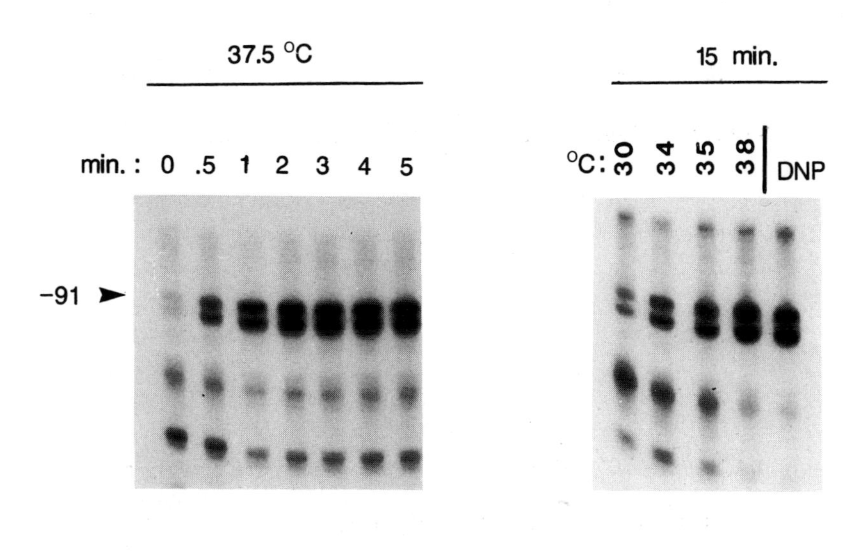

Figure 4.3 Detection of HSF binding to the HSE 91 bases upstream (−91) of the start site for transcription in the *Drosophila hsp82* gene using the exonuclease III assay (see Figure 4.1). Note the increased binding of HSF with increasing time of exposure to heat shock or increased severity of heat shock. HSF binding is also induced by exposure to 2,4-dinitrophenol (DNP), which is known to induce transcription of the heat-shock genes.

833-amino-acid protein including a region of 118 amino acids which mediates its ability to bind to the HSE (Sorger and Pelham, 1988; Wiederrecht *et al.*, 1988).

Hence prior to heat shock, the heat-shock genes are poised for transcription. The binding of a factor (presumably TFIID — see Chapter 3) to the TATA box has resulted in the displacement of the histone-containing nucleosomes from the promoter region and rendered this region exquisitely sensitive to digestion with the enzyme DNase I (Figure 4.4). Although such a DNase I hypersensitive site marks a gene as poised for transcription (review: Gross and Garrard, 1988) it is not in itself sufficient for transcription. This is achieved following heat shock by the binding of the HSE to the HSF. This factor then interacts with TFIID and other components of the basal transcription complex, resulting in the activation of transcription.

The critical role of the HSF in this process obviously begs the question of how this factor is activated in response to heat.

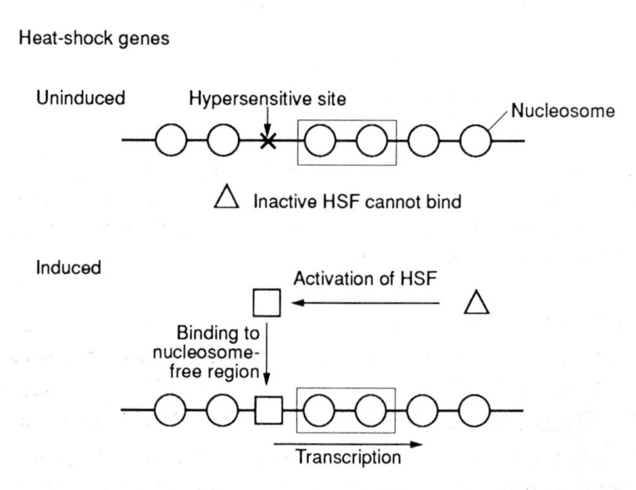

Figure 4.4 Activation of HSF by heat is followed by its binding to a pre-existing nucleosome-free region in the heat-shock gene promoters which is marked by a DNase I hypersensitive site. Binding of HSF then results in the activation of heat-shock gene transcription.

4.2.2 Activation of HSF by heat

If cells are heat-treated in the presence of cycloheximide, which is an inhibitor of protein synthesis, increased binding of HSF to the HSE is observed exactly as in cells treated in the absence of the drug (Zimarino and Wu, 1987). This indicates that the observed binding of HSF following heat shock does not require *de novo* protein synthesis. Rather, this factor must pre-exist in non-heat-treated cells in an inactive form whose ability to bind to the HSE sequence in DNA is activated post-translationally by heat. In agreement with this, activation of HSF can also be observed following heat treatment of cell extracts *in vitro*, when new protein synthesis would not be possible (Parker and Topol, 1984; Larson *et al.*, 1988).

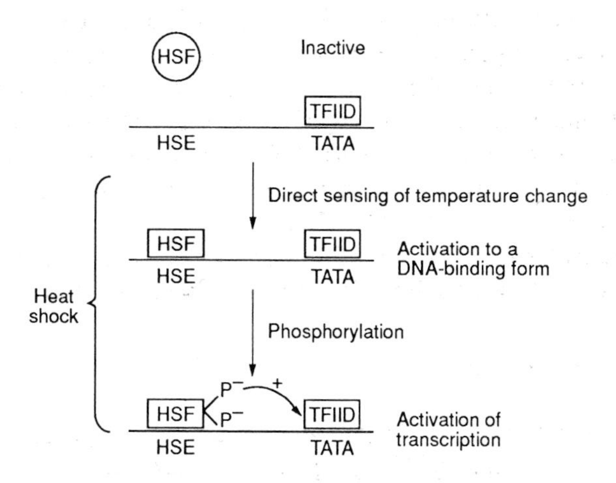

Figure 4.5 Stages in the activation of HSF in mammalian and *Drosophila* cells. Initial activation of HSF to a DNA-binding form following elevated temperature is followed by its phosphorylation, which converts it to a form capable of activating transcription.

Analysis of the activation process using *in vitro* systems from human cells (Larson *et al.*, 1988) has indicated that it is a two-stage process (Figure 4.5). In the first stage, the HSF is activated to a form which can bind to DNA by an ATP-independent mechanism which is directly dependent on elevated temperature. Subsequently this protein is further modified by phosphorylation allowing it to activate transcription. Interestingly, the second of these two stages appears to be disrupted in murine erythroleukaemia (MEL) cells, in which heat shock results in increased binding of HSF to DNA but transcriptional activation of the heat-shock genes is not observed (Hensold *et al.*, 1990).

The two-stage process described above represents a common mechanism for the activation of HSF in higher eukaryotes such as *Drosophila* and mammals. In contrast, however, although the *Saccharomyces cerevisiae* (budding yeast) HSF is indistinguishable from that of higher eukaryotes such as *Drosophila* in size and DNA-binding properties (Wiederrecht *et al.*, 1987), its activation appears to occur by a much simpler mechanism. Thus, unlike *Drosophila* or mammalian HSF, the budding yeast protein can be observed bound to the HSE even in non-heat-shocked cells (Sorger *et al.*, 1987). HSF can activate transcription, however, only following heat treatment, when the protein becomes phosphorylated. Interestingly, in *Schizosaccharomyces pombe* (fission yeast) HSF regulation follows the *Drosophila* and mammalian system, with HSF becoming bound to DNA only following heat shock (Gallo *et al.*, 1991).

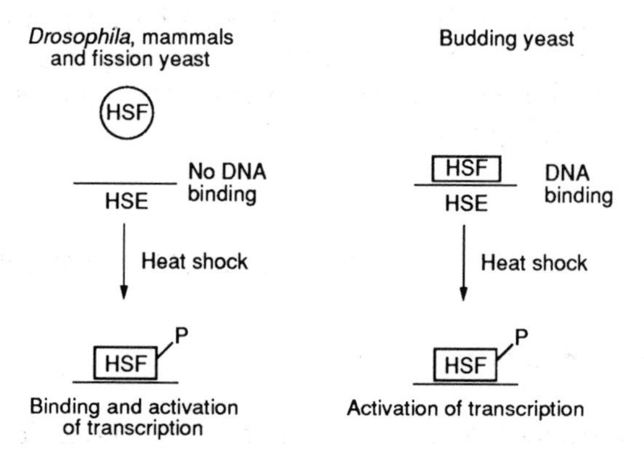

Figure 4.6 HSF activation in *Drosophila*, mammals and fission yeast compared to that in budding yeast. Note that in budding yeast HSF is already bound to DNA prior to heat shock and hence its activation by heat involves only the second of the two stages seen in other organisms, namely its phosphorylation, allowing it to activate transcription.

Hence in mammals, *Drosophila* and fission yeast, activation of HSF is more complex than in budding yeast, involving an initial stage activating the DNA-binding ability of HSF in response to heat as well as the stage, common to all organisms, in which the ability to activate transcription is stimulated by phosphorylation (Figure 4.6). Such phosphorylation is suggested to result in the creation of a highly negatively charged region on the HSF. This region would be analogous to the negatively charged acidic activation domains of other transcription factors (see Section 9.2.2). It would therefore allow HSF to interact with components of the constitutive transcriptional apparatus such as TFIID and activate transcription of the heat-shock genes. In agreement with this idea, HSF has been shown to possess a heat-inducible activation domain (Nieto-Sotelo *et al.*, 1990); for discussion see Parker (1989). Interestingly, this domain can be converted into a form which activates transcription at any temperature by deletion of a seven-amino-acid peptide which is located adjacent to the activation domain within HSF, indicating that this peptide normally inhibits the activity of the activation domain in a temperature-dependent manner (Jakobsen and Pelham, 1991).

In summary, therefore, heat shock induces the increased transcription of a small number of cellular genes via the post-translational modification of a pre-existing transcription factor, HSF. This factor binds to a sequence known as the HSE which is located in a region of the heat-shock gene promoters which is free of nucleosomes and contains a DNase I hypersensitive site even prior to heat shock. The activated form of HSF is capable, following binding, of interacting with components of the constitutive transcriptional apparatus which are bound at other sites in the promoter region and thereby stimulating transcription.

4.3 STEROID-INDUCIBLE TRANSCRIPTION

4.3.1 Steroid receptors

The steroid hormones are a group of substances derived from cholesterol which exert a very wide range of effects on biological processes such as growth, metabolism and sexual differentiation (review: King and Mainwaring, 1974). Early studies using radioactively labelled hormones showed that they act by interacting with specific receptor proteins (review: Jensen and De Sombre, 1972). This binding of hormone to its receptor activates the receptor and allows it to bind to a limited number of specific sites in chromatin. In turn this DNA binding activates transcription of genes carrying the receptor-binding site. Hence, as with the heat-shock factor, these receptor proteins are transcription factors becoming activated in response to a specific signal and in turn activating specific genes (reviews: Yamamoto, 1985; Beato, 1989). These receptor proteins were therefore amongst the earliest transcription factors to be identified, well before the techniques described in Chapter 2 were in routine use, simply on the basis of their ability to bind radioactively labelled steroid ligand.

As with heat shock, genes which are induced by a particular steroid hormone contain a specific binding site for the receptor–hormone complex. The responses to different hormones such as glucocorticoids and oestrogen are mediated by distinct palindromic sequences which are related to one another. In turn such sequences are related to the sequence which mediates induction by the related substances thyroid hormone and retinoic acid (Table 4.2) (review: Beato, 1989).

Table 4.2 Relationship of concensus sequences conferring responsivity to various hormones

Glucocorticoid/progesterone	GGTACANNNTGTTCT
Oestrogen	AGGTCANNNTGACCT
Thyroid hormone/retinoic acid	TCAGGTCA----TGACCTGA

N indicates that any base can be present at this position; a dash indicates that no base is present, the gap having been introduced to align the sequence with the other sequences.

The basis of this binding site relationship was revealed when the genes encoding the receptor proteins were cloned. They were found to constitute a family of genes encoding closely related proteins of similar structure with particular regions being involved in DNA binding, hormone binding and transcriptional activation (Figure 4.7). This has led to the idea that these receptors are encoded by an

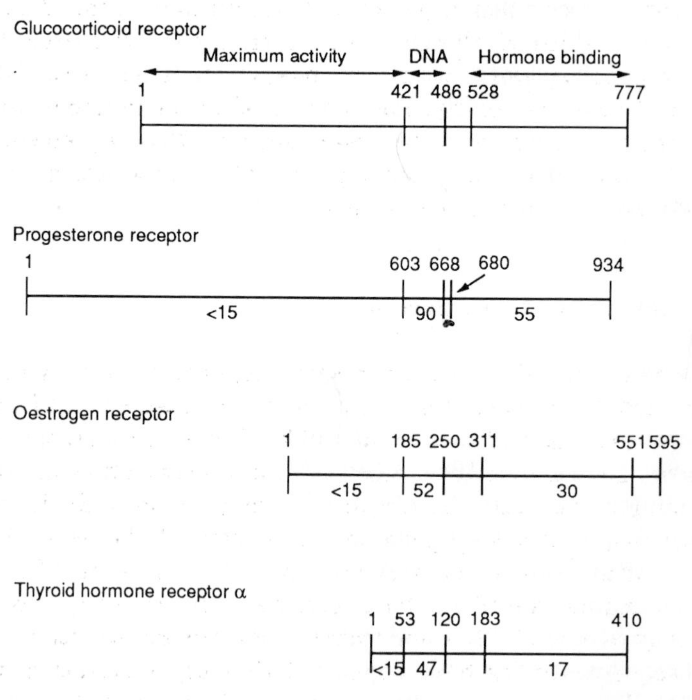

Figure 4.7 Domain structure of individual members of the steroid–thyroid hormone receptor superfamily. The proteins are aligned on the DNA-binding domain, which shows the most conservation between different receptors. The percentage homologies in each domain of the receptors with that of the glucocorticoid receptor are indicated.

evolutionarily related gene family which is known as the steroid–thyroid hormone receptor gene superfamily (reviews: Evans, 1988; Ham and Parker, 1989).

As shown in Figure 4.7, the most conserved region between the different receptors is the DNA-binding domain, explaining the ability of the receptors to bind to similar DNA sequences. Clearly, the differences in amino acid sequences between the DNA-binding motifs in the different receptors will determine the precise DNA sequence which each receptor recognizes. The replacement of particular amino acids in the DNA-binding domain with their equivalents in other receptors has therefore provided considerable information on the features which determine the precise sequence recognized or the optimal spacing between the two halves of the palindromic sequence. These experiments and the so-called 'zinc finger' motifs in the different receptors which mediate their DNA binding are discussed in Section 8.3.2. Interestingly, both DNase I protection and methylation studies support the idea that the receptor binds to DNA as a dimer, each receptor molecule binding to one half of the palindromic recognition sequence (Scheidereit and Beato, 1984).

Having established, therefore, that steroid hormones exert their effect on gene expression via specific receptor proteins, two questions remain. These are, firstly, how does binding of hormone to the receptor activate its ability to bind to specific DNA sequences, and secondly, how does such binding activate transcription? These questions will be considered in turn.

4.3.2 Activation of the receptor

Following identification of the hormone receptors, it was very rapidly shown that the receptors were only found associated with DNA after hormone treatment (Jensen *et al.*, 1968). These early studies were subsequently confirmed by using DNase I footprinting on whole chromatin to show that the receptor was only bound to the hormone response sequence following hormone treatment (Becker *et al.*, 1986). These studies were therefore consistent with a model in which the hormone induces a conformational change in the receptor, activating its ability to bind to DNA and thereby activate transcription.

Subsequent studies have suggested that the situation is more complex, however. Thus, although in the intact cell the receptor binds to DNA only in the presence of the hormone, purified receptor can bind to DNA *in vitro* in a band shift or footprinting assay regardless of whether hormone is present or not (Wilmann and Beato, 1986; Figures 4.8 and 4.9).

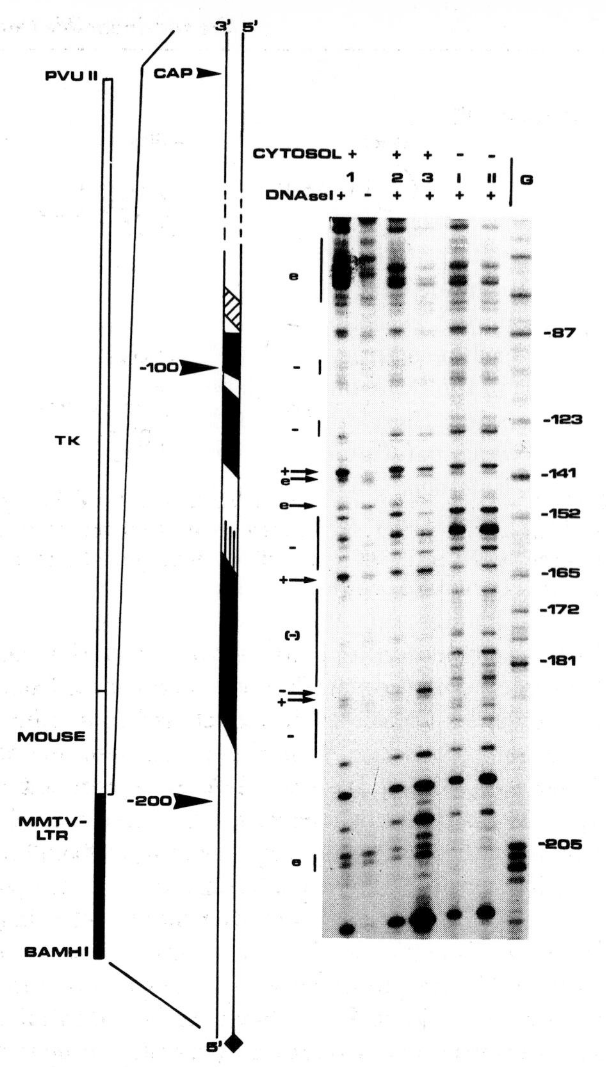

Figure 4.8 DNase I footprint analysis of the binding of the glucocorticoid receptor to the glucocorticoid-inducible mouse mammary tumour virus long terminal repeat promoter (MMTV-LTR). In tracks I and II the DNase I digestion has been carried out without any added receptor. In tracks 1–3, glucocorticoid receptor has been added prior to DNase I digestion, either alone (track 1+), with the glucocorticoid hormone corticosterone (track 2) or with the anti-hormone RU486, which inhibits steroid-induced activation of the receptor (track 3). Track 1− shows the result of adding receptor to the DNA in the absence of DNase I addition in which some cleavage by endogenous nucleases (e) occurs, whilst track G is a marker track produced by cleaving the same DNA at each guanine residue. Minus signs indicate footprinted regions protected by receptor; plus signs are hypersensitive sites at which cleavage is increased by the presence of the receptor. The DNA fragment used and position of the radioactive label (diamond) are shown together with the distances upstream from the initiation site for transcription. Note that the identical footprint is produced by the receptor either alone or in the presence of hormone or anti-hormone. Hence *in vitro* the receptor can bind to DNA in the absence of hormone.

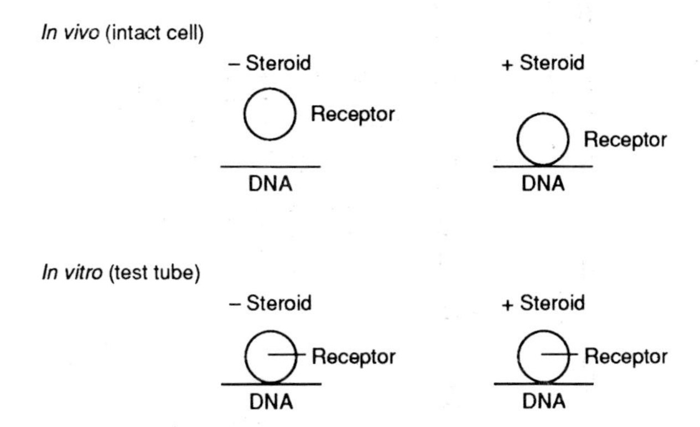

Figure 4.9 Comparison of steroid receptor binding to DNA in the presence or absence of hormone *in vivo* and *in vitro*. Note that whilst *in vivo* DNA binding can occur only in the presence of hormone, *in vitro* it can occur in the presence or absence of hormone.

This discrepancy led to the suggestion that the receptor is inherently capable of binding to DNA but is prevented from doing so in the absence of steroid because it is anchored to another protein. The hormone acts to release it from this association and allow it to fulfil its inherent ability to bind to DNA. In agreement with this possibility, in the absence of hormone the glucocorticoid receptor protein is found in the cytoplasm complexed to a 90 000 molecular weight heat-inducible protein (hsp90) in an 8*S* complex. This complex is dissociated upon steroid treatment, releasing the 4*S* receptor protein (Housley *et al.*, 1985; Denis *et al.*, 1988). The released receptor is free to dimerize and move into the nucleus. Since these processes have been shown to be essential for DNA binding and transcriptional activation by steroid hormone receptors (Tsai *et al.*, 1988; Fawell *et al.*, 1990), dissociation of the receptor from hsp90 is essential if gene activation is to occur. In agreement with this, antiglucocorticoids which inhibit the positive action of glucocorticoids have been shown to stabilize the 8*S* complex of hsp90 and the receptor (Groyer *et al.*, 1987). Similar complexes with hsp90 have also been reported for the other steroid hormone receptors (Joab *et al.*, 1984). Thus the activation of the different steroid receptors by their specific hormones is likely to involve disruption of the protein–protein interaction with hsp90 (Figure 4.10). The manner in which the steroid achieves this effect is discussed further in Section 10.3.3.

In addition to the steroid-induced dissociation of the receptors from hsp90, recent *in vitro* studies have suggested that a second step following dissociation from hsp90 may also be required for receptor

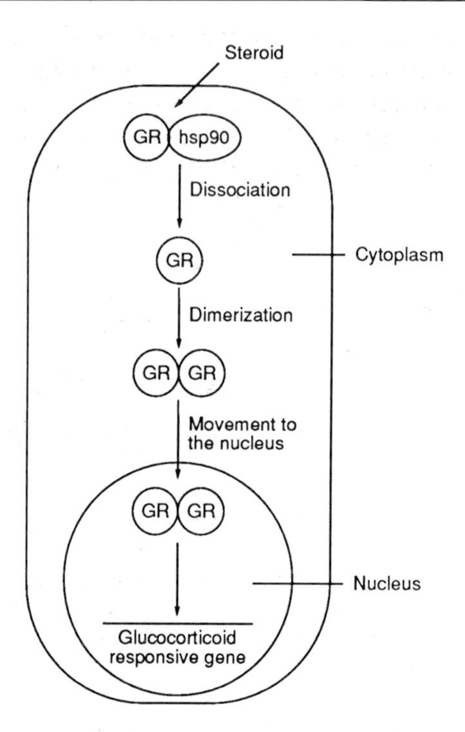

Figure 4.10 Activation of the glucocorticoid receptor (GR) by steroid involves dissociation of hsp90, allowing dimerization and movement to the nucleus.

activation. Thus in a cell-free system in which the progesterone receptor exists in a 4*S* form, free of bound hsp90, the addition of progesterone is still required for the activation of progesterone-responsive genes (Bagchi *et al.*, 1990). This indicates that the hormone has an additional effect on the receptor apart from dissociating it from hsp90. This effect is likely to involve the unmasking of a previously inactive transcriptional activation domain in the receptor, allowing it to activate gene expression in a hormone-dependent manner following DNA binding. Thus, domain-swopping experiments (see Section 2.3.3) have identified regions in both the glucocorticoid and oestrogen receptors which, when linked to the DNA-binding domain of another factor, can activate transcription only following hormone addition. These regions hence constitute hormone-dependent activation domains (Webster *et al.*, 1988). Moreover, in the case of the oestrogen receptor, it has been shown that the oestrogen antagonist 4–hydroxytamoxifen induces the receptor to bind to DNA (presumably by promoting dissociation from hsp90 and dimerization) but does not induce gene activation (Kumar and Chambon, 1988), suggesting that it fails to activate the oestrogen-responsive trans-activation domain. Hence the mechanism by which the receptors are activated is now thought to involve both dissociation from hsp90 and a change in their transcriptional activation ability (Figure 4.11). The various means by which steroid mediates receptor activation are discussed further in Section 10.3.3.

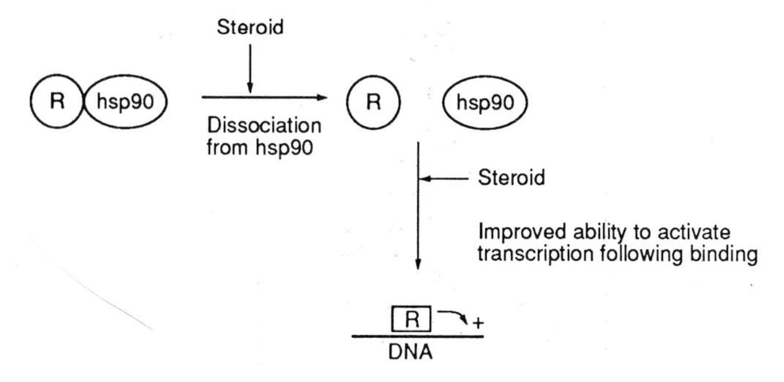

Figure 4.11 Activation of the steroid receptors by treatment with steroid. As well as inducing dissociation of the receptor from hsp90, steroid treatment also increases the ability of the receptor to activate transcription following DNA binding.

The mechanisms of hormone-induced receptor activation are distinct from the mechanism of activation of HSF discussed in Section 4.2.2. Thus in *Drosophila*, mammals and fission yeast, although HSF is present prior to heat shock, in contrast to the steroid receptors, it does not show any DNA-binding ability whether examined *in vivo* or tested *in vitro*. Heat shock acts therefore by activating the ability to bind to DNA of a factor which totally lacks this ability before heat shock. Conversely, in budding yeast HSF is actually bound to the response element prior to heat shock and heat shock has no apparent effect on its DNA binding but only switches on its ability to activate transcription. This is clearly in contrast to the steroid receptors, which are not associated with DNA prior to steroid treatment.

4.3.3 Activation of transcription

Having considered the activation of the receptor by hormone, it is necessary to discuss how binding of the activated receptor in turn activates transcription of the genes bearing its target site.

As discussed in Section 4.2.2. the heat-shock factor binds to a region of chromatin which is free of histone-containing nucleosomes and is hypersensitive to DNase I digestion. This hypersensitive site exists prior to heat shock, marking the potential binding site for the transcription factor. In contrast, in a number of cases, steroid hormone treatment has been shown to cause the induction of a DNase I hypersensitive site located at the DNA sequence to which the receptor binds (Burch and Weintraub, 1983; Becker *et al.*, 1984). Hence the binding of the receptor may activate transcription by displacing a nucleosome from the promoter of the gene, creating the hypersensitive site. In turn this would facilitate the binding of other transcription factors necessary for gene activation. These factors would be present in the cell in an active form prior to steroid treatment but could not bind to the gene because their binding sites were masked by a nucleosome (Figure 4.12). In agreement with this idea, the binding sites for TFIID and CTF/NFI (see Sections 3.2.3 and 3.3.3) in the glucocorticoid-responsive mouse mammary tumour virus promoter are occupied only following hormone treatment although these factors are present in an active DNA-binding form at a similar level in treated and untreated cells (Cordingley *et al.*, 1987).

This mechanism, in which the receptor acts by displacing a nucleosome, allowing constitutive factors access to their binding

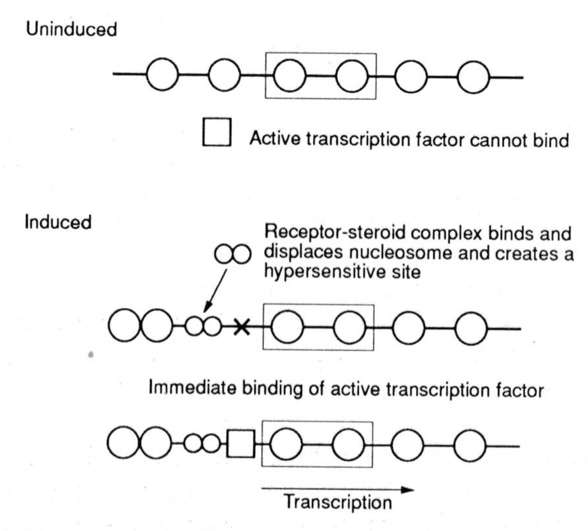

Figure 4.12 Binding of the steroid receptor–steroid hormone complex to the promoter of a steroid-inducible gene results in the displacement of a nucleosome, creating a hypersensitive site and allowing pre-existing constitutively expressed transcription factors to bind and activate transcription. Compare with the binding of HSF to a pre-existing nucleosome-free region illustrated in Figure 4.4.

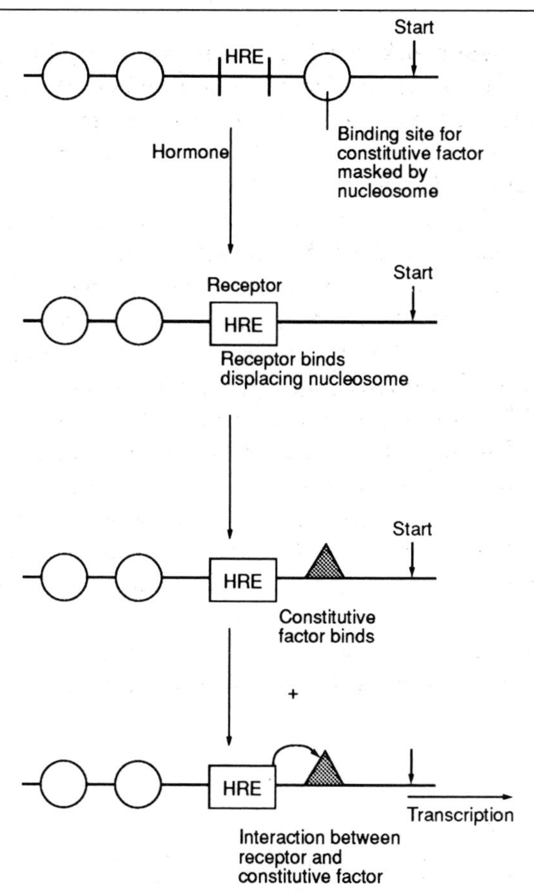

Figure 4.13 Activation of steroid-inducible genes by steroid receptors. As well as displacing a nucleosome and allowing constitutive factors to bind, the hormone receptor is also able to interact directly with constitutive factors and increase their ability to activate transcription.

sites, is clearly in contrast to the binding of HSF to a promoter which already lacks a nucleosome and contains bound TFIID. In this latter case, activation of transcription must occur not via alteration in chromatin structure but via interaction with the components of the constitutive transcriptional apparatus. It should be noted, however, that these two mechanisms are not exclusive. Thus as discussed above (Section 4.3.2) the steroid receptors contain specific activation domains which can activate transcription when linked to the DNA-binding regions of other factors. As discussed in Section 1.3 such activation domains are known to act by interacting with other bound transcription factors to increase transcription. This finding indicates, therefore, that the steroid receptors promote transcription both by altering chromatin structure to allow constitutive factors to bind and also by interacting directly with these other factors (Figure 4.13).

4.3.4 Inhibition of gene expression by steroid hormones

In the preceding sections we have discussed gene activation mediated by steroid hormones complexed to their corresponding receptors. In addition, however, these hormones can inhibit the expression of specific genes, glucocorticoid, for example, inhibiting expression of the genes encoding bovine prolactin (Sakai *et al.*, 1988) and human pro-opiomelanocortin (Israel and Cohen, 1985). The inhibitory effect observed in these cases is mediated by binding to DNA of the identical receptor–hormone complex which activates glucocorticoid-inducible genes. However, the DNA sequence element to which the complex binds when mediating its negative effect (nGRE) is distinct from the glucocorticoid response element (GRE) to which it binds when inducing gene expression, although the two are related (Figure 4.14).

Binding site for positive regulation G G T A C A N N N T G T T C T

Binding site for negative regulation A T Y A C N N N N T G A T C W

Figure 4.14 Relationship of the sites in DNA which mediate gene activation or repression by binding the glucocorticoid receptor. Note that the sites are related but distinct.

This has led to the suggestion that the sequence difference causes the receptor–hormone complex to bind to the nGRE in a configuration in which its activation domain cannot interact with other transcription factors to activate transcription as occurs following binding to the positive element (Sakai *et al.*, 1988; Figure 4.15). The receptor bound in this configuration to the negative element may exert a direct negative influence, resulting in decreased transcription of the gene. More probably, however, it simply acts by preventing binding of a positively acting factor to this or an adjacent site, thereby preventing gene induction. In agreement with this idea, the nGRE in the human glycoprotein hormone α-subunit gene which overlaps a cyclic AMP response element (CRE) is only able to inhibit gene expression when the CRE is left intact. Hence it is likely that receptor bound at the negative element prevents binding of a transcriptional activator to the CRE and thereby inhibits gene expression (Akerblum *et al.*, 1988; Figure 4.16).

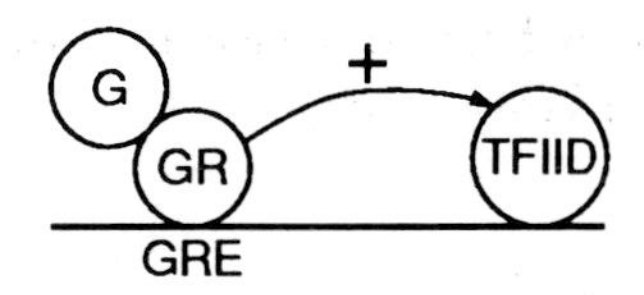

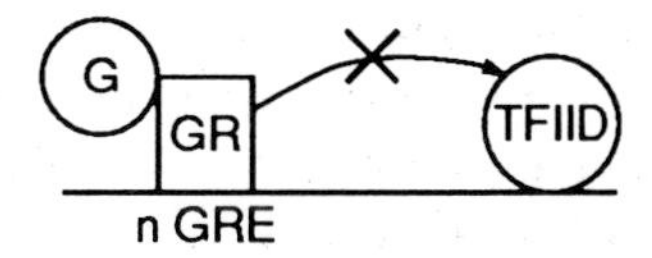

Figure 4.15 Consequences of glucocorticoid receptor binding to the DNA-binding sites which mediate gene activation (GRE) or repression (nGRE). Note that the receptor is likely to bind in a different configuration to the two different sequences, resulting in its ability to activate transcription only following binding to the GRE.

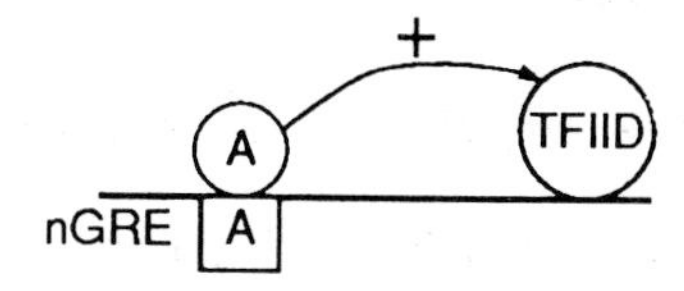

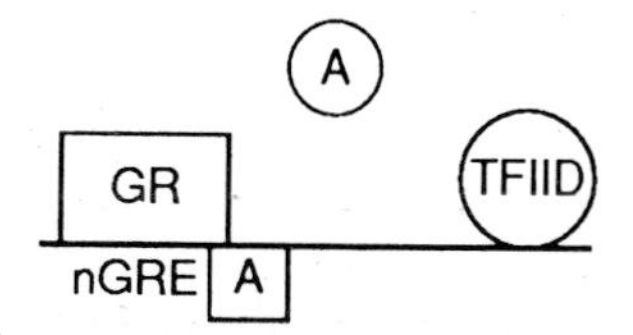

Figure 4.16 Inhibition of gene expression by glucocorticoid receptor binding to an nGRE is likely to be mediated by preventing the binding of a positively acting activator protein (A) to a site adjacent to or overlapping the nGRE.

This mechanism, in which binding of inactive receptor inhibits gene activation by an unrelated transcription factor, can be extended to the interaction between different members of the steroid–thyroid hormone receptor family. For example, as shown in Table 4.2, the binding sites mediating responsiveness to oestrogen (ERE) and thyroid hormone (TRE) are identical in sequence and differ only in the spacing between the two halves of the palindrome which are adjacent in the TRE and separated by three bases in the ERE.

Because of this sequence similarity, the thyroid hormone receptor can bind to an ERE, but, presumably due to a difference in its configuration when bound to a gapped site, it cannot activate transcription. It does, however, inhibit gene activation by oestrogen since it prevents binding of the oestrogen receptor–hormone complex to the ERE (Glass *et al.*, 1988; Figure 4.17). Similarly, the identical structure and sequence of the sites mediating thyroid hormone and retinoic acid responsiveness (Table 4.2) allows the thyroid hormone receptor to bind to a TRE in the absence of thyroid hormone and thereby inhibit gene induction by retinoic acid and its receptor (Graupner *et al.*, 1989; Figure 4.18).

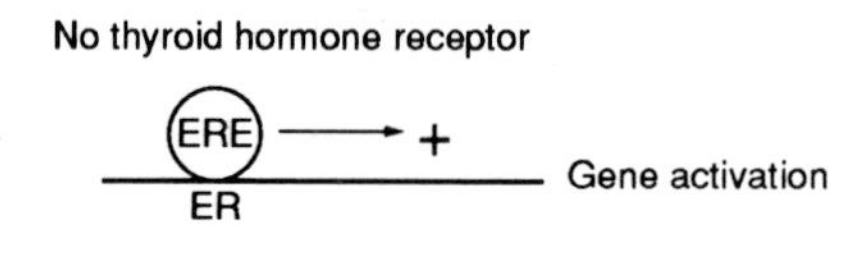

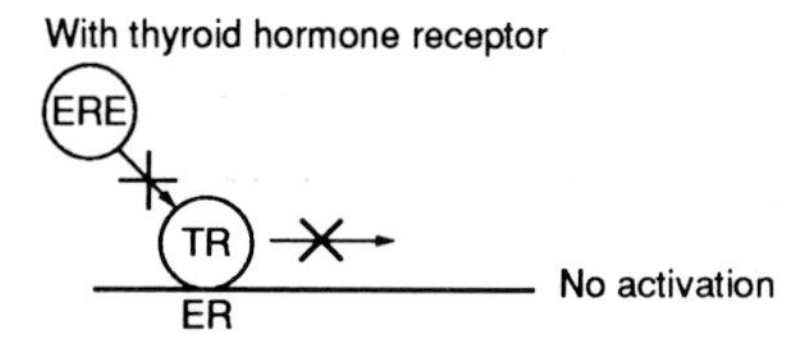

Figure 4.17 The thyroid hormone receptor can bind to a site mediating gene activation by the oestrogen receptor (ERE) but cannot activate transcription. As this binding inhibits binding of the oestrogen receptor itself to the ERE, it results in the inhibition of oestrogen receptor-mediated gene activation.

These examples illustrate, therefore, how interacting networks of transcription factors, each capable of being activated by hormones or other signals, can produce complex patterns of gene activation and repression. Interestingly, however, these interactions can also take place between different forms of the same hormone receptor. Thus, the thyroid hormone receptor exists in two alternatively spliced forms. These two forms differ in their ability to bind thyroid hormone with one form ($\alpha 2$) lacking a part of the hormone-binding domain and, therefore, being unable to bind hormone (Koenig *et al.*, 1989; Figure 4.19a). Both the $\alpha 2$ form and the hormone-binding $\alpha 1$ form can bind to DNA, however. Binding of $\alpha 2$ to the TRE sequence prevents binding of $\alpha 1$ and thereby prevents gene induction in response to thyroid hormone (Figure 4.19b). As discussed in Section 7.3, a similar non-hormone-binding form of the thyroid hormone

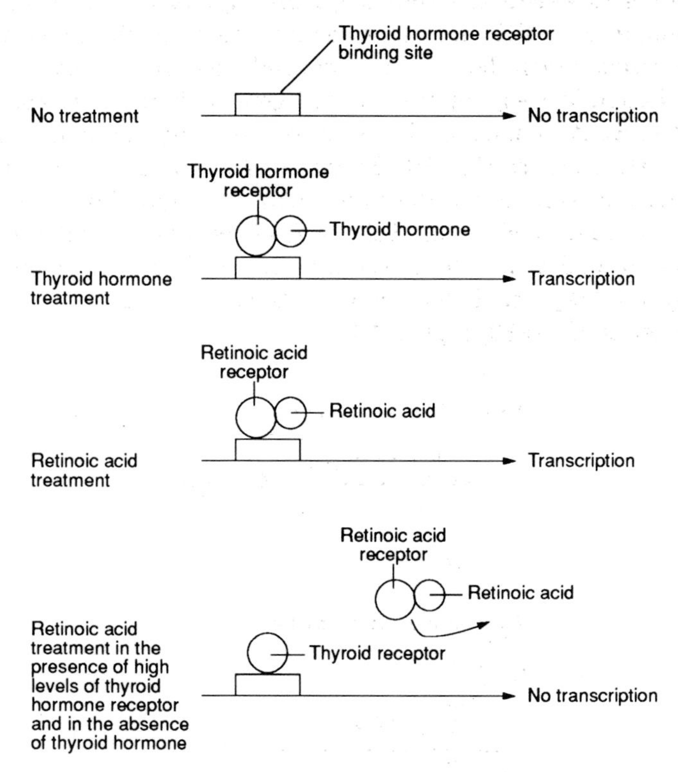

Figure 4.18 The thyroid hormone receptor can bind to its DNA binding site in the absence of thyroid hormone and prevent binding of the retinoic acid–retinoic acid receptor complex to the identical binding site, thereby preventing gene activation in response to retinoic acid.

receptor is encoded by the v-*erbA* oncogene, which produces cancer by inhibiting the expression of thyroid hormone-responsive genes involved in erythroid differentiation.

It should be noted that all the examples of repression by members of the steroid–thyroid receptor family discussed in this section involve their binding to specific sequences in the DNA such as the nGRE or the specific binding sites for other members of the family. It is also possible, however, for receptors such as the glucocorticoid receptor to produce repression without binding to DNA by forming a protein–protein complex with another transcription factor. Cases of this type are discussed in Section 9.3.

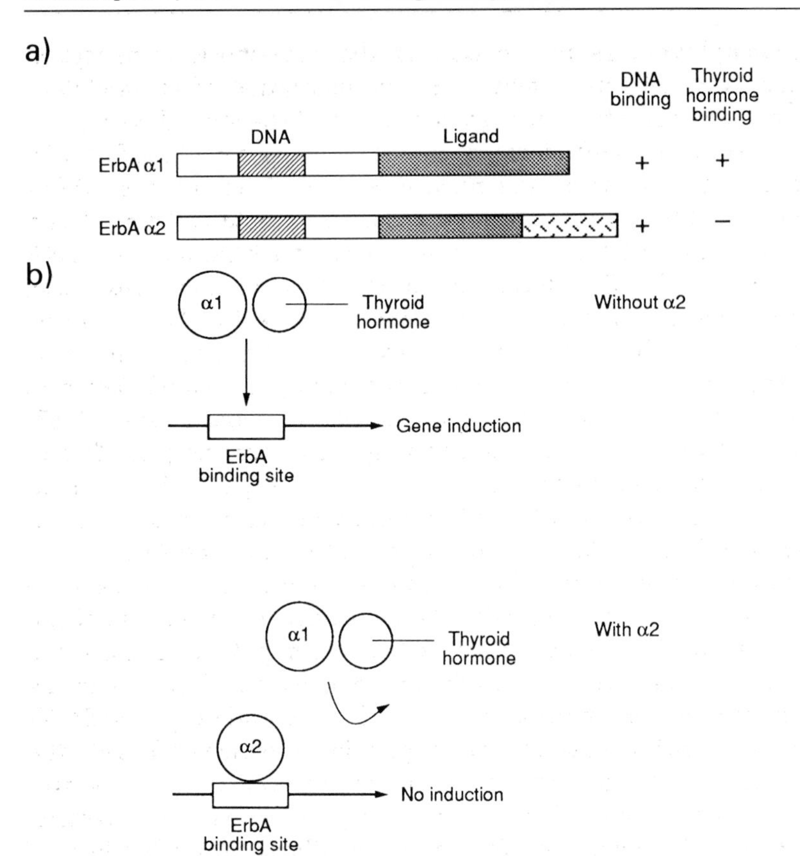

Figure 4.19 (a) Relationship of the ErbA α1 and α2 proteins. Note that only the α1 protein has a functional thyroid hormone-binding domain. (b) Inhibition of ErbA α1 binding and of gene activation in the presence of the α2 protein.

4.4 CONCLUSIONS

This chapter has discussed in detail two systems in which specific genes are switched on in response to a particular inducer. As indicated in Table 4.1, many other inducing agents can activate particular genes containing specific DNA sequences. The examples described here, however, illustrate the general principles by which such induction takes place.

Thus the induction event normally involves the activation of a particular transcription factor which is present in the cell in an inactive form prior to induction. This activation of the transcription factor may take place via a post-translational modification such as phosphorylation, as in the case of the heat-shock transcription factor. Alternatively it may occur via disruption of an inhibitory protein–protein interaction, as in the case of the steroid receptors.

In turn such modifications may produce activation of the factor by allowing it to move to the nucleus and fulfil its inherent DNA-binding ability as in the case of the steroid receptors. Alternatively they may convert the protein from a form which cannot bind DNA into a DNA-binding form, as in the case of the *Drosophila*, mammalian and fission yeast heat-shock factors. Lastly the conversion to an active form may not involve any alteration in DNA binding at all but may enhance the ability of bound factor to enhance transcription, as in the case of the budding yeast HSF. These activating effects of inducing agents on transcription factors will be discussed further in Chapter 10 as part of a general discussion on the activation of transcription factors in both tissue-specific and inducible gene expression and in development.

Once the transcription factor has been activated, it induces gene expression by interacting with the constitutive components of the transcriptional apparatus discussed in Chapter 2 to enhance the stability of the stable transcription complex and hence increase transcription. This process, which is characteristic of virtually all factors mediating inducible gene expression, is observed for both the heat-shock factor and the steroid receptors, which contain activation domains capable of interacting with other transcription factors. In addition, however, the steroid receptors illustrate another facet of gene activation not seen in the heat-shock factor. Thus the binding of the receptor to DNA results in an alteration of chromatin structure involving the displacement of a nucleosome. This exposes the DNA binding sites for constitutive factors such as CTF/NF1, allowing these factors to bind to their exposed binding sites in DNA and activate transcription.

The activation of the various transcription factors in response to the inducing stimulus and the consequent effect of the activated factors on the constitutive transcriptional apparatus, therefore, results in the observed inducible pattern of gene expression.

REFERENCES

Akerblum, I. W., Slater, E. P., Beato, M., Baxter, J. D. and Mellon, P. L. (1988). Negative regulation by glucocorticoids through interference with a cAMP responsive enhancer. *Science* **241**, 350–353.

Bagchi, M. K., Tsai, S. Y., Tsai, M.-J. and O'Malley, B. W. (1990).

Identification of a functional intermediate in receptor activation in progesterone dependent cell-free transcription. *Nature* **345**, 547–550.

Beato, M. (1989). Gene regulation by steroid hormones. *Cell* **56**, 335–344.

Becker, P. B., Gloss, B., Schmid, W., Strahle, U. and Schutz, G. (1986). In vivo protein–DNA interactions in a glucocorticoid response element require the presence of the hormone. *Nature* **324**, 686–688.

Becker, P., Renkawitz, P. and Schutz, G. (1984). Tissue specific DNase I hypersensitive sites in the 5′ flanking sequences of the tryptophan oxygenase and tyrosine aminotransferase genes. *EMBO Journal* **3**, 2015–2020.

Bienz, M. and Pelham, H. R. B. (1987). Mechanism of heat-shock gene activation in higher eukaryotes. *Advances in Genetics* **24**, 31–32.

Burch, J. B. E. and Weintraub, H. (1983). Temporal order of chromatin structural changes associated with activation of the major chicken vitellogenin gene. *Cell* **33**, 64–76.

Cordingley, M. G., Riegel, A. T. and Hager, G. L. (1987). Steroid dependent interaction of transcription factors with the inducible promoter of mouse mammary tumour virus *in vivo*. *Cell* **48**, 261–270.

Davidson, E. H., Jacobs, H. T. and Britten, R. J. (1983). Very short repeats and coordinate induction of genes. *Nature* **301**, 468–470.

Denis, M., Poellinger, L., Wikstom, A.-C. and Gustafsson, J.-A. (1988). Requirement of hormone for thermal conversion of the glucocorticoid receptor to a DNA-binding state. *Nature* **333**, 686–688.

Evans, R. M. (1988). The steroid and thyroid hormone receptor gene superfamily. *Science* **240**, 889–895.

Fawell, S. E., Lees, J. A., White, R. and Parker, M. G. (1990). Characterization and colocalization of steroid binding and dimerization activities in the mouse oestrogen receptor. *Cell* **60**, 953–962.

Gallo, G. J., Schuetz, T. J. and Kingston, R. E. (1991). Regulation of heat shock factor in *Schizosaccharomyces pombe* more closely resembles regulation in mammals than in *Saccharomyces cerevisiae*. *Molecular and Cellular Biology* **11**, 281–288.

Glass, C. K., Holloway, I. M., Devary, O. V. and Rosenfeld, M. G. (1988). The thyroid hormone receptor binds with opposite transcriptional effect to a common sequence motif in thyroid hormone and estrogen response elements. *Cell* **54**, 313–323.

Graupner, G., Wills, K. N., Tzukerman, M., Zhang, X.-K. and Pfahl, M. (1989). Dual regulatory role for thyroid-hormone receptors allows control of retinoic acid receptor activity. *Nature* **340**, 653–656.

Gross, D. S. and Garrard, W. T. (1988). Nuclease hypersensitive sites in chromatin. *Annual Review of Biochemistry* **57**, 159–197.

Groyer, A., Schweizer-Groyer, G., Cadepond, F., Mariller, M. and Baullieu, E. E. (1987). Antiglucocorticosteroid effects suggest why steroid hormone is required for receptors to bind DNA *in vivo* but not *in vitro*. *Nature* **328**, 624–626.

Ham, J. and Parker, M. G. (1989). Regulation of gene expression by hormone receptors. *Current Opinion in Cell Biology* **1**, 503–511.

Hensold, J. O., Hunt, C. R., Calderwood, S. K., Housman, D. E. and Kingston, R. E. (1990). DNA binding of heat shock factor to the heat shock element is insufficient for transcriptional activation in murine erythroleukemia cells. *Molecular and Cellular Biology* **10**, 1600–1608.

Housley, P. R., Sanchez, E. R., Westphal, H. M., Beato, M. and Pratt, W. B. (1985). The molybdate-stabilized L-cell glucocorticoid receptor isolated by affinity chromatography or with a monoclonal antibody is associated with a 90–92 kDa non-steroid binding phosphoprotein. *Journal of Biological Chemistry* **260**, 13810–13817.

Israel, A. and Cohen, S. N. (1985). Hormonally mediated negative regulation of human pro-opiomelanocortin gene expression after transfection into mouse L cells. *Molecular and Cellular Biology* **5**, 2443–2453.

Jakobsen, B. K. and Pelham, H. R. B. (1991). A conserved heptapeptide restrains the activity of the yeast heat shock transcription factor. *EMBO Journal* **10**, 369–375.

Jensen, E. V. and De Sombre, E. R. (1972). Mechanism of action of female sex hormones. *Annual Review of Biochemistry* **41**, 203–230.

Jensen, E. V., Suzuki, T., Kawashima, T., Stumpf, W. E., Jungblut, P. W. and De Sombre, E. R. (1968). A two step mechanism for the interaction of estradiol with rat uterus. *Proceedings of the National Academy of Sciences, USA* **59**, 632–637.

Joab, I., Radanyi, C., Renoir, M., Buchan, T., Catelli, M-G., Binart, N., Mester, J. and Baulieu, E.-E. (1984). Common non-hormone binding component in non-transformed chick oviduct receptors of four steroid hormones. *Nature* **308**, 850–853.

Jones, N. C., Rigby, P. W. J. and Ziff, E. B. (1988). Trans-acting protein factors and the regulation of eukaryotic transcription: lessons from studies on DNA tumour viruses. *Genes and Development* **2**, 267–281.

King, R. J. B. and Mainwaring, W. I. P. (1974). *Steroid Cell Interactions*, London: Butterworths.

Koenig, R. G., Lazar, M. A., Hodin, R. A., Brent, G. A., Larsen, P. R., Chin, W. W. and Moore, D. D. (1989). Inhibition of thyroid hormone action by a non-hormone binding c-erbA protein generated by alternative RNA splicing. *Nature* **337**, 659–661.

Kumar, V. and Chambon, P. (1988). The estrogen receptor binds tightly to its responsive element as a ligand-induced homodimer. *Cell* **55**, 145–156.

Larson, J. S., Schuetz, T. J. and Kingston, R. E. (1988). Activation *in vitro* of sequence-specific DNA binding by a human regulatory factor. *Nature* **335**, 372–375.

Lindquist, S. (1986). The heat-shock response. *Annual Review of Biochemistry* **55**, 1151–1191.

Lindquist, S. and Craig, E. A. (1988). The heat shock proteins. *Annual Review of Genetics* **22**, 631–677.

Maniatis, T., Goodbown, S. and Fischer, J. A. (1987). Regulation of inducible and tissue-specific gene expression. *Science* **236**, 1237–1245.

Nieto-Sotelo, J., Wiederrecht, G., Okuda, A. and Parker, C. S. (1990). The yeast heat shock transcription factor contains a transcriptional activation domain whose activity is repressed under non shock conditions. *Cell* **62**, 807–817.

Parker, C. S. (1989). Transcription factors. *Current Opinion in Cell Biology* **1**, 512–518.

Parker, C. S. and Topol, J. (1984). A *Drosophila* RNA polymerase II transcription factor binds to the regulatory site of an hsp70 gene. *Cell* **37**, 273–283.

Sakai, D. D., Helms, S., Carlstedt-Duke, J., Gustafsson, J.-A., Rottman, F. M. and Yamamoto, K. R. (1988). Hormone-mediated repression: a negative glucocorticoid response element from the bovine prolactin gene. *Genes and Development* **2**, 1144–1154.

Scheidereit, C. and Beato, M. (1984). Contacts between receptor and DNA double helix within a glucocorticoid regulatory element of mouse mammary tumour virus. *Proceedings of the National Academy of Sciences, USA* **81**, 3029–3033.

Sorger, P. K., Lewis, M. J. and Pelham, H. R. B. (1987). Heat shock factor is regulated differently in yeast and HeLa cell. *Nature* **329**, 81–84.

Sorger, P. K. and Pelham, H. R. B. (1988). Yeast heat shock factor is an essential DNA-binding protein that exhibits temperature-dependent phosphorylation. *Cell* **54**, 855–864.

Tsai, S. Y., Carlstedt-Duke, J., Weigel, N. L., Dalman, K., Gustafsson, J. A., Tsai, M.-J. and O'Malley, B. W. (1988). Molecular interactions of steroid hormone receptor with its enhancer element: evidence for receptor dimer formation. *Cell* **55**, 361–369.

Webster, N. J. G., Green, S., Jin, J. R. and Chambon, P. (1988). The hormone-binding domains of the estrogen and glucocorticoid receptors contain an inducible transcription activation function. *Cell* **54**, 199–207.

Wiederrecht, G., Shuey, D. J., Kibbe, W. A. and Parker, C. S. (1987). The *Saccharomyces* and *Drosophila* heat shock transcription factors are identical in size and DNA binding properties. *Cell* **48**, 507–515.

Wiederrecht, G., Seto, D. and Parker, G. (1988). Isolation of the gene encoding the *S. cerevisiae* heat shock transcription factor. *Cell* **54**, 841–853.

Wilmann, T. and Beato, M. (1986). Steroid-free glucocorticoid receptor binds specifically to mouse mammary tumour DNA. *Nature* **324**, 688–691.

Wu, C. (1984). Two protein-binding sites implicated in the activation of heat shock genes. *Nature* **309**, 229–234.

Wu, C. (1985). An exonuclease protection assay reveals heat shock element and TATA-box binding proteins in crude nuclear extracts. *Nature* **317**, 84–87.

Yamamoto, K. R. (1985). Steroid receptor regulated transcription of specific genes and gene networks. *Annual Review of Genetics* **19**, 209–252.

Zimarino, V. and Wu, C. (1987). Induction of sequence-specific binding of *Drosophila* heat-shock activator protein without protein synthesis. *Nature* **327**, 727–730.

CHAPTER FIVE

Transcription factors and cell type-specific transcription

5.1 CELL TYPE-SPECIFIC GENE EXPRESSION

As discussed in Chapter 4, both prokaryotes and eukaryotes respond to specific stimuli by inducing the expression of particular genes, this process being mediated by transcription factors which are activated in response to the stimulus. In eukaryotes, however, transcription factors also play a critical role in processes which have no parallel in prokaryotes. Thus, the higher eukaryote contains a vast range of different cell types, each of which expresses specific genes encoding particular products necessary for the specialized function of that cell type. The role of transcription factors in controlling the cell type-specific expression of particular genes is the subject of this chapter.

A number of different factors which are involved in mediating gene expression in specific cell types have been defined and a selection of these is listed in Table 5.1. These factors are normally synthesized or activated only in one specific tissue, resulting in the cell type-specific transcription of genes whose expression is dependent upon them.

In general such factors have been defined by studying the regulation of a specific, well-characterized gene which is expressed only in a particular cell type and identifying the cell type-specific transcription factor responsible for this pattern of expression. Subsequently, once this factor has been identified, its role in the regulation of other, less well characterized genes expressed in the particular cell type can be assessed. For example, the transcription factor EF-1 (NF-E1), which is expressed at very high levels in erythroid cells,

Table 5.1 Transcription factors regulating cell type-specific gene expression in mammals

Factor	Cell type	Gene regulated	References
EF1	Erythroid	Haemoglobin, porphobillinogen deaminase	Evans *et al.* (1988) Tsai *et al.* (1989)
Isl-1	Islet cells of the pancreas	Insulin	Ohlsson *et al.* (1988) Karlsson *et al.* (1990)
LFB1	Liver	Albumin, α_1-antitrypsin, fibrinogen	Courtois *et al.* (1987) Nicosia *et al.* (1990)
MyoD1	Skeletal muscle	Myosin, creatine Kinase	Davis *et al.* (1987)
NFκB	B cells, activated T cells	Immunoglobulin κ light chain, IL2 α-receptor	Sen and Baltimore (1986a) Nabel and Baltimore (1987)
NFATI	Activated T cells	Interleukin-2	Shaw *et al.* (1988)
Oct-2	B cells	Immunoglobulin heavy and light chains	Singh *et al.* (1986) Muller *et al.* (1988)
Pit-1	Anterior pituitary	Growth hormone	Ingraham *et al.* (1988)

was originally identified on the basis of its binding to the promoter or enhancer regions of chicken and mammalian globin genes (Plumb *et al.*, 1986; Evans *et al.*, 1988) and was shown to be essential for their erythroid-specific pattern of gene expression, deletion or mutation of EF-1 binding sites resulting in the abolition of such expression (Reitman and Felsenfeld, 1988). Subsequently this protein has also been implicated in the erythroid-specific gene expression of the porphobillinogen deaminase gene, which also contains a binding site for the factor within its promoter (Mignotte *et al.*, 1989; Plumb *et al.*, 1989). This factor is likely, therefore, to play a critical role in the regulation of a number of different erythroid-specific genes.

In Section 5.2 we will consider one example of the role of transcription factors in tissue-specific gene expression which has been defined in this way. This involves the B cell-specific expression of the genes encoding the heavy and light chains of the immunoglobulin molecule which is regulated by two transcription factors that are specifically active in B cells, namely NFκB and Oct-2.

However, not all tissue-specific transcription factors have been identified in this manner. Thus, the factor MyoD which activates muscle-specific genes was not identified on the basis of studying genes of this type. Rather it was isolated on the basis of the finding that its artificial expression within undifferentiated fibroblast-like cells was sufficient to transform them into muscle precursor cells

(Davis *et al.*, 1987). Hence, this factor is not only capable of switching on muscle-specific genes but by doing so actually plays a central role in the production of the differentiated cell type itself. This role of MyoD in determining the muscle cell type is discussed in Section 5.3.

The role of transcription factors in producing specific cell types has been particularly well characterized in homothallic yeasts. In this case, whether a cell is a or α in mating type is determined by which transcription factor gene is expressed at the mating type locus. This system, which is discussed in Section 5.4, not only offers insights into cell type-specific gene expression but also provides a possible model system for studies on the developmental regulation of gene expression by transcription factors which will be discussed in Chapter 6.

5.2 TRANSCRIPTION FACTORS AND IMMUNOGLOBULIN GENE EXPRESSION

5.2.1 B cell-specific expression of the immunoglobulin genes

The B lymphocytes which circulate in the blood of higher vertebrates have as their primary role the production of antibodies to defend the body against foreign organisms etc. These antibodies are produced by the association of two immunoglobulin heavy chains and two immunoglobulin light chains to produce a functional antibody molecule (for a recent review of the immune system, emphasizing molecular aspects, see Watson *et al.* (1987)). Consistent with the need to produce antibodies only in B cells, the separate genes encoding the heavy and light chains are expressed in a B cell-specific manner.

This B cell specificity is produced by the combined action of a promoter element located upstream of the transcribed region and an enhancer element located within the intervening sequence (intron) separating the exons encoding the variable and constant regions of the molecule (Figure 5.1). Both these elements, if tested in isolation by linkage to a marker gene, are capable of driving high-level expression in B cells and are hence both B cell-specific (Gillies *et al.*, 1983; Mason *et al.*, 1985). In combination, however, the promoter and the enhancer which acts on it constitute a very powerful B cell-specific regulatory element and produce the high-level expression of the immunoglobulin genes observed in B cells (Garcia *et al.*, 1986).

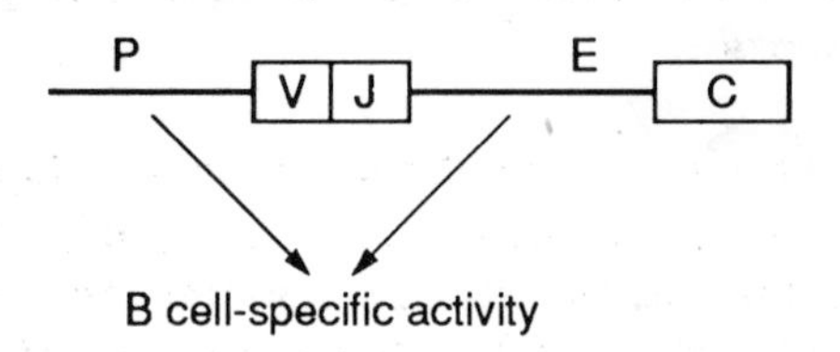

Figure 5.1 Structure of the rearranged immunoglobulin genes in a B cell. The B cell-specific promoter (P) is located adjacent to the DNA segments encoding the variable (V) and joining (J) regions of the molecule. These DNA segments are separated from the DNA segment encoding the constant (C) region of the immunoglobulin molecule by an intervening sequence (intron) which contains the B cell-specific enhancer (E).

Interestingly, the promoter and enhancer are separated by a large distance in gene-line DNA and are brought together by a DNA rearrangement event which allows the enhancer to act on the promoter, producing maximal gene expression (Figure 5.2). The role of this rearrangement in gene regulation is entirely secondary, however, to its primary role in producing antibody diversity by the combination of different DNA sequences encoding different variable, joining and constant regions to produce different antibody molecules (review: Tonegawa, 1983).

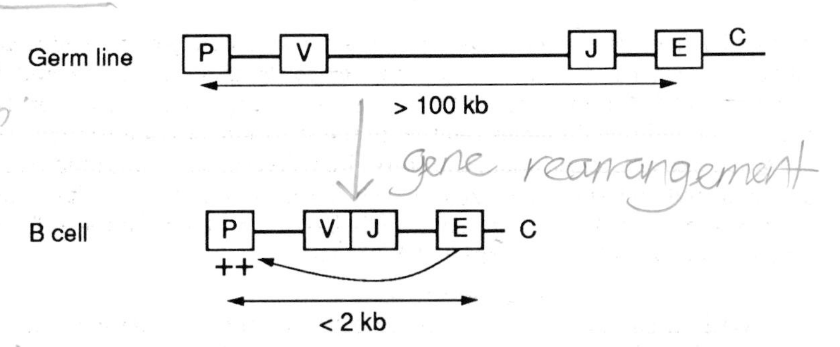

Figure 5.2 The rearrangement of the heavy-chain gene brings an enhancer (E) in the intervening sequence between J and C close to the promoter (P) adjacent to V, and results in the activation of the promoter and gene transcription.

Hence the B cell-specific expression of the immunoglobulin genes is controlled by the inherent B cell specificity of both their promoters and enhancers. An understanding of this B cell specificity requires an understanding, therefore, of the transcription factors binding to these promoters and enhancers and their role in gene expression.

5.2.2 Structure and activity of immunoglobulin promoter and enhancer elements

The promoters of the immunoglobulin heavy- and light-chain genes have a relatively simple structure containing only one identifiable transcription factor-binding site in addition to the TATA box (Parslow *et al.*, 1984; Figure 5.3). This octamer motif, which has the consensus sequence ATGCAAAT, is found approximately 70 bases upstream of the transcription initiation site in both light- and heavy-chain genes. Interestingly, the octamer is oriented in the opposite direction relative to the start site of transcription in the heavy- and light-chain genes.

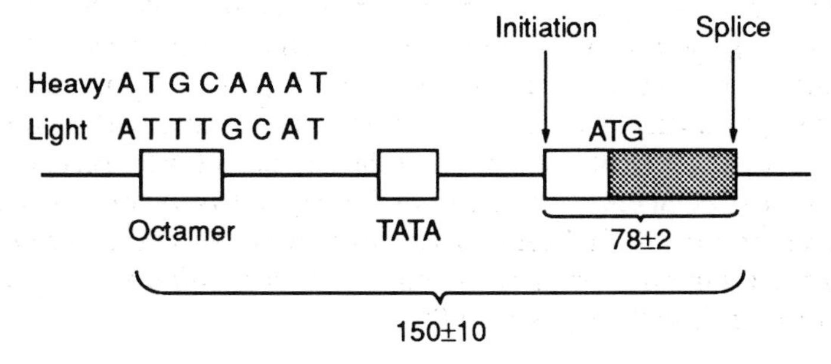

Figure 5.3 The immunoglobulin heavy- and light-chain gene promoter and adjacent transcribed region. The promoter contains a TATA box and an octamer motif. Note the opposite orientation of the octamer motif in the heavy- and light-chain genes relative to the start site of transcription. The initiation site of transcription, the start site (ATG) of the translated region (shaded) and the splice site at the end of the first exon are illustrated. Numbers indicate the average distance (in base pairs) between the different elements in the different immunoglobulin genes.

In both these genes the octamer element plays a critical role in the B cell-specific activity of the promoter. Thus its deletion or mutation, leaving the rest of the promoter intact, abolishes B cell-specific expression (Falkner and Zachau, 1984; Mason *et al.*, 1985). Similarly, linkage of the octamer motif to a non-immunoglobulin promoter results in a B cell-specific expression pattern of the heterologous promoter (Wirth *et al.*, 1987).

Hence the B cell-specific activity of the immunoglobulin promoters is produced simply by their possession of the octamer motif. In DNA mobility shift assays the octamer motif binds two major proteins in B cell extracts (see Figure 2.2) which are known as Oct-1 and Oct-2 (Singh *et al.*, 1986; Staudt *et al.*, 1986). One of these proteins, Oct-1, is present in virtually all cell types, whereas the smaller Oct-2 protein is found in B cells which express the immunoglobulin genes and not in most other cell types (Figure 2.2).

This expression pattern immediately suggests that it is the binding of Oct-2 to the octamer motif in the immunoglobulin promoters which determines their B cell-specific expression. This idea was confirmed when the gene encoding Oct-2 was isolated by using the octamer DNA sequence to screen a B cell cDNA expression library. Thus introduction of an immunoglobulin promoter into non-expressing fibroblast cells with a DNA construct expressing Oct-2 resulted in activation of the immunoglobulin promoter (Muller *et al.*, 1988). This directly confirms that Oct-2 is capable of activating the immunoglobulin promoter. Hence in B cells where Oct-2 is expressed, the immunoglobulin promoter is activated, whereas in other cell types which express only the constitutive protein Oct-1, the promoter is inactive. The reasons for this difference in activity between Oct-1 and Oct-2 are discussed further in Section 9.2.1.

The specific expression of Oct-2 in B cells therefore controls the B cell specificity of the immunoglobulin promoter. This tissue-specific expression of Oct-2 is paralleled by the presence of its corresponding mRNA at high levels in B cells whilst it is absent in most other cell types (Clerc *et al.*, 1988; Muller *et al.*, 1988). Hence, unlike the heat-shock factor or the steroid receptors, Oct-2 is not activated from a pre-existing inactive form when it is required; rather, the Oct-2 mRNA and protein are only synthesized in B cells. Oct-2-dependent expression of the immunoglobulin genes is therefore controlled by the tissue-specific synthesis of Oct-2 rather than by its tissue-specific activation.

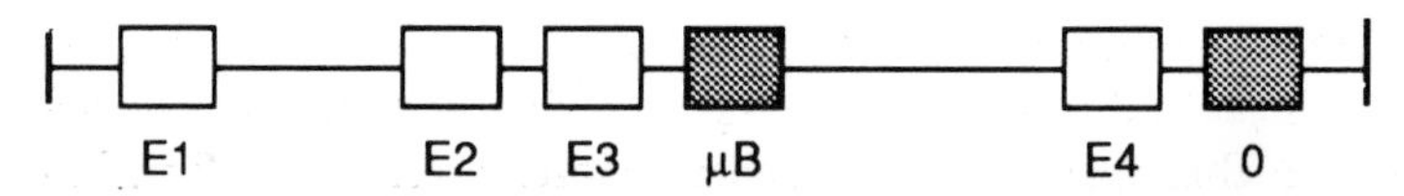

Figure 5.4 Structure of the immunoglobulin heavy-chain gene enhancer. Note the binding sites (shaded) for the B cell-specific factors μB and Oct-2 (0) and the E1–E4 sites (unshaded) which bind constitutive factors.

Most interestingly, the octamer motif is also found within the enhancer element of the immunoglobulin heavy-chain gene (Sen and Baltimore, 1986a). In this situation also, deletion of the octamer diminishes the activity of the enhancer in B cells (Gerster *et al.*, 1987), indicating that the B cell specificity of both immunoglobulin promoters and enhancers involves the octamer motif. Unlike the promoter, however, the enhancer also contains binding sites for several other factors (Sen and Baltimore, 1986a; Figure 5.4). One of these (μB) binds a factor which is found only in B cells (Libermann *et al.*, 1990; Nelson *et al.*, 1990) whilst the others (E1–E4) bind ubiquitous factors present in all cell types (Sen and Baltimore, 1986a). As with the octamer, however, deletion or mutation of the binding sites for these ubiquitous factors reduces the activity of the enhancer in B cells even though the octamer motif is intact (Lenardo *et al.*, 1987). Hence in the heavy-chain gene enhancer, B cell-specific activity is achieved by the interaction of B cell-specific factors and ubiquitous factors rather than, as in the promoter, by the B cell-specific Oct-2 protein alone. Such interactions presumably increase the strength of the enhancer, allowing it to act at a distance from the promoter (see Section 1.2.4).

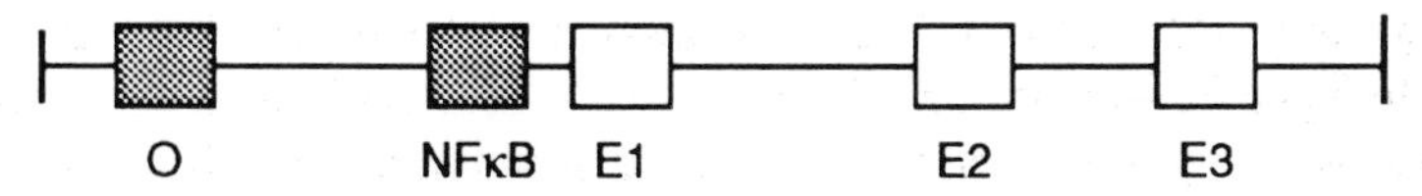

Figure 5.5 Structure of the immunoglobulin light-chain gene enhancer. Note the binding sites (shaded) for the B cell-specific factors NFκB and Oct-2 and the E1–E3 sites (unshaded) which bind constitutive factors.

A similar mixture of binding sites for ubiquitous and B cell-specific proteins is found in the enhancer element of the immunoglobulin κ light-chain genes (Sen and Baltimore, 1986a; Figure 5.5). Thus, for example, the enhancer contains a motif known as E2 which serves as a binding site for two ubiquitously expressed proteins E12 and E47 (Murre *et al.*, 1989a) as well as two other binding sites (E1 and E3) for constitutively expressed factors. In addition, the enhancer contains a binding site for Oct-2 (Nelms and van Ness, 1990) and a binding site for another B cell-specific protein, NFκB (review: Lenardo and Baltimore, 1989). The binding of NFκB to its binding site in the enhancer plays an especially critical role in the expression of the light-chain gene. Thus mutation or deletion of the NFκB binding site abolishes the B cell-specific activity of the light-chain enhancer (Lenardo *et al.*, 1987) whilst, like the octamer motif, it can render a heterologous promoter B cell specific (Pierce *et al.*, 1988).

This suggests, therefore, that binding of NFκB to its site in the enhancer controls the B cell-specific activity of the enhancer. In agreement with this, binding activity can be detected only in B cell lines which express the immunoglobulin κ gene and not in pre-B cells or other cell types which do not. Moreover, treatment of pre-B cells with lipopolysaccharide results in the appearance of NFκB binding activity and concomitantly activates κ gene expression (Atchison and Perry, 1987).

This activation of NFκB by lipopolysaccharide treatment can occur in the presence of the protein synthesis activator cycloheximide (Sen and Baltimore, 1986b). Hence NFκB is present in B cells in an inactive form and can be activated post-translationally without new protein synthesis. Interestingly, this inactive form of NFκB is widely distributed in different cell types and can be activated in both T cells and HeLa cells by treatment with phorbol ester (Sen and Baltimore, 1986b; Nabel and Baltimore, 1987). Although in these cases NFκB activation does not result in immunoglobulin light-chain gene expression, since the gene has not rearranged and is tightly packed within inactive chromatin, it does play a role in gene regulation. Thus the activation of NFκB by agents which activate T cells results in the active transcription factor inducing increased expression of cellular genes such as that encoding the interleukin-2 α-receptor (Bohnlein *et al.*, 1988) and is also responsible for the increased activity of the human immunodeficiency virus promoter in activated T cells (Nabel and Baltimore, 1987). NFκB therefore plays a role not only in B cell-specific gene activity but also in gene activity specific to activated T cells.

Unlike Oct-2, therefore, NFκB is present in all tissues and is activated post-translationally. This occurs, as with HSF, by increased phosphorylation. However, the protein phosphorylated is not NFκB but an inhibitory protein associated with it known as IκB. Phosphorylation of IκB results in its dissociation from NFκB, which is free to dimerize and move to the nucleus in a manner similar to the dimerization and nuclear movement of hsp90-free glucocorticoid receptor following steroid treatment (Baeurele and Baltimore, 1988a,b; Figure 5.6). For further discussion of the mechanisms activating NFκB, see Section 10.3.3.

In summary, therefore, B cell-specific expression of the immunoglobulin genes is controlled by the B cell-specific activity of their promoters and enhancers, which is in turn controlled by the B cell-specific synthesis of Oct-2 and the B cell-specific activation of NFκB.

5.3 MYOD AND THE CONTROL OF MUSCLE-SPECIFIC GENE EXPRESSION

5.3.1 Identification of MyoD

Probably the most novel approach to the cloning of the gene encoding a transcription factor was taken by Davis *et al.* (1987), who

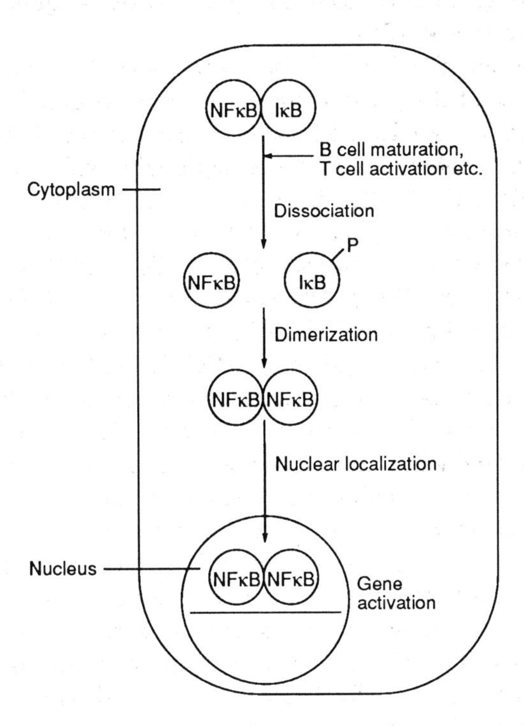

Figure 5.6 Activation of NFκB by dissociation of the inhibitory protein IκB, allowing NFκB to dimerize, move to the nucleus and switch on gene expression. Note that dissociation of IκB from NFκB is caused by its phosphorylation (P).

isolated cDNA clones encoding MyoD, a factor which plays a critical role in skeletal muscle-specific gene regulation. They used an embryonic muscle fibroblast cell line known as C3H 10T1/2. Although these cells do not exhibit any differentiated characteristics, they can be induced to differentiate into myoblast cells expressing a number of muscle-lineage genes upon treatment with 5-azacytidine (Constantinides *et al.*, 1977). This agent is a cytidine analogue having a nitrogen instead of a carbon atom at position 5 on the pyrimidine ring and is incorporated into DNA instead of cytidine. Unlike cytidine, however, it cannot be methylated at this position and hence its incorporation results in demethylation of DNA. As methylation of DNA at C residues is thought to play a critical role in transcriptional silencing of gene expression (reviews: Cedar, 1988; Razin and Riggs, 1980), this artificial demethylation can result in the expression of particular genes which were previously silent.

In the case of 10T1/2 cells, therefore, this demethylation was thought to result in the expression of, previously silent, regulatory loci which are necessary for differentiation into muscle myoblasts. Several experiments also suggested that the activation of only one key regulatory locus might be involved. Thus 5–azacytidine induces myoblasts at very high frequency, consistent with only the demethylation of one gene being required (Konieczny and Emerson, 1984), whilst DNA prepared from differentiated cells can also induce differentiation in untreated cells at a frequency consistent with the transfer of only one activated locus (Lassar *et al.*, 1986).

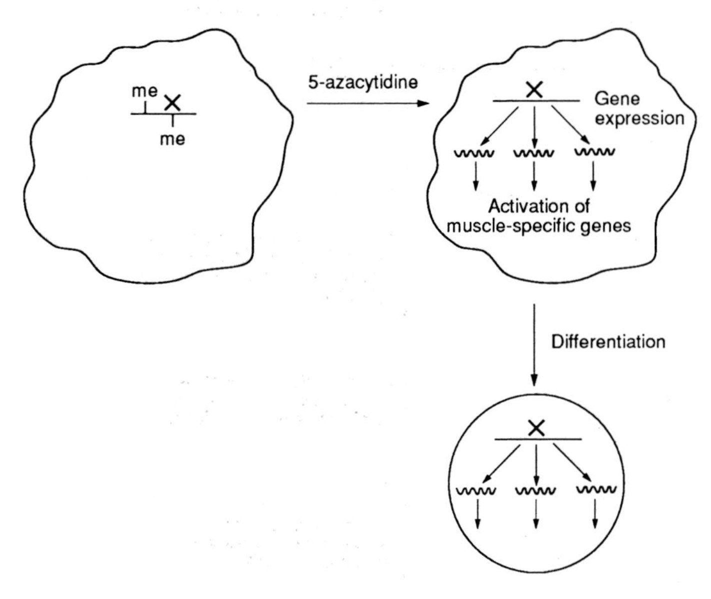

Figure 5.7 Model for differentiation of 10T1/2 cells in response to 5-azacytidine. Activation of a master locus (x) by demethylation allows its product to activate the expression of muscle-specific genes, thereby producing differentiation.

Hence differentiation is thought to occur via the activation of one regulatory locus (gene X in Figure 5.7) whose expression in turn switches on the expression of genes encoding muscle-lineage markers which is observed in the differentiated 10T1/2 cells and thereby induces their differentiation. This suggested that the regulatory locus might encode a transcription factor which switched on muscle-specific gene expression.

To isolate the gene encoding this factor, Davis *et al.* (1987) reasoned that it would continue to be expressed in the myoblast cells but would evidently not be expressed in the undifferentiated cells. They therefore prepared RNA from the differentiated cells and

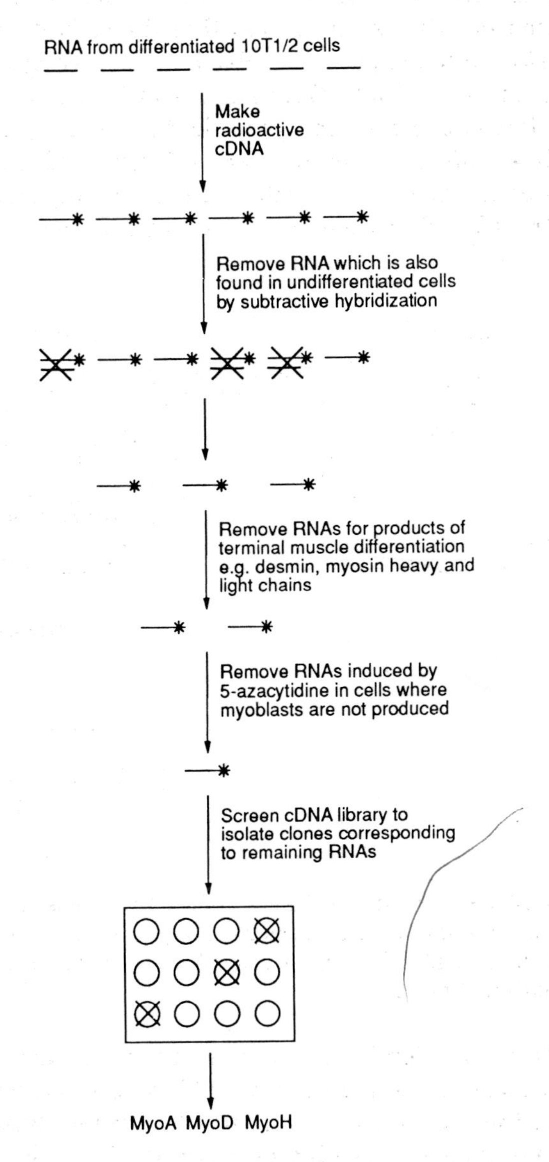

Figure 5.8 Strategy for isolating the master regulatory locus expressed in 10T1/2 cells after but not before treatment with 5-azacytidine. Subtractive hybridization was used to isolate all RNA molecules which are present in 10T1/2 cells only following treatment with 5-azacytidine. After removal of RNAs for terminal differentiation products of muscle and RNAs induced in non-muscle-producing cells by 5-azacytidine, the remaining RNAs were used to a screen a cDNA library. Three candidates for the master regulatory locus, MyoA, MyoD and MyoH, were isolated in this way.

removed from it by subtractive hybridization all the RNAs which were also expressed in the undifferentiated cells. After various further manipulations to exclude RNAs characteristic of terminal muscle differentiation, such as myosin and others induced non-specifically in all cells by 5-azacytidine, the enriched probe was used to screen a cDNA library prepared from differentiated 10T1/2 cells.

This procedure (Figure 5.8) resulted in the isolation of three clones, *MyoA, MyoD* and *MyoH* whose expression was specifically activated when 10T1/2 cells were induced to form myoblasts with 5-azacytidine. When each of these genes was artificially expressed in

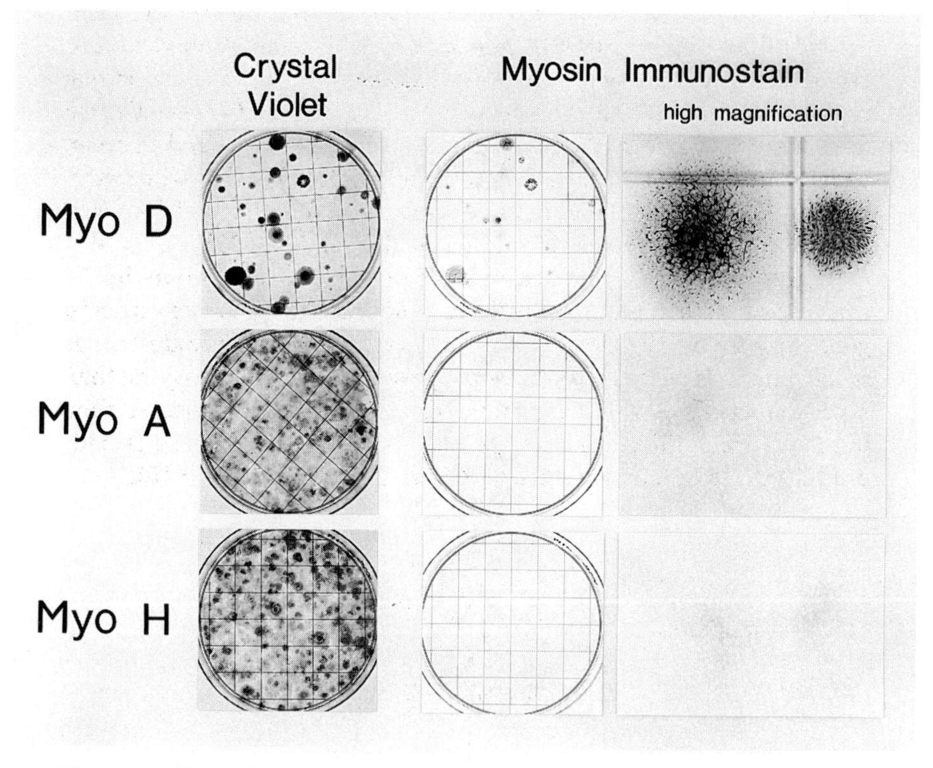

Figure 5.9 Test of each of the putative master regulatory loci *MyoA, MyoD* and *MyoH*. Each of the genes was introduced into 10T1/2 cells and tested for the ability to induce the cells to differentiate into muscle cells. Note that whilst *MyoA* and *MyoH* have no effect, introduction of MyoD results in the production of muscle cells which contain the muscle protein myosin. The differentiated muscle cells induced by *MyoD* cease to divide on differentiation, resulting in less cells being detectable by staining with crystal violet compared to the *MyoA*- and *MyoH*-treated cells, which continue to proliferate. Hence only *MyoD* has the capacity to cause 10T1/2 cells to differentiate into non-proliferating muscle cells producing myosin, identifying it as a master regulatory locus for muscle differentiation.

10T1/2 cells, *MyoA* and *MyoH* had no effect. However, artificial expression of *MyoD* was able to convert undifferentiated 10T1/2 cells into myoblasts (Figure 5.9). Hence expression of *MyoD* alone can induce differentiation of 10T1/2 cells.

The differentiated 10T1/2 cells produced in this manner, like those induced by 5-azacytidine, express a variety of muscle-lineage markers and indeed also switch on both *MyoA* and *MyoH* as well as the endogenous *MyoD* gene itself (Thayer *et al.*, 1989). This suggests that MyoD is a transcription factor which switches on genes expressed in muscle cells. In agreement with this, MyoD was shown to bind to a region of the creatine kinase gene upstream enhancer which was known to be necessary for its muscle-specific gene activity (Lassar *et al.*, 1989). This binding is dependent upon a basic region of the protein which binds directly to the DNA and an adjacent region which can form a helix–loop–helix structure and is essential for dimerization of MyoD. These elements, which are also found in other transcription factors such as the E12 and E47 proteins that bind to the E2 site in immunoglobulin enhancers (Murre *et al.*, 1989a: see Section 5.2.2), are discussed further in Section 8.4.

Hence MyoD is a transcription factor whose activation by 5-azacytidine treatment results in the activation of muscle-specific gene expression. Interestingly, the observation that the introduction of *MyoD* into cells switches on the endogenous *MyoD* gene suggests that a positive feedback loop normally regulates *MyoD* expression so that once the gene is initially expressed, expression is maintained, producing commitment to the myogenic lineage (Figure 5.10).

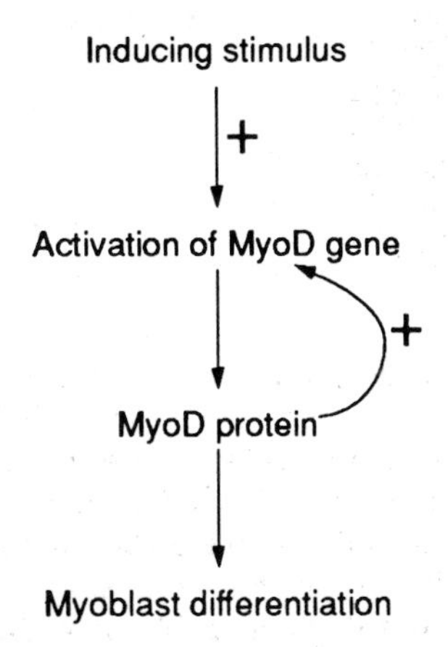

Figure 5.10 Ability of MyoD protein to activate expression of its own gene, creating a positive feedback loop which ensures that, following an initial stimulus, the MyoD protein is continuously produced and hence maintains myoblast differentiation.

Following the identification of *MyoD*, several other genes which can induce the differentiation of 10T1/2 cells into myoblasts, such as *Myf-5* and the myogenin gene (Wright *et al.*, 1989), have been identified. Like MyoD the proteins encoded by these genes are transcription factors containing the helix–loop–helix motif and are all clearly part of a complex regulatory pathway controlling muscle differentiation (review: Olson, 1990). Thus whilst expression of MyoD in cells activates the myogenin gene, suggesting that MyoD may act by inducing myogenin, expression of myogenin or Myf-5 activates the *MyoD* gene, indicating that the regulatory relationships of these factors are complex.

Whatever its relationship to these other factors, it is clear that the MyoD protein is critical for muscle differentiation and that it

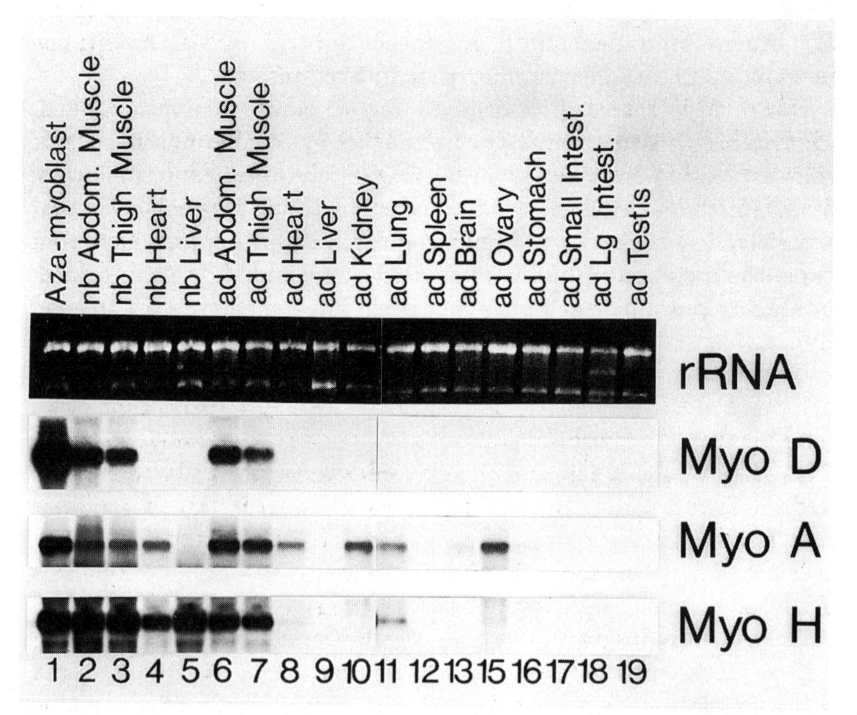

Figure 5.11 Northern blotting experiment to detect the mRNAs encoding MyoA, MyoD and MyoH in different muscle and non-muscle tissues. Note that the MyoD mRNA is present only in skeletal muscle, as expected in view of its ability to produce muscle differentiation, whereas the MyoA and MyoH mRNAs are more widely distributed. nb, new born; ad, adult; rRNA, the ribosomal RNA control used to show that all samples contain intact RNA.

acts by switching on muscle-specific gene expression. This has been established, however, via expression of MyoD in cells which do not normally express it, and it is necessary, therefore, to investigate how MyoD activity is normally regulated in muscle differentiation.

5.3.2 Regulation of MyoD

The ability of MyoD to induce cells to form myoblasts when artificially expressed within them suggests that activation of the endogenous *MyoD* gene during myogenesis *in vivo* plays a critical role in the development of muscle cells. In agreement with this idea, the MyoD mRNA is present in skeletal muscle tissue taken from a variety of different sites in the body but is absent in all other tissues, including cardiac muscle (Davis *et al.*, 1987; Figure 5.11). Hence, like Oct-2, the MyoD mRNA and protein accumulate only in a specific cell type where they are required, and the activation of the *MyoD* gene during myogenesis is likely to be of central importance in switching on the expression of muscle-specific genes.

However, in addition to its control by regulation of its synthesis, MyoD activity also appears to be regulated in another manner. Thus in 10T1/2 cells, increased expression of MyoD, whether produced by 5-azacytidine treatment or by introduction of the *MyoD* gene, results in the formation of myoblasts. In these cells the increased expression of MyoD in turn results in the activation of muscle-lineage genes and of the *MyoD* gene itself as discussed above. Interestingly, however, the products of other genes which can be activated by MyoD *in vitro*, such as creatine kinase, are not activated in these cells. The activation of these products of terminal muscle cell differentiation requires the removal of growth factors from the culture medium of the myoblasts, resulting in cell fusion and the formation of differentiated myotubes. Although this process results in switching on of MyoD-dependent genes such as the creatine kinase gene, no further increase in the level of MyoD is observed during this differentiation (Figure 5.12).

Hence a paradox exists, with the MyoD-dependent genes encoding muscle differentiation markers only becoming activated when myotubes are formed, even though the MyoD level is the same as that in myoblasts. The explanation of this paradox was provided by the identification of the Id protein (Benezra *et al.*, 1990) which, like MyoD, contains a helix–loop–helix motif but lacks the basic domain mediating DNA binding. Because the helix–loop–helix motif mediates dimerization of proteins containing it, Id can dimerize with other proteins like MyoD which contain it and inhibit

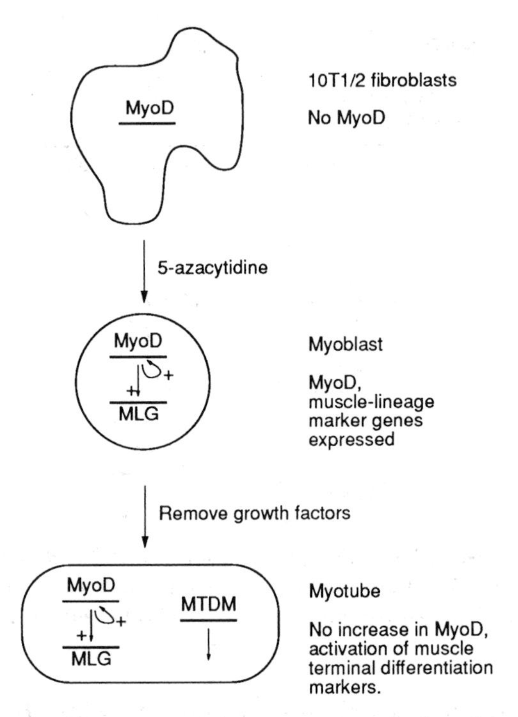

Figure 5.12 Differentiation of 10T1/2 cells into myoblasts by 5-azacytidine and then into myotubes by removal of growth factors. Note that the activation of muscle-lineage genes (MLG) and of the gene encoding MyoD itself in myoblasts occurs in response to a rise in MyoD concentration. In contrast, the MyoD-dependent induction of genes encoding terminal differentiation markers (MTDM) which occurs in myotubes occurs without an increase in MyoD concentration.

their DNA binding since the resulting dimer lacks the necessary pair of DNA-binding motifs (Figure 5.13). As expected from this model, Id can inhibit the activation of muscle-specific genes, such as the creatine kinase gene, by MyoD.

Interestingly, when 10T1/2–derived myoblasts are induced to form myotubes, Id levels decline, indicating that this second stage of myogenesis is mediated by a decline in the inhibitory protein rather than an increase in the activator, MyoD. The induction of muscle-specific genes during this process may simply be dependent on the increased levels of free MyoD produced by its release from Id, these genes requiring a higher level of free MyoD for their activation than the muscle-lineage genes or *MyoD* itself (Figure 5.14a). Alternatively, the reduction in Id may release another helix–loop–helix protein whose dimerization with MyoD is necessary for activation of these muscle-specific genes (Figure 5.14b).

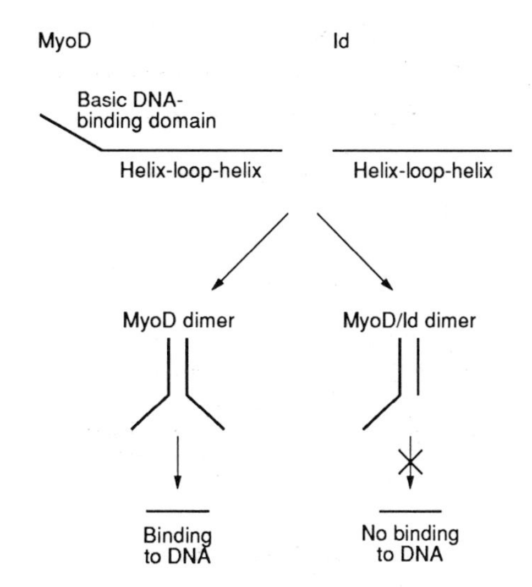

Figure 5.13 Dimerization of MyoD with itself or with Id. Note that whilst Id can dimerize with MyoD via the helix–loop–helix domain, it lacks the basic DNA domain and hence the heterodimer of MyoD and Id cannot bind to DNA.

The immunoglobulin enhancer-binding proteins, E12 and E47, which are ubiquitously expressed, provide possible candidates for this other helix–loop–helix protein. Thus in addition to binding to the E2 motif in the immunoglobulin enhancer (Section 5.2.2), these proteins also bind, like MyoD, to the creatine kinase enhancer. Most interestingly, a dimer of MyoD and E12 or E47 binds to this site with much higher affinity than either homodimer, as expected if such a heterodimer were required for gene activation (Murre *et al.*, 1989b). Moreover, both E12 and E47 bind to Id with much higher affinity than does MyoD, as predicted for proteins whose activity is regulated by association with Id (Benezra *et al.*, 1990).

Hence, as with the immunoglobulin enhancer, the activation of myotube-specific genes may be controlled by the interplay of tissue-specific and constitutive factors, although in the myotube case such factors associate together prior to DNA binding rather than binding to distinct sites in the immunoglobulin enhancer.

Although the possible role of inhibitory helix–loop–helix proteins has only recently been identified, it is clear that it is not confined to myogenesis. Thus the level of Id also declines during differentiation of early embryonic cells and of erythroid cells, whilst the product of the *emc* gene which regulates neurogenesis in *Drosophila* also contains a helix–loop–helix motif and lacks a basic DNA-binding domain (review: Jones, 1990).

a)

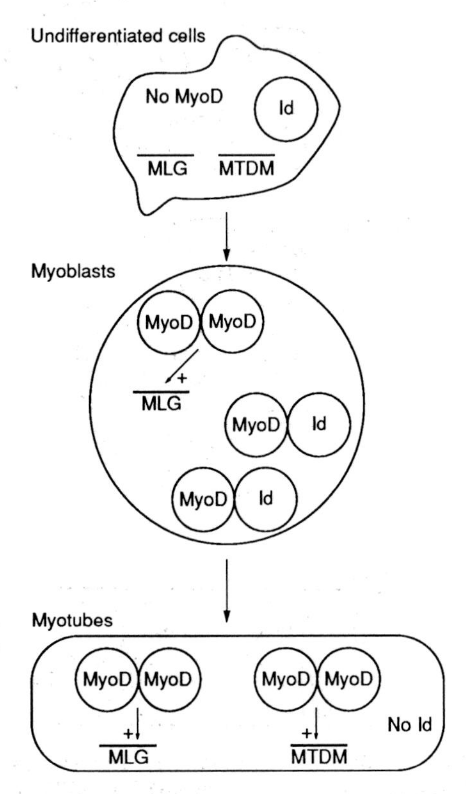

Figure 5.14a Potential mechanisms for the activation of muscle terminal differentiation markers (MTDM) following differentiation of myoblasts to myotubes. (a) The fall in Id concentration results in an increase in MyoD concentration and the activation of the MTDM genes, which are postulated to require a higher level of MyoD for their activation than that required by the muscle-lineage genes (MLG) which are active in myotubes.

In summary, therefore, regulation of MyoD appears to combine the types of regulation we have noted in Oct-2 and NFκB, its activation of lineage markers in myoblasts being controlled by increased synthesis of the protein whilst its subsequent activation of further genes during formation of myotubes is controlled by its increased activity mediated by changes in protein–protein interaction.

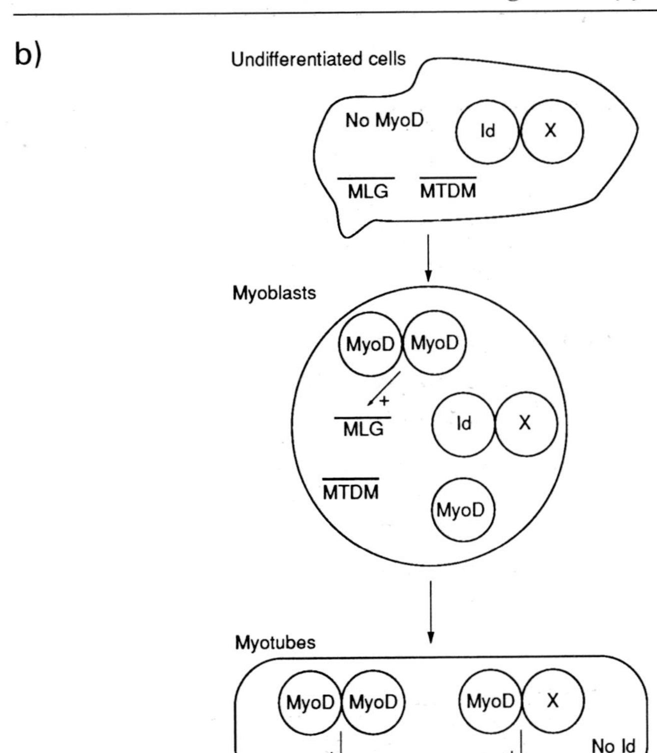

Figure 5.14b (b) The fall in Id concentration releases another helix–loop–helix protein (X) which dimerizes with MyoD. The MTDM genes can only be activated by a MyoD–X heterodimer and are hence induced only in myotubes, whereas the muscle-lineage genes are induced by a MyoD homodimer.

5.4 REGULATION OF YEAST MATING TYPE

5.4.1 Yeast mating type

In the preceding sections, we have discussed some of the control mechanisms which regulate the expression of different genes in the vast array of differentiated cells found in higher eukaryotes. Clearly, such expression of different genes within different cells of the same organism does not occur in single-celled eukaryotes such as yeast. Nonetheless, similar regulatory processes operate in these organisms. Thus, in yeast, two distinct mating types exist, known as a and α, which must fuse together to create a diploid cell. These two mating types each express distinct genes encoding products such as pheromones or pheromone receptors which are involved in mating. Individual yeasts of different mating type can thus be regarded as analogous to the different types of differentiated cell in a higher eukaryote, although they are in fact distinct individual organisms of different phenotype.

In some yeast strains, known as heterothallic strains, the two mating types are entirely separate, as in higher organisms. We shall discuss, however, the situation in homothallic yeast strains, where each cell is capable of switching its mating type from a to α or vice versa and behaves following switching exactly as does any other cell with its new mating type (reviews: Nasmyth and Shore, 1987; Herskowitz, 1988).

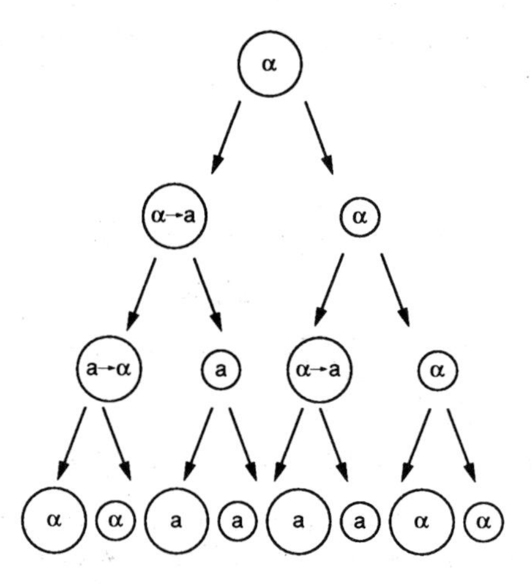

Figure 5.15 Mating type switching in yeast. In every generation the larger mother cell, which has produced a smaller daughter, switches its genotype from a to α or vice versa.

This switching process is a very precise one (Figure 5.15), occurring only in the G1 phase of the cell cycle prior to cell division and only in a mother cell which has already produced a daughter by budding. The daughter itself does not switch until it has grown and produced a daughter itself. This process is thought to be a consequence of the need to rapidly produce diploid progeny, switching from a to α or vice versa, allowing this to occur amongst the progeny of a single cell without the need for contact with another strain of different mating type. This process is of interest in terms of gene regulation from two points of view. Firstly, how does a cell switch from a to α or vice versa, and secondly, how is the expression of genes expressed only in a or α cells regulated? These two aspects will be discussed in turn.

5.4.2 Control of mating type switching

Genetic analysis of homothallic yeast strains allowed the demonstration that whether a cell was a or α in phenotype was controlled by a single gene locus on chromosome 3, known as MAT (mating type). If this transcriptionally active locus contained an a gene, the cell was of a mating type; if it contained an α gene, then the cell was of α mating type. In addition, however, yeast cells also contain transcriptionally silent copies of both the a and α genes elsewhere on chromosome 3 at the HML and HMR loci. The change in mating type occurs via a cassette mechanism in which one of the silent copies replaces the active gene at the MAT locus, changing the mating type (Figure 5.16). This process is controlled by an endonuclease which is the product of the HO (homothallism) gene and which makes a double-stranded cut at the MAT locus initiating switching.

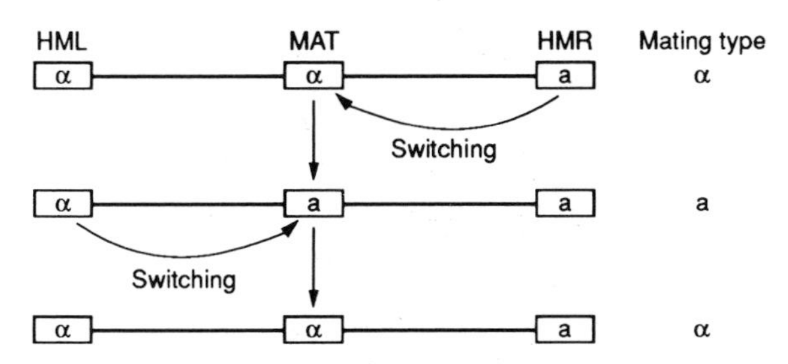

Figure 5.16 Mechanism of mating type switching involving the movement of an a or α gene from the inactive HML or HMR loci to the active MAT locus.

At first sight this process involving DNA cutting by an endonuclease appears to have little relevance to the study of gene regulation by transcription factors. A more detailed consideration of this process reveals, however, a central role for transcription factors. Thus the activity of the HO endonuclease is controlled by transcriptional regulation of its corresponding gene. A number of regulatory processes operate on this gene, ensuring that it is transcribed only in the G1 phase of the cell cycle and only in mother cells and not in daughter cells, resulting in the observed restriction of the switching process to the G1 phase in mother cells (Nasmyth, 1983).

Genetic analysis has allowed the identification of a number of different loci which regulate HO expression. Thus five SWI genes necessary for the expression of HO have been defined whilst six SIN loci which inhibit HO expression have also been identified. It is likely that these genes encode transcription factors which activate or repress HO gene expression in particular situations. Thus the SWI4 and SWI6 gene products have recently been shown to be necessary for the formation of a protein complex on DNA sequences upstream of the HO gene (Andrews and Herskowitz, 1989). This complex binds to a DNA sequence in HO which is necessary for its expression in G1 and which can confer this expression pattern on an unrelated gene. It is likely, therefore, that the SWI4 and SWI6 gene products play a critical role in the limitation of HO expression and hence of switching to the G1 phase of the cell cycle. Similarly, the activation of the HO gene only in mother cells and not in daughter cells is controlled by the opposing actions of the SIN3 gene product which inhibits HO expression, and SWI5 which antagonizes the action of SIN3 and activates HO transcription (Sternberg *et al.*, 1987; Figure 5.17). The SWI5 protein contains zinc finger motifs characteristic of DNA-binding transcription factors (see Section 8.3.1) and binds specifically to sequences in the HO promoter involved in mother/daughter control (Stillman *et al.*, 1988).

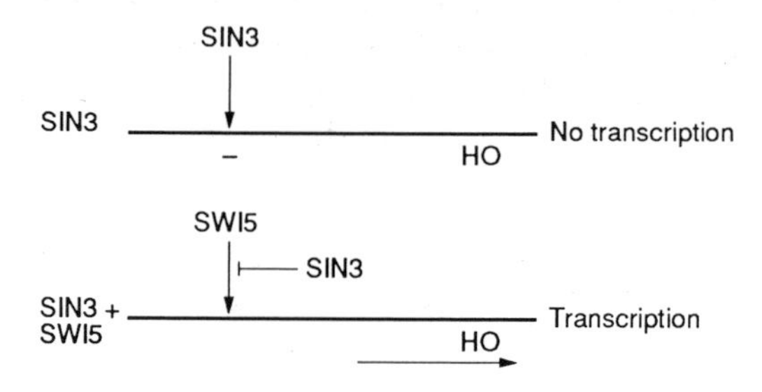

Figure 5.17 Antagonistic action of the SWI5 and SIN3 gene products on HO transcription. Note that when SWI5 is present it antagonizes the inhibitory effect of SIN3 and HO transcription occurs.

Therefore, even though mating type switching is actually brought about by the action of an endonuclease, the activity of the endonuclease and hence switching is regulated at the level of gene transcription by several interacting control mechanisms.

5.4.3 Gene regulation by the a and α gene products

Once mating type switching has determined whether the mating type locus contains an α or a gene, the expression of this gene in turn regulates the activation or repression of genes specific to a or α cells. This is achieved at the level of transcription, the a or α genes encoding transcription factors which can activate or repress gene

Ftz	Ser	Leu	Ser	Glu	Arg	Gln	Ile	Lys	Ile	Trp	Phe	Gln	Asn	Arg	Arg	Met	Lys	Ser	Lys
α 2	Ser	Leu	Ser	Arg	Ile	Gln	Ile	Lys	Asn	Trp	Val	Ser	Asn	Arg	Arg	Arg	Lys	Glu	Lys

Figure 5.18 Relationship of the yeast mating type protein α2 and the homeobox of the *Drosophila* Fushi-Tarazu protein. Boxes indicate identical amino acids.

expression. Interestingly, DNA sequence analysis of these genes (Shepherd *et al.*, 1984) has revealed that they contain a motif with strong homology to the homeobox which is shared by a number of different transcription factors involved in regulating embryonic development in *Drosophila* and other higher eukaryotes (Figure 5.18; see Section 6.2) and which mediates DNA binding by the factors which contain it (see Section 8.2).

Hence the a and α products control mating type by modulating the transcription of genes whose products are required for the a or α phenotype. This is not achieved, however, by the a product switching on a-specific genes and the α product switching on α-specific genes. Rather, the products of the a and α loci interact together to regulate the a- and α-specific genes as well as the expression of genes whose products are required in both mating types but not in diploid cells (Figure 5.19) (review: Nasmyth and Shore, 1987).

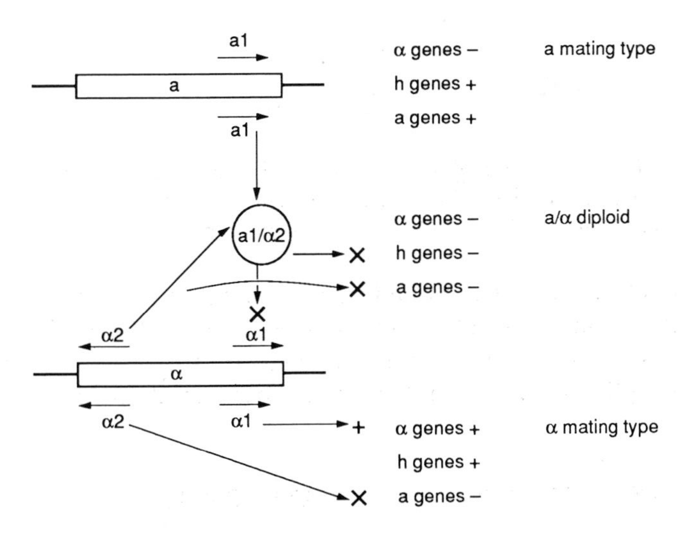

Figure 5.19 Regulatory actions of the a and α gene products. Note that whilst the α1 product is required for the activation of the α-specific genes, the α2 product inhibits the constitutively active a-specific genes. The interaction of the a1 and α2 proteins which occurs in a–α diploids where both are present inhibits the expression of the haploid-specific genes (h) as well as that of α1, thereby indirectly inhibiting the α-specific genes.

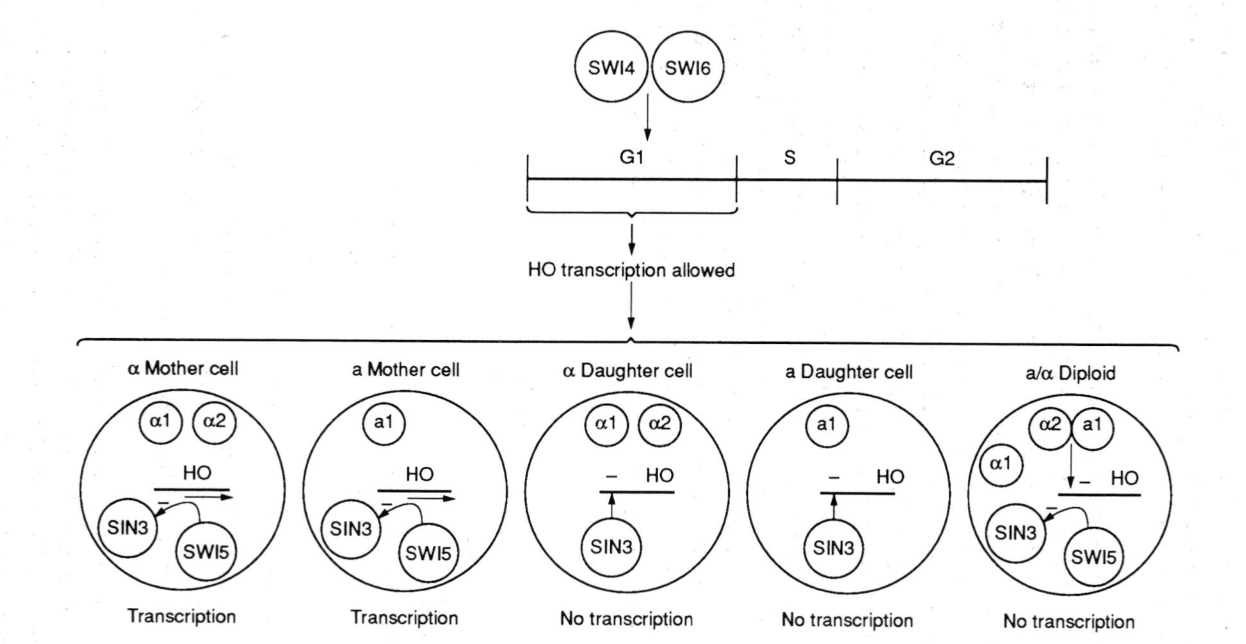

Figure 5.20 Transcription of the HO gene. Note that transcription is only allowed in the G1 phase of the cell cycle when the SWI4 and SWI6 gene products are present and in mother cells where SWI5 inhibits the action of SIN3. Transcription of HO is also inhibited in a–α diploids by the action of the a1 and α2 proteins.

Genetic analysis of this process has indicated that the a-specific genes are constitutively active but are repressed by one of the products of the α gene, α2. Hence they are active in a cells where α2 is absent but not in α cells where it is present. Similarly they are repressed in the presence of α2 in a–α diploid cells where the mating genes must be repressed, since the cells have already mated and fused to give a diploid.

In contrast, the α-specific genes are inactive in the absence of the other product of the α gene, α1, and are hence inactive in a cells where it is absent but are active in α cells where it is present. These genes are also inactive, however, in diploids where α1 is present. This paradox is explained by the fact that when both α2 and the product of the a gene, a1, are present, they co-operate to repress both the α1 gene (hence indirectly preventing α gene expression) and also the genes whose expression is required in both mating types but not in non-mating diploids. Interestingly, one of these genes is that encoding HO which, although required for switching in both a and α cells, is not required in diploids. This gene is therefore repressed by the a1 and α2 complex and its G1-specific expression is hence subject to multiple controls, preventing expression in both haploid daughter cells and in diploids (Figure 5.20). A summary of the regulatory processes acting on the HO gene is given in Figure 5.21.

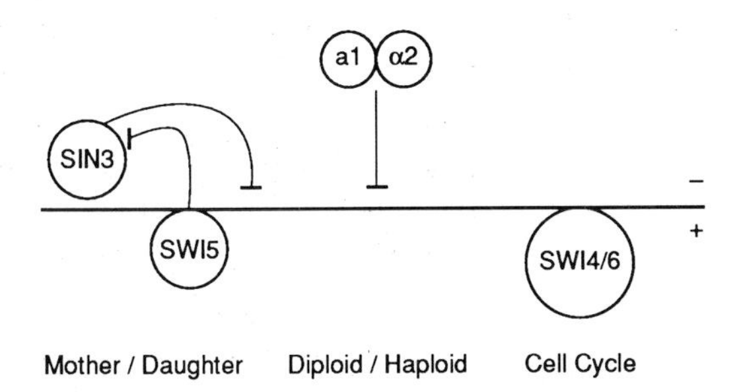

Figure 5.21 Summary of the mechanisms regulating HO transcription involving the negative actions of SIN3 and the a1/α2 complex and the positive actions of SWI5 and the SWI4/6 complex.

Hence genetic studies have indicated that gene expression is regulated by the action of the α1 activator protein and the a1 and α2 repressor proteins. These studies have now been supplemented by a more detailed molecular analysis of the a and α proteins and their interaction with the promoters of the genes which they regulate. Thus it has been shown that the constitutive activity of the a-specific genes is controlled by the binding of a constitutively expressed transcription factor PRTF (pheromone receptor transcription factor, also known as GRM, general regulator of mating type) to a sequence upstream of the a-specific genes known as the P box (Bender and Sprague, 1987). Interestingly, PRTF is also required for transcription of the α-specific genes and is capable of binding to a sequence upstream of these genes known as the P′ box which is related to but distinct from the P box. Following binding to this P′ box, however, transcriptional activation does not occur unless the α1 protein is present (Jarvis *et al.*, 1988). This protein forms a complex with PRTF and transcriptional activation occurs (Figure 5.22).

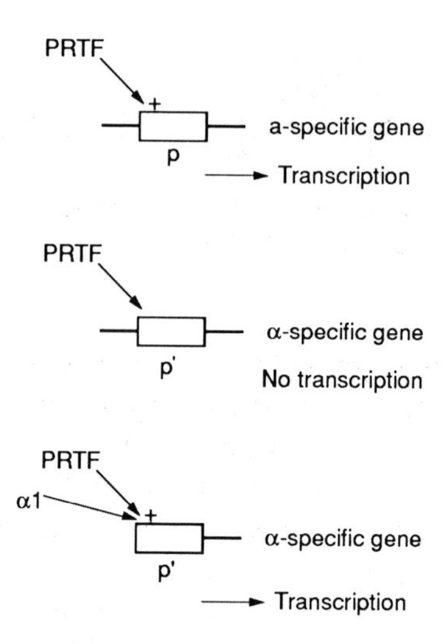

Figure 5.22 Stimulation of the a-specific genes by binding of PRTF to the P box and of the α-specific genes by binding of PRTF and α1 to the P′ box. Note that PRTF cannot activate the α-specific genes via P′ in the absence of α1.

Therefore the distinction between the a- and α-specific genes is controlled by sequences in their promoters which allow transcriptional activation of the a-specific genes following binding of PRTF alone with transcriptional activation of the α-specific genes occurring only following binding of the PRTF–α1 complex and not following binding of PRTF alone. A possible mechanism for this effect has been suggested by a recent study (Tan and Richmond, 1990) which showed that binding of PRTF to the P box produces a conformational change in the structure of the protein which does not occur following binding to the P′ box. A similar conformational change in PRTF is observed, however, following binding to the P′ box in the presence of the α1 protein. This conformational change is postulated to expose a transcriptional activation domain on PRTF, allowing it to activate transcription. Hence the difference in gene activation following binding to the P and P′ boxes is controlled by the fact that PRTF alone can undergo the necessary conformational change following binding to the P box but requires α1 to do so following binding to the P′ box.

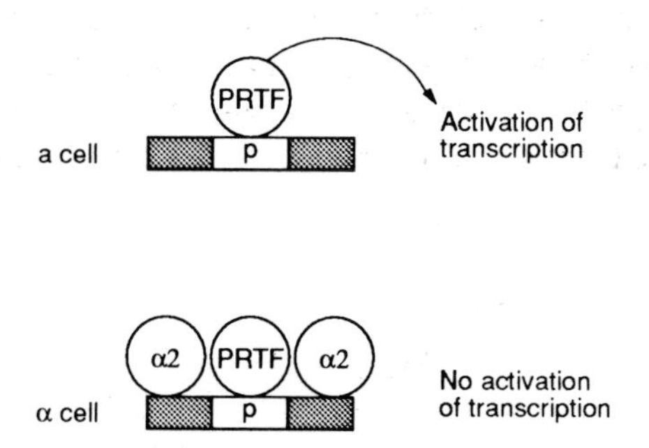

Figure 5.23 Inhibition of the a-specific genes in α cells by binding of α2 to sites flanking the binding site for PRTF.

This model therefore explains the dependence of the α-specific genes on the α1 factor for their expression. A further aspect of the model is required to explain how the constitutively active a-specific genes are repressed in α mating type cells. This occurs because the P box in these genes is flanked by binding sites for the α2 protein. One molecule of the α2 protein binds on either side of the PRTF protein bound to the P box and prevents the transcription of the a-specific genes (Figure 5.23; Keleher *et al.*, 1988). This inhibition of the activity of PRTF does not take place by α2 preventing the binding of PRTF to the P box, since the binding of PRTF is actually enhanced by the presence of α2. Rather, α2 may act by inhibiting or masking the transcriptional activation domain of PRTF (Figure 5.24a) or by forming a tightly locked inactive complex at the promoter which prevents transcription (Figure 5.24b). This latter possibility is supported by the finding (Keleher *et al.*, 1988) that the presence of the α2–PRTF complex prevents transcription from the promoter directed by any other unrelated activation sequences placed upstream of the α2–PRTF binding site, as would be expected if constitutive factors bound at the promoter are locked in a tight, inactive complex.

The differential activity of the a- and α-specific genes is therefore controlled by the constitutive factor PRTF and its interaction with the cell type-specific factors α1 and α2 (Figure 5.25).

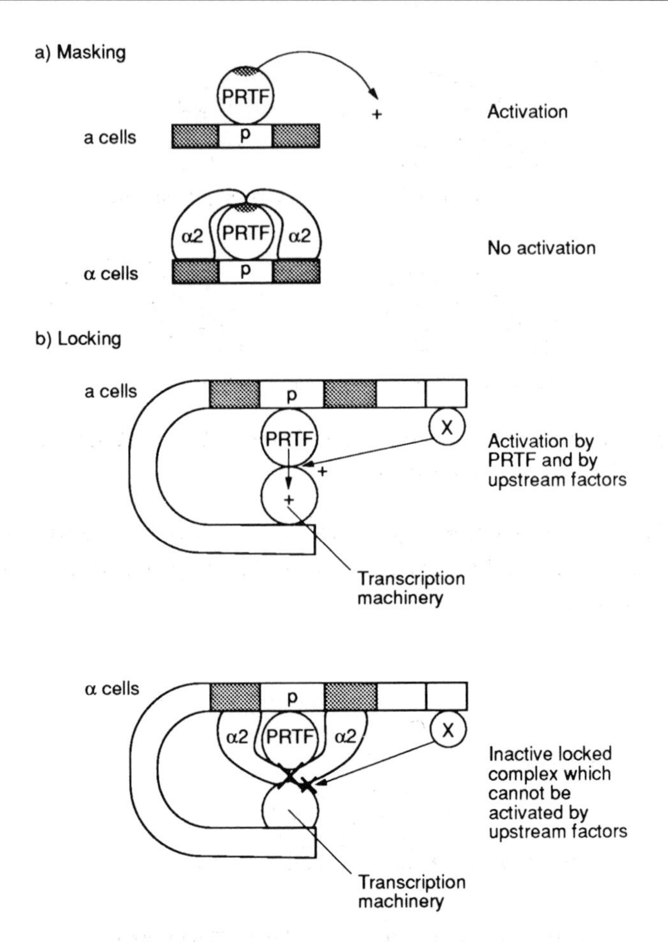

Figure 5.24 Possible mechanisms for the inhibition of the α-specific genes by α2. In (a) the α2 molecules mask the activation domain of PRTF, preventing it from activating transcription. In (b) the α2 molecules lock PRTF and components of the basic transcription machinery into a complex which is inactive transcriptionally and cannot respond to other stimulatory transcription factors (X) binding to other sites in the promoter.

In addition to its role as a repressor of the a-specific genes, α2 also plays a key role in the repression in diploid cells of genes such as HO, whose activity is required for mating in general and which are therefore expressed in both a and α cells but not in non-mating a–α diploids. This repression is dependent upon the association of the α2 protein with the a1 protein. It has been shown (Goutte and Johnson, 1988) that this association alters the DNA-binding specificity of α2, allowing it to bind upstream of the haploid-specific gene promoters and repress their expression (Figure 5.26). a1 and α2 bind as a heterodimer to the palindromic binding site in the

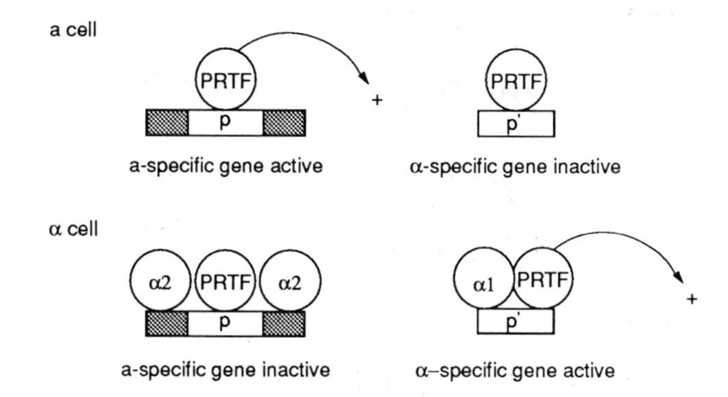

Figure 5.25 Regulation of a- and α-specific genes by the constitutive factor PRTF and the mating type factors α1 and α2.

haploid-specific genes (Dranginis, 1990), providing the first example of the regulation of homeobox-containing transcription factors by dimerization.

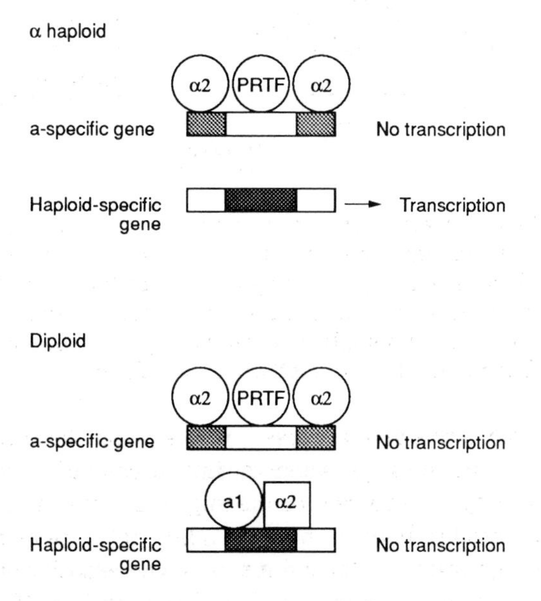

Figure 5.26 Inhibition by the α2 gene product. Note that in α haploid cells, α2 can repress the a-specific genes but cannot bind to haploid-specific genes, which are therefore transcribed. In diploid cells, however, α2 forms a heterodimer with the a1 factor which can bind to the haploid-specific genes and repress their transcription.

It is clear, therefore, that the products of the a and α mating type genes act to regulate gene expression via complex interactions both with themselves and with constitutive factors. A detailed analysis of this process and the control of switching itself has been made possible by the combination of the genetic techniques available in yeast and the molecular techniques described in Chapter 2. The possibility that these processes which are understood in such detail may serve as a model for similar regulatory processes in higher organisms is discussed in the next section.

5.4.4 Relevance of the yeast mating type system to higher organisms

At first sight, a process such as mating type switching, which involves DNA rearrangement, is apparently of little relevance to higher organisms, where, with very few exceptions such as the immunoglobulin genes (see Section 5.2.1), DNA rearrangements have not been shown to be involved in developmental processes. The more detailed discussion of this process given above illustrates, however, a number of possible events in this process which are applicable to higher organisms.

Thus both the action of the a and α gene products and the regulation of HO transcription involve the complex interaction of different antagonistic or synergistic transcription factors. Many of these interactions, such as the difference in binding specificity of homo- and heterodimers or the inhibition of gene activation by one factor which is mediated via a second factor, both of which have been extensively studied in this system, have now been shown to have counterparts in higher organisms. Similarly, the yeast transcription factors themselves show considerable similarity in structure to factors present in higher organisms. Thus PRTF shows considerable homology to the mammalian serum response transcription factor and can bind to sites within mammalian genes which bind this factor (Passmore *et al.*, 1989) whilst, as discussed above (Section 5.4.3), the *a* and α gene products contain the homeobox also found in many developmentally regulated transcription factors in higher eukaryotes.

Hence the ability to study transcription factors and their interactions both genetically and biochemically in yeast has led to the accumulation of considerable information of potential relevance to gene control in higher eukaryotes.

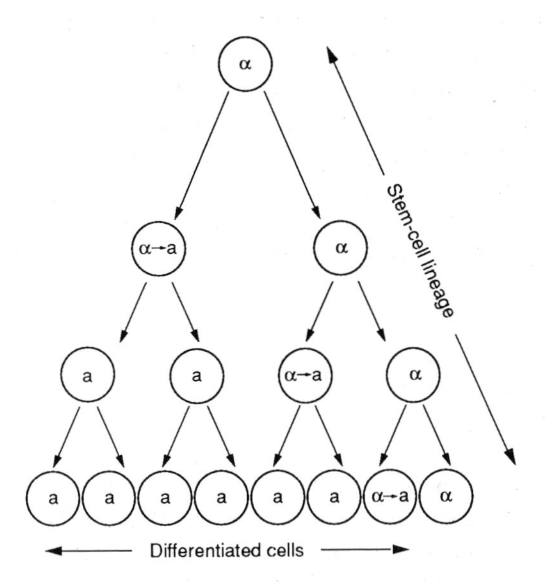

Figure 5.27 Model for the generation of a stem-cell lineage, producing differentiated cells by a system based on α to a mating type switching. In this model α represents the stem cell and a the differentiated cell, and it is assumed that, unlike the yeast mating type system, the α to a switch is irreversible.

Interestingly, Herskowitz (1985) has drawn the analogy between mating type switching and lineage in higher eukaryotes. Thus in the switching system illustrated in Figure 5.15, if α represents a stem cell and a is a differentiated cell derived from it, then, if it is assumed that, unlike the yeast situation, switching from α to a is irreversible, the model lineage illustrated in Figure 5.27 is obtained. In this model, the stem cell is continually dividing, with one daughter maintaining the stem cell lineage whilst the other differentiates. This type of system is commonly found in higher organisms, being used in the development of a number of different cell types and organs.

Even though such a lineage is unlikely to have a DNA rearrangement as its basis, the simple model of change in expression from one state to another provided by the yeast system can be used to generate models of how differentiation might occur in higher organisms.

5.5 CONCLUSIONS

In this chapter we have discussed the regulation of cell type-specific transcription via the activity of specific transcription factors which are synthesized or are present in an active form only in the particular cell type where the gene(s) they regulate are active. By interacting both with each other and constitutively expressed factors, these cell type-specific factors control the specific transcription pattern of the genes which are dependent upon them.

In some cases, such as Oct-2 or NFκB, these factors simply regulate one or more genes which are active in the cell types where they are present. In other cases, such as MyoD or the yeast mating type proteins, however, the factor is capable of directly or indirectly switching on all the genes whose expression is characteristic of the differentiated cell and its expression is therefore sufficient to produce the differentiated cell phenotype.

In lower eukaryotes such as yeast, therefore, the switching on of such a transcription factor in a particular cell will result in the appropriate differentiated phenotype. Although this is the case in higher eukaryotes also, such activation must occur at a particular stage of development and in a particular group of adjacent cells so that a tissue or organ forms at the appropriate time and in the appropriate place. The role of transcription factors in this regulation of gene expression during development is discussed in the next chapter.

REFERENCES

Andrews, B. J. and Herskowitz, I. (1989). Identification of a DNA binding factor involved in cell-cycle control of the yeast HO gene. *Cell* **57**, 21–29.

Atchison, M. L. and Perry, R. P. (1987). The role of the kappa enhancer and its binding factor NF-kappa B in the developmental regulation of kappa gene transcription. *Cell* **48**, 121–128.

Baeuerle, P. A. and Baltimore, D. A. (1988a). Activation of DNA-binding activity in an apparently cytoplasmic precursor of the NF-kappa B transcription factor. *Cell* **53**, 211–217.

Baeurele, P. A. and Baltimore, D. (1988b). I kappa B: a specific inhibitor of the NF kappa B transcription factor. *Science* **242**, 540–546.

Bender, A. and Sprague, G. F. Jr (1987). Mat alpha 1 protein, a yeast transcription activator, binds synergistically with a second protein to a set of cell-type specific genes. *Cell* **50**, 681–691.

Benezra, R., Davis, R. L., Lockshon, D., Turner, D. L. and Weintraub, H. (1990). The protein Id: a negative regulator of helix–loop–helix DNA binding proteins. *Cell* **61**, 49–59.

Bohnlein, E., Lowerthal, J. W., Siekevitz, M., Ballard, D. W., Franza, B. R. and Greene, W. C. (1988). The same inducible nuclear protein regulates mitogen activation of both the interleukin-2 receptor-alpha gene and type 1 HIV. *Cell* **53**, 827–836.

Cedar, H. (1988). DNA methylation and gene activity. *Cell* **53**, 3–4.

Clerc, R. G., Corcoran, L. M., LeBowitz, J. H., Baltimore, D. and Sharp, P. A. (1988). The B cell specific Oct-2 protein contains POU box and homeo box type domains. *Genes and Development* **2**, 1570–1581.

Constantinides, P. G., Jones, P. A. and Gevers, W. (1977). Functional striated muscle cells from non-myoblast precursors following 5-azacytidine treatment. *Nature* **267**, 364–366.

Courtois, G., Morgan, J. G., Campbell, L. A., Faurel, G. and Crabtree, G. R. (1987). Interaction of a liver-specific nuclear factor with the fibrinogen and alpha-1–antitrypsin promoters. *Science* **238**, 688–692.

Davis, H. L., Weintraub, H. and Lassar, A. B. (1987). Expression of a single transfected cDNA converts fibroblasts to myoblasts. *Cell* **51**, 987–1000.

Dranginis, A. M. (1990). Binding of yeast a1 and alpha 2 as a heterodimer to the operator DNA of a haploid specific gene. *Nature* **347**, 682–685.

Evans, T., Reitman, M. and Felsenfeld, G. (1988). An erythrocyte specific DNA-binding factor recognizes a regulatory sequence common to all chicken globin genes. *Proceedings of the National Academy of Sciences, USA* **85**, 5976–5980.

Falkner, F. G. and Zachau, H. G. (1984). Correct transcription of an immunoglobulin K gene requires an upstream fragment containing conserved sequence elements. *Nature* **310**, 71–74.

Garcia, J. V., Bich-Thuy, L., Stafford, J. and Queen, C. (1986). Synergism between immunoglobulin enhancers and promoters. *Nature* **322**, 383–385.

Gerster, T., Mathias, P., Thali, M., Jariciny, J. and Schaffner, W. (1987). Cell type-specificity of the immunoglobulin heavy chain gene enhancer. *EMBO Journal* **6**, 1323–1330.

Gillies, S. D., Morrison, S. L., Oi, V. T. and Tonegawa, S. (1983). A tissue specific enhancer is located downstream of the joining region in immunoglobulin heavy chain genes. *Cell* **33**, 729–740.

Goutte, C. and Johnson, A. D. (1988). a1 protein alters the DNA binding specificity of alpha 2 repressor. *Cell* **52**, 875–882.

Herskowitz, L. (1985). Master regulatory loci in yeast and lambda. *Cold Spring Harbor Symposia* **50**, 565–574.

Herskowitz, I. (1988). A regulatory hierachy for cell specialization in yeast. *Nature* **342**, 749–757.

Ingraham, H. A., Chen, R., Mangalam, H. J., Elsholtz, H. P., Flynn, S. E., Linn, C. R., Simmons, D. M., Swanson, L. and Rosenfeld, M. G. (1988). A tissue specific factor containing a homeo domain specifies a pituitary phenotype. *Cell* **55**, 519–529.

Jarvis, E. E., Hagen, D. C. and Sprague, G. F. Jr (1988). Identification of a DNA segment that is necessary and sufficient for alpha-specific gene control in *Saccharomyces cerevisiae*: implications for regulation of a and alpha-specific genes. *Molecular and Cellular Biology* **8**, 309–320.

Jones, N. (1990). Transcriptional regulation by dimerization: two sides to an incestuous relationship. *Cell* **61**, 9–11.

Karlsson, O., Thor, S., Norberg, T., Ohlsson, H. and Edlund, T. (1990). Insulin gene enhancer binding protein Isl-1 is a member of a novel class of proteins containing both a homeo and a Cys-His domain. *Nature* **344**, 879–882.

Keleher, C. A., Gouthe, C. and Johnson, A. D. (1988). The yeast cell-type specific repressor alpha 2 acts co-operatively with a non cell-type-specific protein. *Cell* **53**, 927–936.

Konieczny, S. F. and Emerson, C. P. Jr (1984). 5-Azacytidine induction of stable mesodermal lineages from 10T1/2 cells: evidence for regulatory genes controlling determination. *Cell* **38**, 791–800.

Lassar, A. B., Paterson, B. M. and Weintraub, H. (1986). Transfection of a DNA locus that mediates the conversion of 10T1/2 fibroblasts to myoblasts. *Cell* **47**, 649–656.

Lassar, A. B., Buskin, J. N., Lockshon, D., Davis, R. L., Apone, S., Hauschka, S. D. and Weintraub, H. (1989). MyoD is a sequence-specific DNA binding protein requiring a region of *myc* homology to bind to the muscle creatine kinase enhancer. *Cell* **58**, 823–831.

Lenardo, M. J. and Baltimore, D. (1989). NF-kappa B: a pleiotropic mediator of inducible and tissue-specific gene control. *Cell* **58**, 227–229.

Lenardo, M., Pierce, J. W. and Baltimore, D. (1987). Protein binding sites in Ig gene enhancers determine transcriptional activity and inducibility. *Science* **236**, 1573–1577.

Libermann, T. A., Lenardo, M. and Baltimore, D. (1990). Involvement of a second lymphoid-specific enhancer element in the regulation of immunoglobulin heavy chain gene expression. *Molecular and Cellular Biology* **10**, 3155–3162.

Mason, J. O., Williams, G. T. and Neuberger, M. S. (1985). Transcription cell type specificity is conferred by an immunoglobulin VH gene promoter that includes a functional consensus sequence. *Cell* **41**, 479–487.

Mignotte, V., Wall, L., de Boer, E., Grosveld, F. and Romeo, P. H. (1989). Two tissue-specific factors bind the erythroid promoter of the human porphobillinogen deaminase gene. *Nucleic Acids Research* **17**, 37–54.

Muller, M. M., Ruppert, S., Schaffner, W. and Matthias, P. (1988). A cloned octamer transcription factor stimulates transcription from lymphoid-specific promoters in non-B cells. *Nature* **336**, 544–551.

Murre, C., McCaw, P. S. and Baltimore, D. (1989a). A new DNA binding and dimerization motif in immunoglobulin enhancer binding, *daughterless*, MyoD and myc proteins. *Cell* **56**, 777–783.

Murre, C., McCaw, P. S., Baessin, H., Caudy, M., Jan, L. Y., Jan, Y. N., Cabera, C. V., Buskin, J. N., Hauschkn, S. D., Lassar, A. B., Weintraub, H. and Baltimore, D. (1989b). Interactions between heterologous helix–loop–helix proteins generate complexes that bind specifically to a common DNA sequence. *Cell* **58**, 537–544.

Nabel, G. and Baltimore, D. (1987). An inducible transcription factor activates expression of human immunodeficiency virus in T cells. *Nature* **326**, 711–713.

Nasmyth, K. (1983). Molecular analysis of a cell lineage. *Nature* **302**, 670–676.

Nasmyth, K. and Shore, D. (1987). Transcriptional regulation in the yeast life cycle. *Science* **237**, 1162–1170.

Nasmyth, K., Stillman, D. and Kipling, D. (1987). Both positive and negative regulators of HO transcription are required for mother-cell specific mating type switching in yeast. *Cell* **48**, 579–587.

Nelms, K. and van Ness, B. (1990). Identification of an octamer binding site in the human kappa light chain enhancer. *Molecular and Cellular Biology* **10**, 3843–3846.

Nelson, B., Kadesch, T. and Sen, R. (1990). Complex regulation of the immunoglobulin mu heavy chain gene enhancer: mu B a new determinant of enhancer function. *Molecular and Cellular Biology* **10**, 3145–3154.

Nicosia, A. Monaci, P., Tumei, L., de Francesco, R., Nuzzo, M., Stunnenberg, H. and Cortese, R. (1990). A myosin-like dimerization helix and an extra-large homeodomain are essential elements of the tripartite DNA binding structure of LFB1. *Cell* **61**, 1225–1236.

Ohlsson, H., Karlsson, O. and Edlund, T. (1988). A beta-cell-specific protein binds to the two major sequences of the insulin gene enhancer. *Proceedings of the National Academy of Sciences, USA* **85**, 4228–4231.

Olson, E. N. (1990). MyoD family: a paradigm for development? *Genes and Development* **4**, 1454–1461.

Parslow, T. G., Blair, D. C., Murphy, W. J. and Granner, D. K. (1984). Structure of the 5′ ends of immunoglobulin genes: a novel conserved sequence. *Proceedings of the National Academy of Sciences, USA* **81**, 2650–2654.

Passmore, S., Elble, R. and Tye, B.-K. (1989). A protein involved in minichromosome maintenance in yeast binds a transcriptional enhancer conserved in eukaryotes. *Genes and Development* **3**, 921–935.

Pierce, J. W., Lenardo, M. and Baltimore, D. (1988). Oligonucleotide that binds nuclear factor NF-kappa B acts as a lymphoid specific and inducible enhancer element. *Proceedings of the National Academy of Sciences, USA* **85**, 1482–1486.

Plumb, M. A., Lobenkov, V. V., Nicolas, R. H., Wright, C. A., Zavou, S. and Goodwin, G. H. (1986). Characterization of chicken erythroid nuclear proteins which bind to the nuclease hypersensitive regions upstream of the beta ^{A} and ^{H} globin genes. *Nucleic Acids Research* **14**, 7675–7693.

Plumb, M., Frampton, J., Wainwright, H., Walker, M., Macleod, K., Goodwin, G. and Harrison, P. (1989). GATAAG: a cis control region binding an erythroid-specific nuclear factor with a role in globin and non-globin gene expression. *Nucleic Acids Research* **17**, 73–92.

Razin, A. and Riggs, A. D. (1980). DNA methylation and gene function. *Science* **210**, 604–610.

Reitman, M. and Felsenfeld, G. (1988). Mutational analysis of the chicken beta globin enhancer reveals two positive acting domains. *Proceedings of the National Academy of Sciences, USA* **85**, 6267–6271.

Sen, R. and Baltimore, D. (1986a). Multiple nuclear factors interact with the immunoglobulin enhancer sequences. *Cell* **46**, 705–716.

Sen, R. and Baltimore, D. (1986b). Inducibility of kappa immunoglobulin enhancer binding protein NF-kappa B by a posttranslational mechanism. *Cell* **47**, 921–928.

Shaw, J.-P., Utz, P. J., Durand, D. B., Toole, J., Emmel, E. A. and Crabtree, G. R. (1988). Identification of a putative regulator of early T cell activation genes. *Science* **241**, 202–205.

Shepherd, J. C. W., McGinnis, W., Carrasco, A. E., De Robertis, E. M. and Gehring, W. J. (1984). Fly and frog homoeo domains show homologies with yeast mating type regulatory loci. *Nature* **310**, 70–71.

Singh, H., Sen, R., Baltimore, D. and Sharp, P. A. (1986). A nuclear factor that binds to a conserved sequence motif in transcriptional control elements of immunoglobulin genes. *Nature* **319**, 154–158.

Staudt, L. H., Singh, H., Sen, R. Wirth, T., Sharp, P. A. and Baltimore, D. (1986). A lymphoid specific protein binding to the octamer motif of immunoglobulin genes. *Nature* **323**, 640–643.

Sternberg, P. W., Stern, M. J., Clark, I. and Herskowitz, I. (1987). Activation of the yeast HO gene by release from multiple negative controls. *Cell* **48**, 567–577.

Stillman, D. J., Bonkier, A. T., Seddan, A., Groenheat, G. and Nasmylh, K. A. (1988). Characterization of a mother cell transcription factor involved in mother cell specific transcription of the yeast HO gene. *EMBO Journal* **7**, 485–494.

Tan, S. and Richmond, T. J. (1990). DNA binding-induced conformational change of the yeast transcriptional activator PRTF. *Cell* **62**, 367–377.

Thayer, M. J., Tapscott, S. J., Davis, R. L., Wright, W. E., Lassar, A. B. and Weintraub, H. (1989). Positive autoregulation of the myogenic determination gene MyoD1. *Cell* **58**, 241–248.

Tonegawa, S. (1983). Somatic generation of antibody diversity. *Nature* **302**, 575–581.

Tsai, S.-F., Martin, D. I. K., Zon, L. I., D'Andrea, A. D., Wong, G. G. and Orkin, S. H. (1989). Cloning of cDNA for the major DNA-binding protein of the erythroid lineage through expression in mammalian cells. *Nature* **339**, 446–451.

Watson, J. D., Hopkins, N. H., Roberts, J. W., Steitz, J. A. and Weiner, A. M. (1987). *Molecular Biology of the Gene*, 4th edn, vol. 2, pp. 832–897. Mendo Park, California: Benjamin/Cummings Publishing Co.

Wirth, T., Staudt, L. and Baltimore, D. (1987). An octamer oligonucleotide upstream of a TATA motif is sufficient for lymphoid specific promoter activity. *Nature* **329**, 174–178.

Wright, W. E., Sassoon, D. A. and Lin, U. K. (1989). Myogenin, a factor regulating myogenesis has a domain homologous to MyoD. *Cell* **56**, 607–617.

CHAPTER SIX

Transcription factors and developmentally regulated gene expression

6.1 DEVELOPMENTALLY REGULATED GENE EXPRESSION

As discussed in Chapter 5, the ability to carry out genetic experiments in yeast resulted in the isolation of genes whose mutagenesis affects cell type-specific gene regulation, allowing the role of their protein products in this process to be elucidated. Clearly such an approach could be equally valuable in understanding the regulation of gene expression during development in which the processes of cell type-specific gene expression discussed in Chapter 5 are integrated and co-ordinated so that each cell type and tissue arises in the right place at the right time. It is evident, however, that this cannot be done in the single-celled yeast. Rather, these studies have focused on the fruit fly *Drosophila melanogaster*, which is well characterized genetically and has a short generation time, facilitating its study.

A very large number of mutations which affect the development of this organism have been isolated and their corresponding genes named on the basis of the observed phenotype of the mutant fly (reviews: Scott and Carroll, 1987; Ingham, 1988). Thus mutations in the so-called homeotic genes result in the transformation of one particular segment of the body into another; mutations in the *Antennapedia* gene, for example, causing the transformation of the segment which normally produces the antenna into one which produces a middle leg (Figure 6.1). Similarly, mutations in genes of the gap class result in the total absence of particular segments; mutations in the *Knirps* gene, for example, resulting in the absence of most of the abdominal segments, although the head and thorax develop normally.

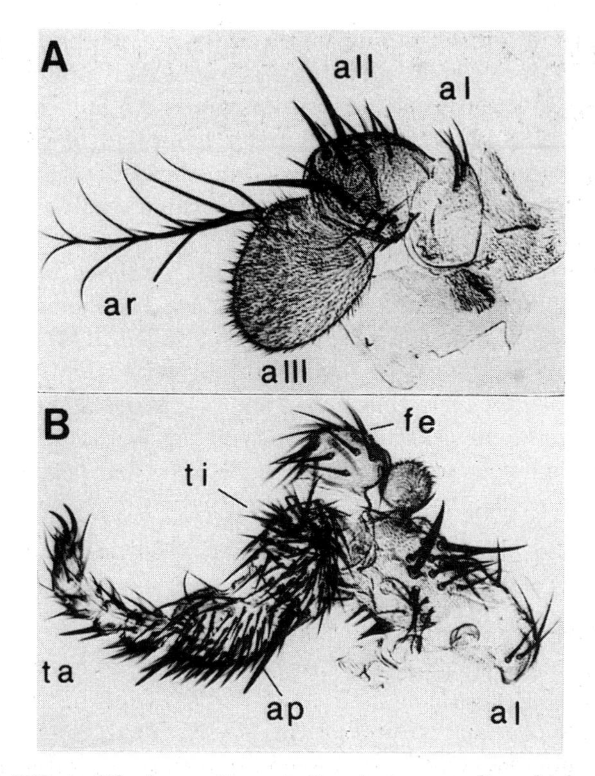

Figure 6.1 Effect of the homeotic mutation *Antennapedia*, which produces a middle leg (B) in the region that would contain the antenna of a normal fly (A). aI, aII, aIII: 1st, 2nd, and 3rd antennal segments. ar: arista. ta: tarsus. ti: tibia. fe: femur. ap: apical bristle.

The products of genes of this type therefore play critical roles in *Drosophila* development. The products of the gap genes, for example, are necessary for the production of particular segments, whilst the homeotic gene products specify the identify of these segments. Given that these processes are likely to require the activation of genes whose protein products are required in the particular segment, it is not surprising that many of these genes have been shown to encode transcription factors. Thus the *Knirps* gene product and that of another gap gene, *Kruppel*, contain multiple zinc finger motifs characteristic of DNA-binding transcription factors and can bind to DNA in a sequence-specific manner (Rosenberg *et al.*, 1986). Similarly, the *tailless* gene whose product plays a key role in defining the anterior and posterior regions of the *Drosophila* embryo has recently been shown to be a member of the steroid receptor super gene family discussed in Section 4.3 (Pignoni *et al.*, 1990).

It is clear, therefore, that the genes identified by mutation as playing a role in *Drosophila* development can encode several different types of transcription factors. However, of the first 25 such genes which were cloned, allowing a study of their protein products, well over half (15) contain a motif known as the homeobox (Levine and Hoey, 1988) which was originally identified in the homeotic genes of *Drosophila*. The evidence that these homeobox proteins modulate transcription in *Drosophila* and the ways in which they do so are discussed in Section 6.2.

Following the successful use of genetic analysis to isolate the *Drosophila* homeobox genes, attempts have been made to isolate homologues of these genes in other organisms, less amenable to genetic analysis, by using hybridization with the DNA of the *Drosophila* genes. Accordingly, Section 6.3 discusses the isolation of homeobox genes in other organisms, including mammals, by this and other means, as well as the evidence that their protein products are involved in developmental gene regulation in these organisms.

6.2 THE HOMEOBOX-CONTAINING GENES OF *DROSOPHILA*

6.2.1 The homeobox

When the first homeotic genes were cloned, it was found that they shared a region of homology approximately 180 base pairs long and therefore capable of encoding 60 amino acids which was flanked on either side by regions which differed dramatically between the different genes. This region was named the homeobox (reviews: Gehring, 1987; Scott *et al.*, 1989). Subsequently, the homeobox was shown to be present in many other *Drosophila* regulatory genes. These include the *Fushi-tarazu* gene (Ftz), which is a member of the pair rule class of regulatory loci whose mutation causes alternate segments to be absent, and the *engrailed* gene (Eng), which is a member of the class of genes whose products regulate segment polarity. The close similarity of the homeoboxes encoded by the homeotic genes *Antennapedia* and *Ultrabithorax* and that encoded by the Ftz gene is shown in Figure 6.2.

The presence of this motif in a large number of different regulatory genes of different classes strongly suggested that it was of importance in their activity. The evidence that the homeobox-containing proteins are transcription factors whose DNA-binding activity is mediated by the homeobox is discussed in the next section (review: Hayashi and Scott, 1990).

Antp	Arg	Lys	Arg	Gly	Arg	Gln	Thr	Tyr	Thr	Arg	Tyr	Gln	Thr	Leu	Glu	Leu	Glu	Lys	Glu	Phe	His	Phe	Asn	Arg	Tyr	Leu	Thr	Arg	Arg	Arg
Ubx		Arg																				Thr		His						
Ftz	Ser			Thr																						Ile				

Helix (Arg–Leu) | Turn (Cys–Leu) | Recognition helix (Thr–Phe)

Antp	Arg	Ile	Glu	Ile	Ala	His	Ala	Leu	Cys	Leu	Thr	Glu	Arg	Gln	Ile	Lys	Ile	Trp	Phe	Gln	Asn	Arg	Arg	Met	Lys	Trp	Lys	Lys	Glu	Asn
Ubx				Met		Tyr										Glu										Leu				Ile
Ftz			Asp			Asn			Ser		Ser															Ser			Asp	Arg

Figure 6.2 Amino acid sequences of several *Drosophila* homeodomains, showing the conserved helical motifs. Differences between the sequences of the Ubx and Ftz homeodomains from that of Antp are indicated; a blank denotes identity in the sequence. The helix–turn–helix region is indicated.

6.2.2 The homeobox proteins as transcription factors

The first indication that the homeobox proteins were indeed transcription factors came from the finding that the homeobox was also present in the yeast mating type a and α gene products (Shepherd *et al.*, 1984). Thus, as discussed in Section 5.4, these proteins are known to be transcription factors which regulate the activity of a and α-specific genes, hence suggesting, by analogy, that the *Drosophila* proteins also fulfilled such a role.

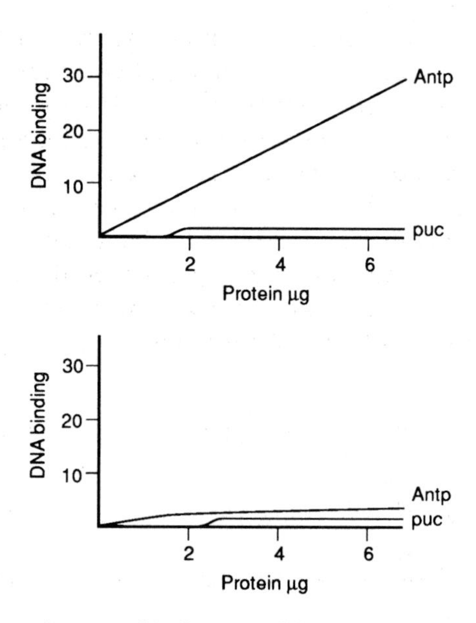

Figure 6.3 Assay of protein binding to a DNA fragment from the *Antennapedia* gene promoter (Antp) or a control fragment of plasmid DNA (pUC) using protein extracts from *E. coli* which have been genetically engineered to express the *Drosophila* Ubx protein (top panel) or protein extracts from control *E. coli* not expressing Ubx (bottom panel). Note the specific binding of Ubx protein to the *Antennapedia* DNA fragment.

Direct evidence that this is the case is now available from a number of different approaches. Thus it has been shown that many of these proteins bind to DNA in a sequence-specific manner as expected for transcription factors (Hoey and Levine, 1988). Moreover, binding of a specific homeobox protein to the promoter of a particular gene correlates with the genetic evidence that the protein regulates expression of that particular gene. For example, the Ultrabithorax (Ubx) protein has been shown to bind to specific DNA sequences within its own promoter and in the promoter of the

Antennapedia gene, in agreement with the genetic evidence that Ubx represses *Antennapedia* expression (Beachy *et al.*, 1988; Figure 6.3).

The ability of the homeobox-containing proteins to bind to DNA is directly mediated by the homeobox itself. Thus if the homeobox of the Antennapedia protein is synthesized in isolation either in bacteria (Muller *et al.*, 1988) or by chemical synthesis (Mihara and Kaiser, 1988), it is capable of binding to DNA in the identical sequence-specific manner characteristic of the intact protein. The structural features of the homeodomain which allow it to do this are discussed in Section 8.2.

Although DNA binding is a prerequisite for the modulation of transcription, it is necessary to demonstrate that the homeobox proteins do actually affect transcription following such binding. In the case of the Ubx protein, this was achieved by Krasnow *et al.* (1989), who showed that co-transfection of a plasmid expressing Ubx with a plasmid in which the *Antennapedia* promoter drives a marker gene resulted in the repression of gene expression driven by the *Antennapedia* promoter. Hence the observed binding of Ubx to the Antp promoter (see above) results in down-regulation of its activity, in agreement with the results of genetic experiments.

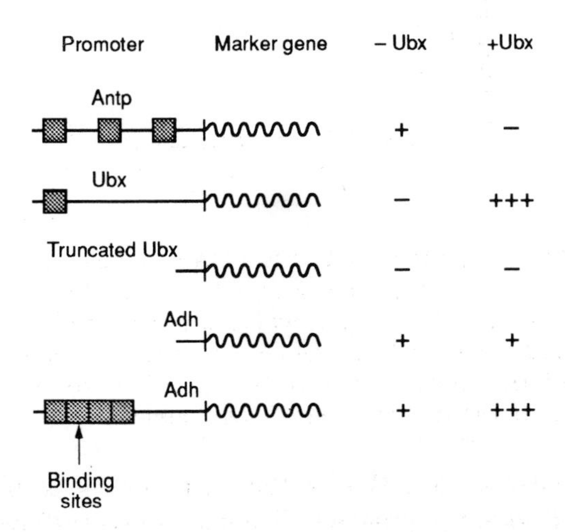

Figure 6.4 Effect of Ubx on various marker genes with or without binding sites (hatched boxes) for the Ubx protein. Note that Ubx can stimulate its own promoter which contains a Ubx-binding site and this effect is abolished by deleting the Ubx-binding site. Similarly the alcohol dehydrogenase (Adh) gene, which is normally unaffected by Ubx, is rendered responsive to Ubx stimulation by addition of Ubx-binding sites. In contrast, the Antennapedia promoter, which also contains Ubx-binding sites, is repressed by Ubx. Hence binding of Ubx can activate or repress different promoters.

Most interestingly, the Ubx expression plasmid was able to up-regulate activity of its own promoter in co-transfection experiments, this ability being dependent on the previously defined binding sites for Ubx within its own promoter. Similarly, although Ubx normally has no effect on expression of the alcohol dehydrogenase (Adh) gene it can stimulate the Adh promoter following linkage of the promoter to a DNA sequence containing multiple binding sites for Ubx (Krasnow *et al.*, 1989). Hence a homeobox protein can produce distinct effects following binding, Ubx activating its own promoter and a hybrid promoter containing Ubx-binding sites but repressing the activity of the Antp promoter (Figure 6.4).

A similar transcriptional activation effect of DNA binding has been demonstrated for the Fushi-tarazu (Ftz) protein. This protein binds specifically to the sequence TCAATTAAATGA. As with Ubx, linkage of this sequence to a marker gene confers responsivity to activation by Ftz, such activation being dependent upon binding of Ftz to its target sequence, a one-base-pair change which abolishes binding, also abolishing the induction of transcription (Jaynes and O'Farrell, 1988; Figure 6.5).

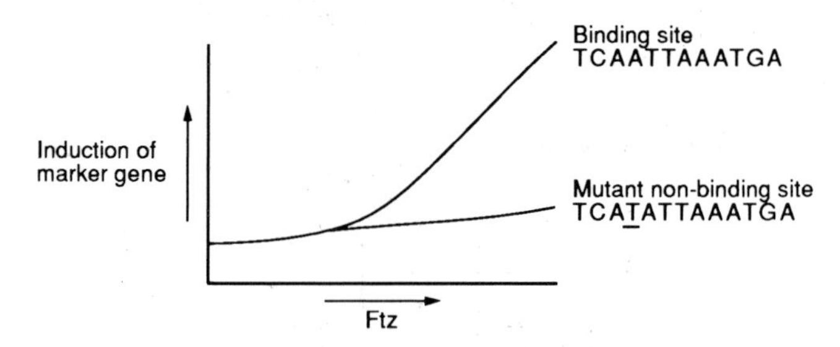

Figure 6.5 Effect of expression of the Ftz protein on the expression of a gene containing its binding site, or a mutated binding site containing a single base pair change which abolishes binding of Ftz.

These co-transfection studies carried out by introducing target DNA into cultured cells along with another DNA molecule expressing the homeobox protein have recently been supplemented by studies using *in vitro* transcription assays. Thus if the Ubx promoter linked to a marker gene is added to a suitable cell-free extract, transcription of the marker gene driven by the promoter can be observed. Addition of the purified homeodomain protein even-skipped (eve) to this extract inhibits Ubx promoter activity, however, and this inhibition is dependent upon binding sites for the eve protein within the Ubx promoter (Biggin and Tjian, 1989). Such findings parallel the ability of a vector expressing eve to repress the

Ubx promoter following co-transfection into cultured cells and the genetic evidence which originally led to the definition of eve as a repressor of Ubx (Figure 6.6). Hence a direct link exists between the genetic data, the results of co-transfections into cells and the *in vitro* data. The fact that the results of experiments in whole organisms or intact cells can be directly reproduced in cell extracts indicates that the homeodomain proteins function directly as transcription factors to control development. Moreover, by showing that these factors can act in a relatively simple *in vitro* system they provide a means of analysing this effect.

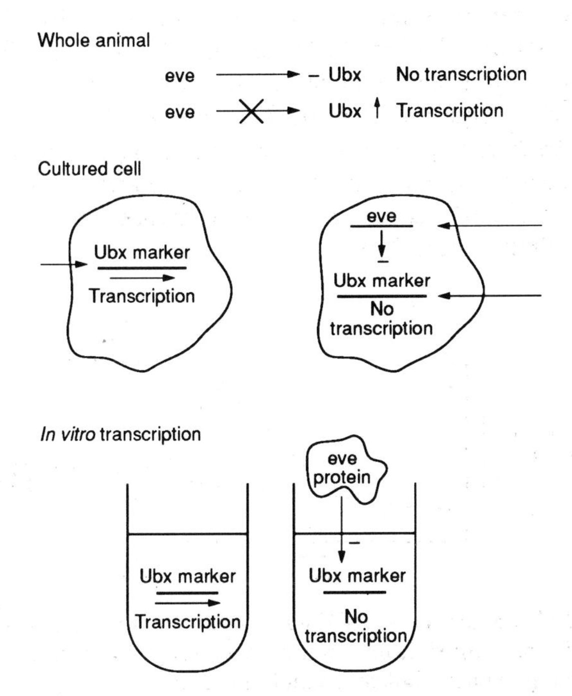

Figure 6.6 Inhibitory effect of the eve protein on expression of the Ubx gene. This inhibitory effect can be observed in the whole animal where mutation of the eve gene enhances Ubx expression (top panel); in cultured cells where introduction of a plasmid expressing the eve gene represses a co-transfected Ubx promoter driving a marker gene (middle panel) and in a test tube *in vitro* transcription system where addition of purified eve protein represses transcription of a marker gene driven by the Ubx promoter (bottom panel).

A variety of evidence is therefore available which allows the conclusion that the homeobox proteins are transcription factors which exert their effects on development by activating or repressing the activity of target genes. What remains to be elucidated is the manner in which such relatively simple effects on gene expression can regulate the enormously complicated process of development. The progress which has been made in this area is discussed in the next section.

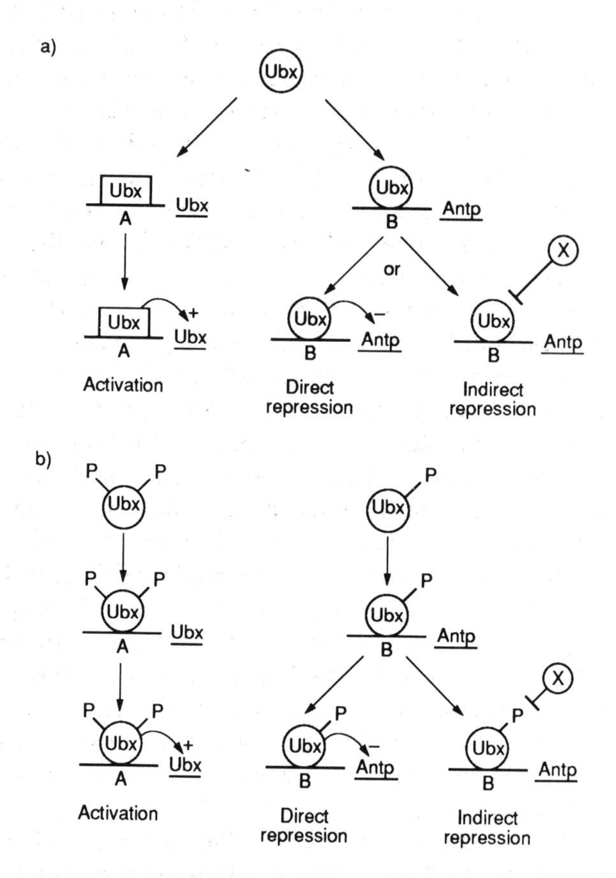

Figure 6.7 Potential mechanisms by which the binding of Ubx to different DNA sequences (A and B) in different promoters could result in activation or repression of gene expression. In (a), binding to the different sequences is suggested to result in a different configuration of the Ubx protein so that it can only activate gene expression following binding to the A sequence. In contrast, this does not occur following binding to the B sequence and gene expression is repressed either by a direct inhibitory effect of the Ubx protein (direct repression) or by bound Ubx preventing the binding of an activating protein (x, indirect repression). In (b), the A and B sequences are suggested to bind different forms of the Ubx protein which differ in the ability to activate gene expression due to differences, for example, in phosphorylation (P). Only the form of the protein bound by the A sequence can activate gene expression whilst the form bound by the B sequence represses by direct or indirect repression as in (a).

6.2.3 Homeobox transcription factors and the regulation of development

At first sight it is remarkably difficult to understand how the relatively simple process of transcriptional regulation by homeobox proteins could in turn control development. Although this process is of course not yet fully understood, a number of findings which indicate ways in which the complex regulatory networks needed to regulate development might be built up are available.

One such finding has already been discussed in the previous section. Thus, in a number of cases, it has already been shown that a single factor can repress some target genes whilst activating others, thereby increasing the range of effects mediated by one factor. Thus the Ubx protein, for example, can activate transcription from its own promoter whilst repressing that of the Antp gene (Krasnow *et al.*, 1989). The mechanism by which this occurs is unclear, although as with the activation and repression effects of the glucocorticoid receptor (Section 4.3.4) it may involve differences in the DNA sequences in the activated and repressed promoters. This could result in the Ubx protein binding to the sequence in the Antp gene promoter in a configuration in which it could not activate gene expression (Figure 6.7a). Alternatively, differences between the two different promoters might result in their binding different forms of the protein which differ, for example, in their level of phosphorylation and only one of which is capable of activating transcription (Figure 6.7b). In either case binding of the form incapable of activating gene expression would result in repression, either by interfering with the binding of another activating factor or by direct repression (see Section 9.3 for discussion of the mechanisms of negative regulation).

Whatever its mechanism, the activation and repression of different promoters by the Ubx protein has important consequences in terms of the control of development. Thus the ability of Ubx to induce its own transcription provides a mechanism for the long-term maintenance of Ubx gene expression during development, since once expression has been switched on and some Ubx protein made, it will induce further transcription of the gene via a simple positive feedback loop even if the factors which originally stimulated its expression are no longer present (Figure 6.8). This long-term maintenance of Ubx expression is essential, since if the Ubx gene is mutated within the larval imaginal disc cells which eventually produce the adult fly, the cells which would normally produce the haltere (balancer) will produce a wing instead (Struhl, 1982). Thus although these cells are known to already be committed to

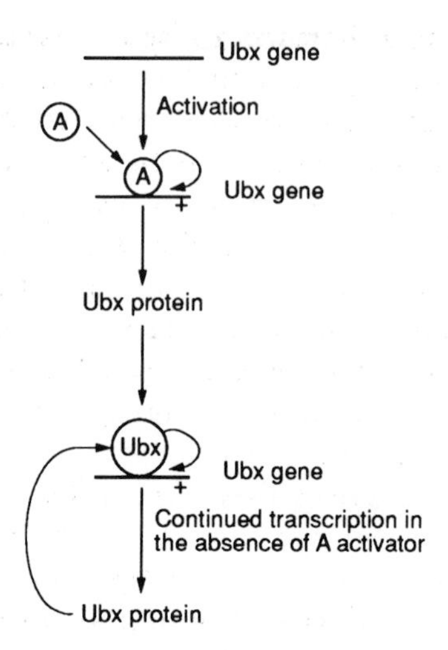

Figure 6.8 The stimulatory effect of the Ubx protein on the transcription of its own gene ensures that once Ubx gene transcription is initially switched on by an activator protein (A), transcription will continue even if the activator protein is removed.

form the adult haltere at the larval stage, the continued expression of the Ubx gene is essential to maintain this commitment and allow eventual overt differentiation (see Hadorn (1968) for a review of imaginal discs and their role in *Drosophila* development).

Similarly, the inhibition of the Antp gene by Ubx indicates that the homeobox factors do not act simply by activating the transcription of genes for structural proteins or enzymes required in particular cell types. Although this must be one of their functions, they can also regulate the transcription of each other, creating the potential for regulatory networks. Thus, we have already seen how the eve protein can repress Ubx gene transcription whilst Ubx in turn represses the Antp promoter. Since Antp has been shown to stimulate both its own promoter and that of Ubx (Winslow *et al.*, 1989), this creates the possibility of complex interactions in which the synthesis of one particular factor at a particular time will create changes in the levels of numerous other factors and ultimately result in the activation or repression of numerous target genes. Hence by both regulating each other's expression and that of non-homeobox target genes both positively and negatively the homeobox transcription factors can create regulatory networks of the type which are necessary for the control of development.

The effects which we have described so far involve the activity of single homeobox factors affecting the expression of other genes. In addition, however, it is also possible for the same target gene to be regulated by multiple homeobox factors, with combinations of factors either synergizing with each other or interfering with one another so that effects are observed with combinations of factors which are not observable with either factor alone.

For example, it has been shown that a number of homeobox proteins, for example eve, engrailed (Eng), Fushi-tarazu (Ftz), paired (prd) and zerknult (zen), can all bind to the sequence TCAATTAAAT (Hoey and Levine, 1988). When plasmids expressing each of these genes are co-transfected with a target promoter carrying multiple copies of this binding site, the Ftz, prd and zen proteins can activate transcription of the target promoter (Jaynes and O'Farrell, 1988; Han *et al.*, 1989). In contrast, the eve and Eng proteins have no effect on the transcription of such a promoter. They do, however, interfere with the ability of the activating proteins to induce transcription, presumably by blocking the binding of the activating factor. Thus, for example, whilst Ftz can stimulate the target promoter when co-transfected with it, it cannot do so in the presence of Eng (Jaynes and O'Farrell, 1988). Hence the expression of Ftz alone in a cell would activate particular genes, whereas its expression in a cell also expressing engrailed would not have any effect (Figure 6.9).

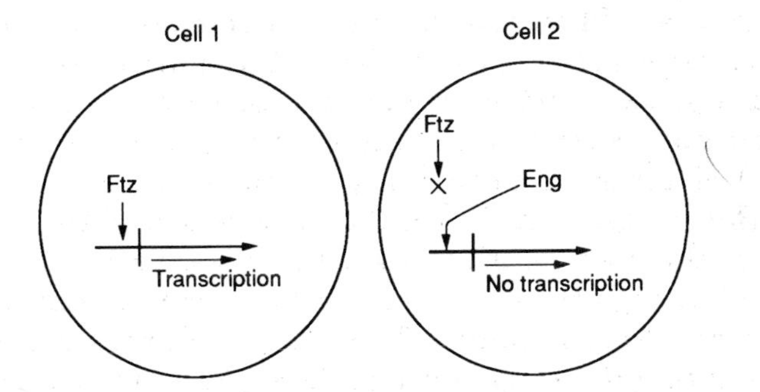

Figure 6.9 Blockage of gene induction by Ftz in cells expressing the engrailed (Eng) protein which binds to the same sequence as Ftz but does not activate transcription.

Similar types of interaction can also take place between different positively acting factors binding to the same site. Thus, although as noted above, the Ftz, prd and zen products can all activate transcription of a target promoter when transfected alone, this effect is relatively small, producing only approximately two-fold activation. In contrast, much larger effects can be obtained by activating the target promoter with two of these proteins in combination, producing 10–20-fold activation, or by all three activators together,

Transfected DNA				Effect on target promoter
Ftz	prd	zen	eng	
–	–	–	–	–
+	–	–	–	+
+	+	–	–	+++
+	–	+	–	+++
+	+	+	–	+++++
+	–	–	+	–

Figure 6.10 Activation of a target promoter containing multiple copies of the identical binding site for the Ftz, prd and zen proteins when it is co-transfected with plasmids expressing each of these proteins. Note the synergistic activation of the promoter when more than one of the activator proteins is co-transfected and the inhibitory effect of the Eng protein which binds to the same site as the other proteins but inhibits gene expression.

producing 400-fold activation of the target (Han *et al.*, 1989; Figure 6.10).

This synergistic stimulation of the target promoter by factors which bind the same sequence can be explained on the basis of the observation that the target promoter contains multiple copies of the binding site for these factors. It has been suggested, therefore (Han *et al.*, 1989), that gene activation proceeds via a multi-switch mechanism in which much stronger activation is produced by binding of different factors at each binding site than by the binding of multiple copies of the same factor (Figure 6.11).

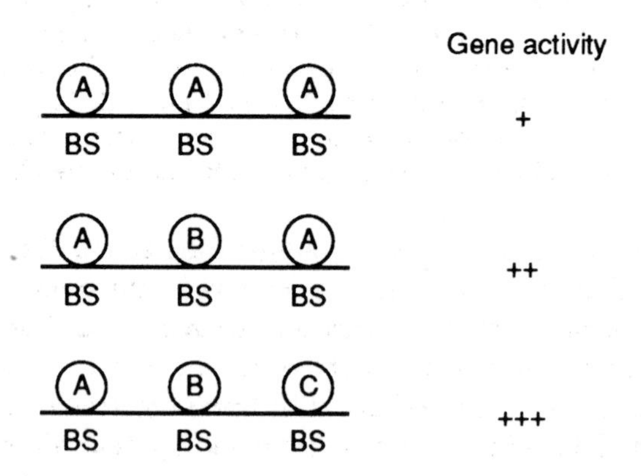

Figure 6.11 Model for the synergistic activation of gene expression by the various homeobox proteins which have the same DNA-binding site as illustrated in Figure 6.10. Gene activation is suggested to be enhanced when different factors (A, B, C) are bound to the multiple copies of their binding site (BS) present in the target promoter compared to the level of activation when the same factor (A) is bound to each of the binding sites.

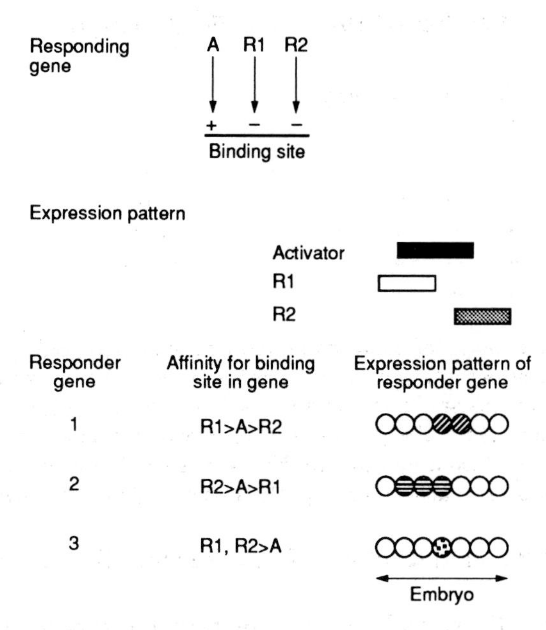

Figure 6.12 Model system producing variation in the spatial expression of three different responder genes along the length of an embryo as indicated by the open or shaded circles. All the responder genes are activated by the activator A and inhibited by the two repressors R1 and R2, each of which is expressed in different but overlapping regions of the embryo indicated by the solid, open or hatched boxes. The three genes differ, however, in their relative affinity for the activators and repressors as indicated. This results in variation between the genes in whether they are activated or repressed in the presence of a particular combination of activator and repressor molecules in each part of the embryo as indicated by the open circles (no expression) or the shaded circles (expression). Thus, for example, in the second circle both A and R1 are present, so genes 1 and 3 are not expressed (since R1 binds more strongly than A to these genes) while gene 2 is expressed (since A binds more strongly than R1 to this gene).

Hence the complex effects of single homeobox factors on the expression of other genes can be rendered still more complex by means of synergistic or inhibitory effects of combinations of factors creating effects which would not be obtained with a single factor alone. Indeed, such interactions of different factors can be used to generate models which predict complex spatial distributions of responder gene activity in response to relatively simple expression patterns of homeobox protein distribution. One such model (Jaynes and O'Farrell, 1988; Figure 6.12) is based on the interaction of activator and repressor molecules which bind to the same binding site in the manner of the Ftz and Eng products. By assuming that target genes vary in the affinity of their binding sites for an activator and two repressor molecules whose areas of expression are overlapping but not identical, it is possible to generate different patterns of responder gene activity in each cell type, depending on which particular factors are present (Figure 6.12).

Hence the activation and repression of target genes by different homeobox factors both alone and in combination can generate complex overlapping patterns of target gene expression of the sort which must occur in development. In the model described above and in our discussion so far, however, it has been assumed that a homeobox factor is either present in a particular cell or is entirely absent. In fact, however, a further level of complexity exists since many homeobox factors are not expressed in a simple on–off manner but rather show a concentration gradient ranging from high levels in one part of the embryo via intermediate levels to low levels in another part. For example, the Bicoid protein (Bcd), whose absence leads to the development of a fly without head and thoracic structures, is found at high levels in the anterior part of the embryo and declines progressively posteriorly, being absent in the posterior one-third of the embryo (Figure 6.13; Driever and Nusslein-Volhard, 1988a).

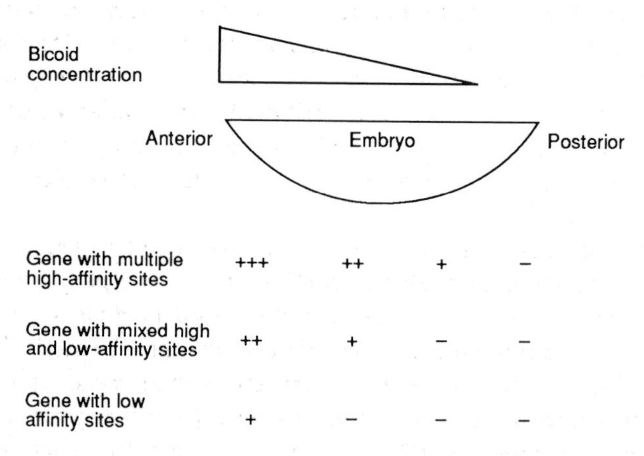

Figure 6.13 The gradient in Bicoid concentration from the anterior to the posterior point of the embryo results in Bicoid-dependent genes with only low-affinity binding sites for the protein being active only at the extreme anterior part of the embryo, whereas genes with high-affinity binding sites are active more posteriorly. Note that in addition to the different posterior boundaries in the expression of genes with high- and low-affinity binding sites, genes with high-affinity binding sites will be expressed at a higher level than genes with low-affinity binding sites at any point in the embryo.

Most interestingly, genes which are activated in response to bicoid contain binding sites in their promoters which have either high affinity or low-affinity for the bicoid protein. If these sites are linked to a marker gene, it can be demonstrated that genes with low affinity binding sites are only activated at high concentrations of bicoid and are therefore expressed only at the extreme anterior end of the embryo. In contrast, genes which have higher-affinity binding sites are active at much lower protein concentrations and will be active both at the anterior end and more posteriorly. Moreover, the greater the number of higher-affinity sites the greater the level of gene expression which will occur at any particular point in the gradient (Driever *et al.*, 1989; Figure 6.13).

The gradient in Bcd expression can be translated, therefore, into the differential expression of various Bcd-dependent genes along the anterior part of the embryo. Each cell in the anterior region will be able to 'sense' its position within the embryo and respond by activating specific genes. Since the genes regulated by Bcd include the homeobox-containing segmentation gene *Hunchback* (Tautz, 1988) which in turn regulates the zinc finger-containing genes *Kruppel* and *Knirps* (Hulskamp *et al.*, 1990), such changes in gene expression lead to the activation of regulatory networks of the type discussed above.

Hence Bcd has all the properties of a morphogen whose concentration gradient determines position in the anterior part of the embryo. This idea is strongly supported by the results of genetic experiments in which the Bcd gradient was artificially manipulated, cells containing artificially increased levels of Bcd assuming a phenotype characteristic of more anterior cells and vice versa (Driever and Nusslein-Volhard, 1988b).

The studies discussed in this section have indicated, therefore, how by various means, such as affecting each other's expression and that of other target genes both positively and negatively, interacting with each other both synergistically and antagonistically, and affecting gene expression in a concentration-dependent manner, the homeobox transcription factors can be used to build up the complex regulatory networks which are necessary for the regulation of *Drosophila* development.

6.3 HOMEOBOX-LIKE GENES IN OTHER ORGANISMS

6.3.1 Homeobox-containing genes

The critical role played by the homeobox genes in the regulation of *Drosophila* development suggests that they may also play a similar role in other organisms. Indeed, we have already seen that the a and α mating type gene products which determine the mating type in yeast contain homeoboxes (Section 5.4). Similarly, in the nematode *C. elegans*, homeoboxes have been identified in several genes whose mutation affects development, such as the *mec-3* gene which controls the terminal differentiation of specific sensory cells (Way and Chalfie, 1988).

As in *Drosophila*, studies in the nematode have been facilitated by the availability of well-characterized mutations affecting development, allowing the corresponding genes to be isolated and the homeobox identified. In higher organisms, where such genetic evidence is unavailable, numerous investigators have used Southern blot hybridization with labelled probes derived from *Drosophila* homeoboxes in an attempt to identify homeobox-containing genes in these species. Thus, for example, Holland and Hogan (1986) used a probe from the *Antennapedia* homeobox to identify homeobox genes in a wide range of species, including not only other invertebrates such as the molluscs, but also chordates such as the sea urchin, and vertebrates such as the mouse (Figure 6.14). Subsequent studies have resulted in the identification of a large number of different homeobox-containing genes from both mouse and human, and many of these genes have been isolated and their DNA sequence obtained (reviews: Dressler and Gruss, 1988; Akam, 1989; Wright *et al.*, 1990).

It is clear from these studies that homeobox-containing genes are not confined to invertebrates such as *Drosophila* but are found also in vertebrates, including mammals such as mouse and human. Interestingly, this evolutionary conservation is not confined to the homeobox portion of these genes. Thus homologues of individual homeobox genes of *Drosophila*, such as *Engrailed* and *Deformed*, have been identified in mouse and human, the fly and mammalian proteins showing extensive sequence homology which extends beyond the homeobox to include other regions of the proteins (Joyner and Martin, 1985; Mavilio *et al.*, 1986).

Moreover, the similarity between the *Drosophila* and mammalian systems extends also to the manner in which the homeobox-containing genes are organized in the genome. Thus in both *Drosophila* and

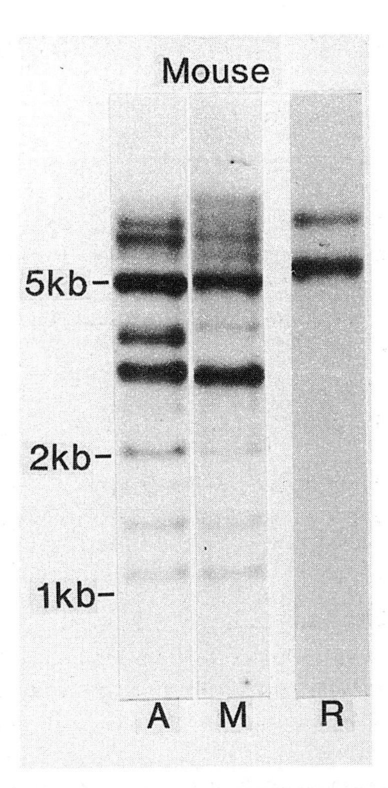

Figure 6.14 Southern blot of mouse DNA hybridized with a probe from the *Drosophila* Antennapedia gene (A), a mouse *Antennapedia*-like gene (M) and mouse ribosomal DNA (R). Note the presence of DNA fragments which hybridize to both *Antennapedia*-like DNAs but not to ribosomal DNA and which represent *Antennapedia*-like sequences in the mouse genome.

mammals these genes are organized into clusters containing several homeobox-containing genes, with homologous genes in the different organisms occupying equivalent positions in the clusters. For example, in a detailed comparison of the genes in the *Drosophila Bithorax* and *Antennapedia* complexes with those of the mouse homeobox gene complex Hox 2, Graham *et al.* (1989) showed that the first gene in the mouse complex, Hox 2.5, was most homologous to the first gene in the *Drosophila Bithorax* complex, Abd-B, and so on across the complex (Figure 6.15). Hence both the homeobox genes and their arrangement are highly conserved in evolution, the common ancestor of mammals and insects having presumably possessed a similar cluster of homeobox-containing genes.

Most interestingly, in both *Drosophila* and mammals, the position of a gene within a cluster is related to its expression pattern during

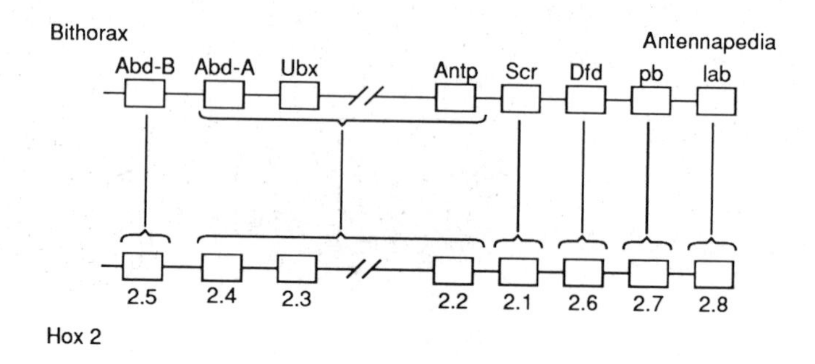

Figure 6.15 Comparison of the *Bithorax/Antennapedia* complex on *Drosophila* chromosome 3 with the Hox 2 complex on mouse chromosome 11. Individual genes are indicated by open boxes. Note that each gene in the *Drosophila* complex is most homologous to the equivalent gene in the mouse complex, as indicated by the vertical lines. The *Drosophila* Abd-A, Ubx and Antp genes are too closely related to each other to be individually related to a particular mouse gene but are most closely related to the Hox 2.4, 2.3 and 2.2 genes which occupy the equivalent positions in the Hox 2 cluster as indicated by the brackets.

embryogenesis. Thus in the mouse Hox 2 cluster discussed above, all the genes are expressed in the developing central nervous system of the embryo. However, in moving from the 5′ to the 3′ end of the cluster (i.e. from Hox 2.5 to Hox 2.8 in Figure 6.15), each successive gene displays a more anterior boundary of expression within the central nervous system (Figures 6.16 and 6.17). Similar expression patterns have also been observed in *Drosophila*, where each successive gene in the *Bithorax* and *Antennapedia* clusters is expressed more anteriorly and affects progressively more anterior segments when it is mutated (Harding *et al.*, 1985; Akam, 1987).

In the case of the mouse genes a possible molecular mechanism for this differential expression pattern is provided by the recent observation (Simeone *et al.*, 1990) that genes in the 3′ half of the Hox 2 cluster are activated in cultured cells by treatment with low levels of retinoic acid whereas genes in the 5′ half of the cluster require much higher levels of retinoic acid for their activation. Considerable evidence exists that retinoic acid can act as a morphogen in vertebrate development and it has been suggested that a gradient of retinoic acid concentration may exist across the developing embryo (review: Brickell and Tickle, 1989). Hence the observed difference in expression of the Hox 2 genes could be controlled by a

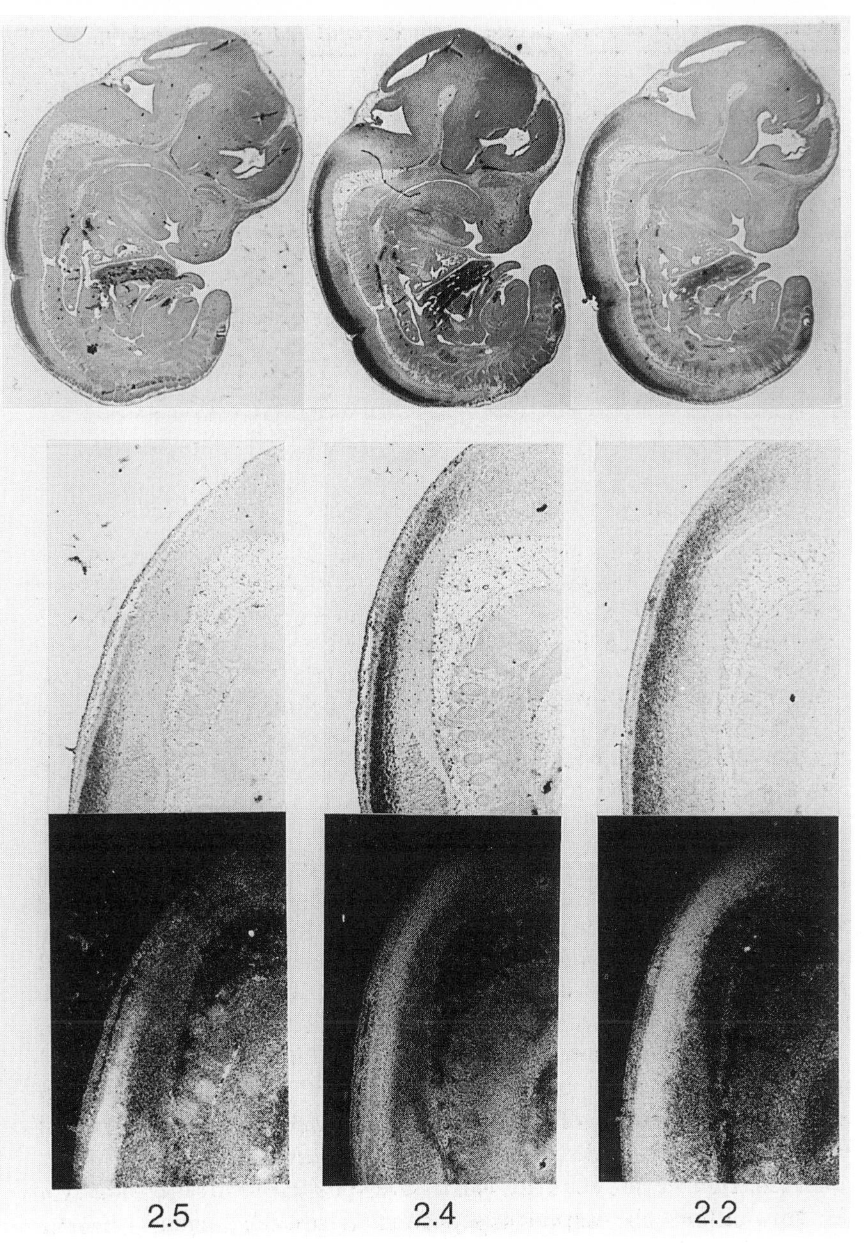

Figure 6.16 Comparison of the expression pattern of the Hox 2.5, 2.4 and 2.2 genes in the 12.5-day mouse embryo. The top panel shows *in situ* hybridization with the appropriate gene probe to a section of the entire embryo, whilst the middle row shows a high-power view of the region in which the anterior limit of gene expression occurs. In these panels, which show the sections in bright field, hybridization of the probe and therefore gene expression is indicated by the dark areas. In the lower panel, which shows the same area in dark field, hybridization is indicated by the bright areas. Note the progressively more anterior boundary of expression of Hox 2.2 compared to Hox 2.4 and to Hox 2.5 and compare with their positions in the Hox 2 complex in Figure 6.15.

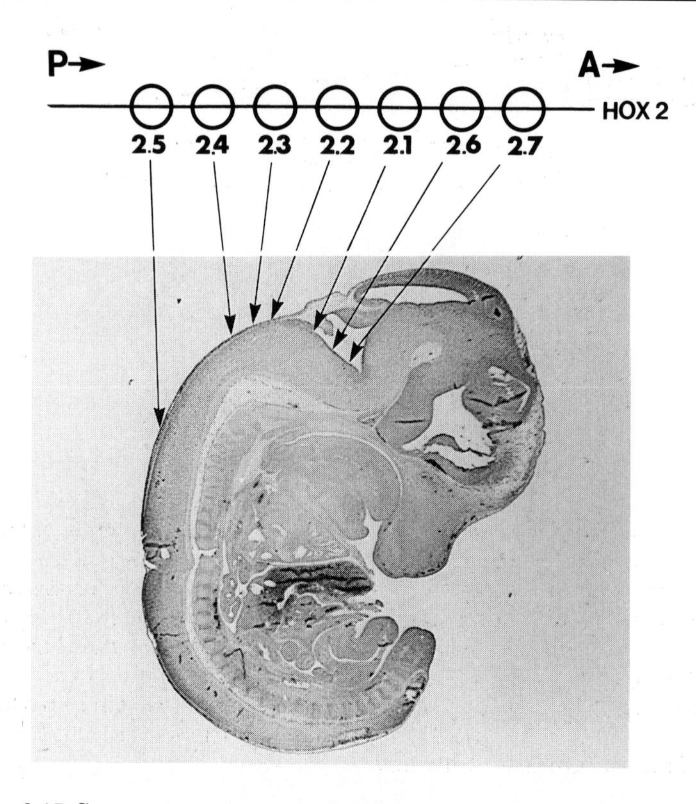

Figure 6.17 Summary of the anterior boundary of expression of the genes in the Hox 2 complex indicated on a section of a 12.5-day mouse embryo and compared to the position of the gene in the Hox 2 cluster. Note the progressively more anterior boundary of expression from the 5′ to the 3′ end of the Hox 2 cluster.

retinoic acid gradient in a manner similar to the differential activation of Bcd-dependent genes in the bicoid concentration gradient observed in *Drosophila* (Figure 6.18). In turn the Hox 2 genes, like their *Drosophila* counterparts, would switch on other genes required in cells at particular positions in the embryo, accounting for the morphogenetic effects of retinoic acid.

This process is, of course, related to the induction of gene expression in response to treatment with steroid hormones discussed in Section 4.3, retinoic acid functioning by binding to its receptor which is a member of the steroid–thyroid hormone receptor superfamily and which in turn binds to specific sequences within retinoic acid-responsive genes, activating their expression (review: Evans, 1988). Hence the activation of regulatory genes and the initiation of a regulatory cascade can be achieved by a modification of the processes mediating inducible gene expression discussed in Chapter

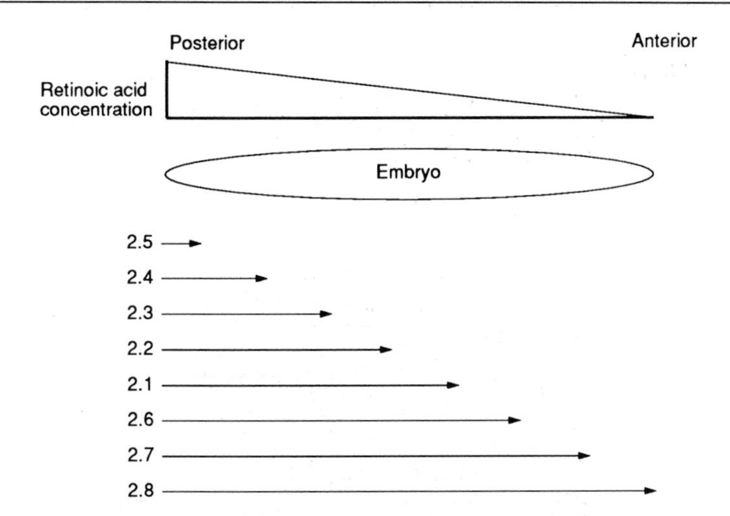

Figure 6.18 Model for the progressively more anterior expression of the genes in the Hox 2 cluster in which expression is controlled by a posterior to anterior gradient in retinoic acid concentration and the increasing sensitivity to induction by retinoic acid which occurs from the 5′ to the 3′ end of the cluster. Thus because genes at the 3′ end of the cluster are inducible by very low levels of retinoic acid they will be expressed in anterior points of the embryo where the retinoic acid level will be too low to induce the genes at the 5′ end of the cluster, which require a much higher level of retinoic acid to be activated.

4. Although this mechanism for regulating expression of the Hox 2 genes is attractive, therefore, no direct evidence is as yet available that retinoic acid is actually responsible for the expression patterns of these genes observed in the developing embryo.

Whatever the case, it is clear that homologues of the *Drosophila* homeobox-containing genes do exist in mammals and other vertebrates and that they are expressed in specific cell types in the early embryo. It is likely, therefore, that they play a similar role to their *Drosophila* counterparts in regulating development by affecting the transcription of other genes. In agreement with this idea, the artificial expression of the mouse Hox 2.2 gene in *Drosophila* has been shown to produce similar effects on development to those produced by the expression of *Antennapedia*, indicating that the mouse gene can substitute functionally for Antp, presumably by regulating the same set of downstream genes (Malicki *et al.*, 1990). Thus far, however, because of the lack of genetic information available in mammals compared to *Drosophila*, no direct evidence on the role of the mammalian factors in regulating mammalian development is available. Such evidence is available, however, for another group of mammalian homeobox-containing proteins which contain the homeobox as part of a much larger conserved domain. These POU proteins are discussed in the next section.

6.3.2 Homeobox-containing POU proteins

As discussed in other chapters (Sections 5.2.2 and 9.2.1) the octamer-binding transcription factors Oct-1 and Oct-2 play an important role in regulating the expression of specific genes such as those encoding histone H2B, the SnRNA molecules and the immunoglobulins. Similarly, the transcription factor Pit-1, which binds to a sequence two bases different from the octamer sequence, plays a critical role in pituitary-specific gene expression (Section 1.2.3).

When the genes encoding these factors were cloned, they were found to share a 150–160-amino-acid sequence which was also found in the protein encoded by the nematode gene *unc*-86, whose mutation affects sensory neuron development (Finney *et al.*, 1988). This common POU (Pit-Oct-Unc) domain contains both a homeobox sequence and a second conserved domain, the POU-specific domain (Figure 6.19) (reviews: Herr *et al.*, 1988; Hoey and Levine, 1988; Ruvkun and Finney, 1991).

Interestingly, whilst the homeoboxes of the different POU proteins are closely related to one another (53 out of 60 homeobox residues are the same in Oct-1 and Oct-2 and 34 out of 60 in Oct-1 and Pit-1) they show less similarity to the homeoboxes of other mammalian genes lacking the POU-specific domain, sharing at best only 21 out of 60 homeobox residues. Hence they may represent a distinct class of homeobox proteins containing both a POU-specific domain and a diverged homeodomain.

As with the *Drosophila* homeobox proteins, however, the isolated homeodomains of the Pit-1 and Oct-1 proteins are capable of mediating sequence-specific DNA binding in the absence of the POU-specific domain (Theill *et al.*, 1989; Kristie and Sharp 1990). The affinity and specificity of binding by such an isolated homeodomain is much lower, however, than that exhibited by the intact POU domain, indicating that the POU-specific domain plays a critical role in producing high-affinity binding to specific DNA sequences (Ingraham *et al.*, 1990; Kristie and Sharp, 1990).

In addition to its role in DNA binding, the POU domain also plays a critical role in several other features of the POU proteins which are not found in the simple homeobox-containing proteins. For example the ability of both Oct-1 and Oct-2 (like NF1—see Section 3.3.2) to stimulate DNA replication as well as transcription is also a property of the isolated POU domains of these factors (Verrijzer *et*

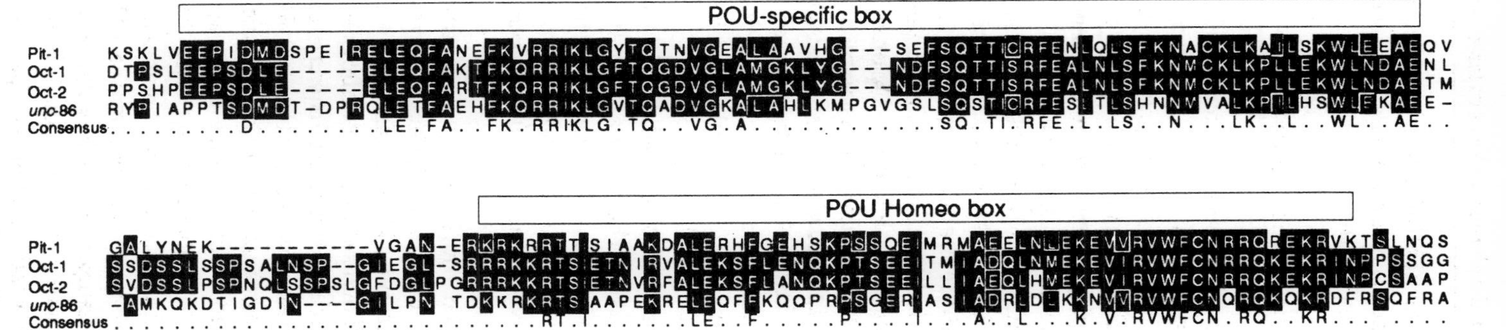

Figure 6.19 Amino acid sequences of the POU proteins. The homeodomain and the POU-specific domain are indicated. Solid boxes indicate regions of identity between the different POU proteins. The final line shows a consensus sequence obtained from the four proteins.

al., 1990). Similarly, the ability of Oct-1 and not Oct-2 to interact with the herpes simplex virus trans-activator protein Vmw65 is controlled by differences in the homeodomain region of the POU domains in the two proteins (Stern *et al.*, 1989). In contrast, however, it should be noted that differences in the ability of Oct-1 and Oct-2 to activate different octamer-containing promoters are controlled by regions outside the POU domain of the molecule which contain the activation domains (Tanaka and Herr, 1990; see Section 9.2.1).

As in the case of the homeobox-containing proteins, the POU proteins appear to play a critical role in the regulation of developmental gene expression and in the development of specific cell types (review: Ruvkun and Finney, 1991). Thus the *unc*-86 mutation in the nematode results, for example, in the lack of touch receptor neurons or male-specific cephalic companion neurons (Desai *et al.*, 1988), indicating that this POU protein is required for the development of these specific neuronal cell types. Similarly, recent studies have shown that the dwarf mutation in the mouse which results in abnormal development of the pituitary gland and consequent failure of normal growth is itself caused by inactivation of the gene encoding Pit-1 (Li *et al.*, 1990). Hence the activity of the Pit-1 gene is required for the normal development of the cells of the anterior pituitary gland.

The role of POU proteins in the regulation of developmental gene expression is not confined to the original four POU proteins (Pit-1, Oct-1, Oct-2 and *unc*-86). Thus, for example, the *Drosophila* POU protein CFla has been shown to play a critical role in the regulation of the dopa decarboxylase gene in selected dopaminergic neurons (Johnson and Hirsh, 1990). Moreover, all the novel POU domain-containing genes isolated by He *et al.* (1989) from the rat, on the basis of their containing a POU domain, are expressed in the embryonic and adult brain, suggesting a similar role for these proteins in the regulation of neuronal-specific gene expression. Such a close connection of POU proteins and the central nervous system is also supported by studies using the original POU domain genes which revealed expression in the embryonic brain even in the case of Oct-2, which had previously thought to be expressed only in B cells (He *et al.*, 1989).

It is clear, therefore, that, like the homeobox proteins, POU proteins occur in a wide variety of organisms and play an important role in the regulation of gene expression in development. Moreover, these proteins may be of particular importance in the development of the central nervous system.

6.4 CONCLUSIONS

The process of development is clearly immensely complex. The modulation of gene expression during this process will clearly involve many of the mechanisms which have been discussed in previous chapters; some genes must become and remain active in all cell types (Chapter 3), others must be induced in response to specific signals (Chapter 4) whilst others must become and remain active in a particular type of differentiated cell (Chapter 5). In turn such processes must be co-ordinated both temporally and spatially so that each organ forms at the right time and in the right place.

The insights provided by the genetic identification of the *Drosophila* transcription factors which control this process has allowed some understanding of how this might be achieved. Thus gradients of transcription factors such as Bcd, which are laid down in the egg, can be translated into differential gene activity and hence produce specific cell types at particular points in the embryo. Similarly, retinoic acid can activate the expression of specific genes such as the mammalian homeobox-containing genes in a similar manner to the induction of gene expression in response to steroid hormones discussed in Chapter 4. Hence a gradient in a chemical such as retinoic acid can be used to determine position and initiate a regulatory cascade via its ability to differentially activate regulatory genes.

It is clear, therefore, that transcription factors play a critical role in the regulation of development and that gradients in these factors or other substances may allow positional information to be determined. Further studies on these factors, their interaction with each other and with their target genes, will be necessary, however, before the process of development is fully understood.

REFERENCES

Akam, M. E. (1987). The molecular basis for metameric pattern in the *Drosophila* embryo. *Development* **101**, 1–22.

Akam, M. (1989). Hox and HOM: homologous gene clusters in insects and vertebrates. *Cell* **57**, 347–349.

Beachy, P. A., Krasnow, M. A., Gavis, E. R. and Hogness, D. S. (1988). An *ultrabithorax* protein binds sequences near its own and the *Antennapedia* P1 promoters. *Cell* **55**, 1069–1081.

Biggin, M. D. and Tjian, R. (1989). A purified *Drosophila* homeodomain protein represses transcription *in vitro*. *Cell* **58**, 433–440.

Brickell, P. M. and Tickle, C. (1989). Morphogens in chick limb development. *BioEssays* **11**, 145–149.

Desai, C., Garriga, G., McIntire, S. L. and Horvitz, H. R. (1988). A genetic pathway for the development of *Caenorhabditis elegans* HSN motor neurons. *Nature* **336**, 638–646.

Dressler, G. R. and Gruss, P. (1988). Do multigene families regulate vertebrate development? *Trends in Genetics* **4**, 214–219.

Driever, W. and Nusslein-Volhard, C. (1988a). A gradient of bicoid protein in *Drosophila* embryos. *Cell* **54**, 83–93.

Driever, W. and Nusslein-Volhard, C. (1988b). The bicoid protein determines position in the *Drosophila* embryo in a concentration-dependent manner. *Cell* **54**, 95–104.

Driever, W., Thoma, G. and Nusslein-Volhard, C. (1989). Determination of spatial domains of zygotic gene expression in the *Drosophila* embryo by the affinity of binding sites for the bicoid morphogen. *Nature* **340**, 363–367.

Evans, R. M. (1988). The steroid and thyroid hormone receptor gene superfamily. *Science* **240**, 889–895.

Finney, M., Ruvkun, G. and Horvitz, H. R. (1988). The *C. elegans* cell lineage and differentiation gene *unc*-86 encodes a protein with a homeodomain and extended similarity to transcription factors. *Cell* **55**, 757–769.

Gehring, W. J. (1987). Homeo boxes in the study of development. *Science* **236**, 1245–1252.

Graham, A., Papalopulu, N. and Krumlauf, R. (1989). The murine and *Drosophila* homeobox gene complexes have common features of organization and expression. *Cell* **57**, 367–378.

Hadorn, E. (1968). Transdetermination in cells. *Scientific American* **219** 110–120.

Han, K., Levine, M. S. and Manley, J. L. (1989). Synergistic activation and repression of transcription by *Drosophila* homeobox proteins. *Cell* **56**, 573–583.

Harding, K., Weeden, C., McGinnis, W. and Levine, M. (1985). Spatially regulated expression of homeotic genes in *Drosophila*. *Science* **229**, 1236–1242.

Hayashi, S. and Scott, M. P. (1990). What determines the specificity of action of *Drosophila* homeodomain proteins. *Cell* **63**, 883–894.

He, X., Treacy, M. N., Simmons, D. M., Ingraham, H. A., Swanson, L. S. and Rosenfeld, M. G. (1989). Expression of a large family of POU-domain regulatory genes in mammalian brain development. *Nature* **340**, 35–42.

Herr, W., Sturm, R. A., Clerc, R. G., Corcoran, L. M., Baltimore, D., Sharp, P. A., Ingraham, H. A., Rosenfeld, M. G., Finney, M., Ruvkun, G. and Horvitz, H. R. (1988). The POU domain: a large conserved region in the mammalian Pit-1, Oct-1, Oct-2 and *Caenorhabditis elegans* unc-86 gene products. *Genes and Development* **2**, 1513–1516.

Hoey, T. and Levine, M. (1988). Divergent homeobox proteins recognize similar DNA sequences in *Drosophila*. *Nature* **332**, 858–861.

Holland, P. W. H. and Hogan, B. L. M. (1986). Phylogenetic distribution of Antennapedia-like homeoboxes. *Nature* **321**, 251–253.

Hulskamp, M., Pfeifle, C. and Tautz, D. (1990). A morphogenetic gradient of hunchback protein organizes the expression of the gap genes Kruppel and Knirps in the early *Drosophila* embryo. *Nature* **346**, 577–580.

Ingham, P. W. (1988). The molecular genetics of embryonic pattern formation in *Drosophila*. *Nature* **335**, 25–34.

Ingraham, H. A., Flynn, S. E., Voss, J. W., Albert, V. R., Kapiloff, M. S., Wilson, L. and Rosenfeld, M. G. (1990). The POU specific domain is essential for sequence-specific high affinity DNA binding and DNA-dependent Pit1–Pit1 interactions. *Cell* **61**, 1021–1033.

Jaynes, J. B. and O'Farrell, P. H. (1988). Activation and repression of transcription by homeodomain-containing proteins that bind a common site. *Nature* **336**, 744–749.

Johnson, W. A. and Hirsh, J. (1990). Binding of a *Drosophila* POU-domain protein to a sequence element regulating gene expression in specific dopaminergic neurons. *Nature* **343**, 467–470.

Joyner, A. L. and Martin, G. R. (1987). En-1 and En-2 two mouse genes with sequence homology to the *Drosophila* Engrailed gene; expression during embryogenesis. *Genes and Development* **1**, 29–38.

Krasnow, M. A., Saffman, E. E., Kornfeld, K. and Hogness, D. S. (1989). Transcriptional activation and repression by Ultrabithorax proteins in cultured *Drosophila* cells. *Cell* **57**, 1031–1043.

Kristie, T. M. and Sharp, P. A. (1990). Interaction of the Oct-1 POU subdomains with specific DNA sequences and with the HSV alpha-transactivator protein. *Genes and Development* **4**, 2383–2396.

Levine, M. and Hoey, T. (1988). Homeobox proteins as sequence-specific transcription factors. *Cell* **55**, 537–540.

Li, S., Crenshaw, E. B., Rawson, E. J., Simmons, D. M., Swanson, L. M. and Rosenfeld, M. G. (1990). Dwarf locus mutants lacking three pituitary cell types result from mutations in the POU-domain gene Pit-1. *Nature* **347**, 528–533.

Malicki, J., Schughart, K. and McGinnis, W. (1990). Mouse Hox 2.2 specifies thoracic segmental identity in *Drosophila* embryos and larvae. *Cell* **63**, 961–967.

Mavilio, F., Simeone, A., Giampado, A., Faiella, A., Zappavigna, V., Acampora, D., Poiana, G., Russo, G., Peschlo, C. and Boncinelli, E. (1986). Differential and stage related expression in embryonic tissue of a new human homeo box gene. *Nature* **324**, 664–667.

Mihara, J. and Kaiser, E. T. (1988). A chemically synthesized Antennapedia homeo domain binds to a specific DNA sequence. *Science* **242**, 925–927.

Muller, M., Affolter, M., Leupin, W., Otting, G., Wathrich, K. and Gehring, W. J. (1988). Isolation and sequence specific DNA binding of the Antennapedia homeodomain. *EMBO Journal* **7**, 4299–4304.

Pignoni, F., Baldarelli, R. M., Steingrimsson, E., Diaz, R. J., Patapoutian, A., Merriam, J. R. and Lengyel, J. A. (1990). The *Drosophila* gene *tailless* is expressed at the embryonic termini and is a member of the steroid receptor superfamily. *Cell* **62**, 151–163.

Rosenberg, U. B., Schroder, C., Preiss, A., Kienlin, A., Cote, S., Riede, I. and Jackle, H. (1986). Structural homology of the *Drosophila Kruppel* gene with *Xenopus* transcription factor III A. *Nature* **319**, 336–339.

Ruvkun, G. and Finney, M. (1991). Regulation of transcription and cell identity by POU domains proteins. *Cell* **64**, 475–482.

Scott, M. P. and Carroll, S. B. (1987). The segmentation and homeotic gene network in early *Drosophila* development. *Cell* **51**, 689–698.

Scott, M. P., Tamkun, J. W. and Hartzell, G. W. (1989). Structure and function of the homeodomain. *Biochimica et Biophysica Acta* **989**, 25–48.

Shepherd, J. C. W., McGinnis, W., Carrasco, A. E., De Robertis, E. M. and Gehring, W. J. (1984). Fly and frog homoeo domains show homologies with yeast mating type regulatory loci. *Nature* **310**, 70–71.

Simeone, A., Acampora, D., Arcioni, L., Andrews, P. W., Doncinelli, E. and Mavilio, F. (1990). Sequential activation of Hox 2 homeobox genes by retinoic acid in human embryonal carcinoma cells. *Nature* **346**, 763–766.

Stern, S., Tanaka, M. and Herr, W. (1989). The Oct-1 homeodomain directs formation of a multiprotein-DNA complex with the HSV transactivator VP16. *Nature* **341**, 624–630.

Struhl, G. (1982). Genes controlling segmental specification in the *Drosophila* thorax. *Proceedings of the National Academy of Sciences, USA* **79**, 7380–7384.

Tanaka, M. and Herr, W. (1990). Differential transcriptional activation by Oct-1 and Oct-2: Interdependent activation domains induce Oct-2 phosphorylation. *Cell* **60**, 375–386.

Tautz, D. (1988). Regulation of the *Drosophila* segmentation gene hunchback by two maternal morphogenetic centres. *Nature* **332**, 281–284.

Theill, L. E., Castrillo, J.-L., Wu, D. and Karin, M. (1989). Dissection of functional domains of the pituitary-specific transcription factor GHF-1. *Nature* **342**, 945–948.

Verrijzer, C. R., Kal, A. J. and van der Vliet, P. C. (1990). The DNA binding domain (POU domain) of transcription factor Oct-1 suffices for stimulation of DNA replication. *EMBO Journal* **9**, 1883–1888.

Way, J. C. and Chalfie, M. (1988). *mec-3*, a homeobox-containing gene that specifies differentiation of the touch receptor neurons in *C. elegans*. *Cell* **54**, 5–16.

Winslow, G. M., Hayashi, S., Krasnow, M., Hogness, D. S. and Scott, M. P. (1989). Transcriptional activation and repression by the *Antennapedia* and *fushi tarazu* proteins in cultured *Drosophila* cells. *Cell* **57**, 1017–1030.

Wright, C. V. E., Cho, K. W. Y., Oliver, G. and De Robertis, E. M. (1990). Vertebrate homeodomain proteins: families of region-specific transcription factors. *Trends in Biochemical Science* **14**, 52–56.

CHAPTER SEVEN

Transcription factors and cancer

7.1 CELLULAR ONCOGENES AND CANCER

In previous chapters we have discussed the involvement of transcription factors in normal cellular regulatory processes, for example constitutive, inducible, cell type-specific or developmentally regulated transcription. It is not surprising that aspects of this complex process can go wrong and that the resulting defects in transcription factors can result in disease. For example, the absence of a specific transcription factor necessary for the transcription of the HLA class II genes results in the lack of transcription of these genes and the absence of their protein products which is characteristic of one form of congenital severe combined immunodeficiency disease (Reith *et al.*, 1988).

A special place in the human diseases which can involve alterations in transcription factors is occupied, however, by cancer. Thus, because this disease results from growth in an inappropriate place or at an inappropriate time, it can be caused not only by deficiencies in particular genes but also by the enhanced expression or activation of specific cellular genes involved in growth regulatory processes which are normally only expressed at low levels or very transiently.

Interestingly, the cancer-causing genes of this type known as oncogenes (reviews: Weinberg, 1985; Bishop, 1987) were originally identified within cancer-causing retroviruses which had picked them up from the cellular genome. Within the virus, the oncogene has become activated either by over-expression or by mutation and is therefore responsible for the ability of the virus to transform cells to a cancerous phenotype. In contrast, the homologous gene within the cellular genome is clearly not always cancer-causing, since all cells are not cancerous. It can be activated, however, into a cancer-causing form either by over-expression or by mutation, and hence these genes can play an important role in the generation of human cancer (Figure 7.1). The forms of the oncogene isolated from the retrovirus and from the normal cellular genome are distinguished by the prefixes v and c respectively, as in v-*onc* and c-*onc*.

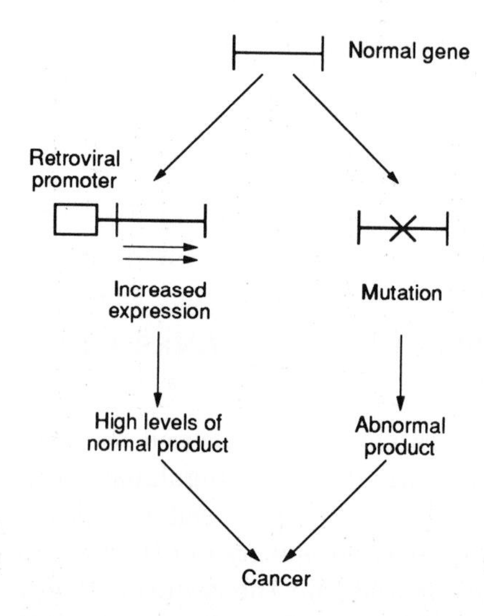

Figure 7.1 A cellular proto-oncogene can be converted into a cancer-causing oncogene by increased expression or by mutation.

Despite this potential to cause cancer, the c-*onc* genes are highly conserved in evolution, being found not only in the species from which the original virus was isolated but also in a wide range of other eukaryotes. This indicates that the products of these oncogenes play a critical role in the regulation of normal cellular growth processes, their malregulation or mutation resulting, therefore, in abnormal growth and cancer. In agreement with this idea, oncogenes identified in this way include genes encoding many different types of protein involved in growth control such as growth factors, growth factor receptors and G proteins. They also include, however, several genes encoding cellular transcription factors which normally regulate specific sets of target genes. This chapter discusses several cases of this type and the insights they have provided into the processes regulating gene expression in normal cells and their malregulation in cancer. A number of these factors, together with their cellular targets, are listed in Table 7.1 (review: Gutman and Wasylyk, 1991).

Table 7.1 Oncogenes and transcription factors

Oncogene	Site bound and DNA sequence	Target genes	Comments	Reference
erbA	Thyroid response element TCAGGTCATGACCTGA	Growth hormone, myosin heavy chain	Mutated form of the thyroid hormone receptor	Sap *et al.* (1986) Sap *et al.* (1989)
ets	PEA 3 site CACTTCCT	Collagenase, stromelysin, interleukin-2, etc.	Often found in association with AP1 site (see below)	Wasylyk *et al.* (1990)
fos	AP1 site TGAGTCAG	Collagenase, stromelysin, interleukin-2, etc.	Can only bind to AP1 site in the presence of Jun as Fos–Jun dimer	Curran and Franza (1988)
jun	AP1 site TGAGTCAG	Collagenase, stromelysin, interleukin-2, etc.	Can bind to AP1 site alone as Jun–Jun homodimer	Curran and Franza (1988)
myb	PyAACG/TG	*mim*-1	*mim*-1 isolated directly as a *myb*-regulated gene	Cole (1990)
myc	CACGTG	Plasminogen activator inhibitor-1, 70-kDa heat-shock protein	Forms DNA binding heterodimer with Max protein	Cole (1986) Cole (1990) Blackwood and Eisenman (1991)
rel	GTGGAGATGGGG AATCCCCA	?	Closely related to transcription factors NFκB and KBF1	Gilmore (1990)
spi-1	GAGGAA	?	Identical to PU.1 transcription factor	Goebl (1990)

7.2 FOS, JUN AND AP1

As noted in Table 4.1, the AP1 binding site is a DNA sequence that renders genes which contain it inducible by treatment with phorbol esters such as TPA (Angel *et al.*, 1987; Lee *et al.*, 1987a). The activity binding to this site is referred to as AP1 (activator protein 1). It is clear, however, that preparations of AP1 purified by affinity chromatography on an AP1-binding site contain several different proteins (Lee *et al.*, 1987b) (review: Curran and Franza, 1988).

DNA binding site

```
GCN4   5' T G A C/G T C A T 3'
 AP1   5' T G A  G  T C A G 3'
```

Figure 7.2 Relationship of the DNA binding sites for the yeast transcription factor GCN4 and the mammalian transcription factor AP1.

A possible clue as to the identity of one of these AP1 binding proteins was provided by the finding that the yeast protein GCN4, which induces transcription of several yeast genes involved in amino acid biosynthesis, does so by binding to a site very similar to the AP1 site (Hill *et al.*, 1986; Figure 7.2). In turn, the DNA-binding region of GCN4 shows strong homology at the amino acid level to v-*jun*, the oncogene of avian sarcoma virus ASV17 (Figure 7.3; Vogt *et al.*, 1987). This suggested, therefore, that the protein encoded by the cellular homologue of this gene, c-*jun*, which was known to be a nuclearly located DNA-binding protein, might be one of the proteins which bind to the AP1 site.

```
Jun   206 P L F P I D M E S Q E R I K A E R K R M R N R I A A S K S R K
GCN4  216 P L S P I V P E S S D P     A A L K R A R N T E A A R R S R A

          R K L E R I A R L E E K V K T L K A Q N S E L A S T A N M L R
          R K L Q R M K Q L E D K V               E E L L S K N Y H L E

          E Q V A Q L K Q K V M N H V N S G C Q L M L T Q Q L G T F 296
          N E V A R L K K K L V G E R 281
```

Figure 7.3 Comparison of the C-terminal amino acid sequences of the chicken Jun protein and the yeast transcription factor GCN4. Boxes indicate identical residues.

In agreement with this, antibodies against the Jun protein react with purified AP1-binding proteins (Bos *et al.*, 1988) whilst Jun protein expressed in bacteria can bind to AP1-binding sites (Angel *et al.*, 1988). Hence the Jun protein is capable of binding to the

AP1-binding site and constitutes one component of purified AP1 preparations which also contain other Jun-related proteins such as Jun B (Figure 7.4). Moreover, co-transfection of a vector expressing the Jun protein with a target promoter resulted in increased transcription if the target gene contained AP1-binding sites but not if it lacked them, indicating that Jun was capable of stimulating transcription via the AP1 site (Figure 7.5; Angel *et al.*, 1988). Hence the Jun oncogene product is a sequence-specific transcription factor capable of stimulating transcription of genes containing its binding site.

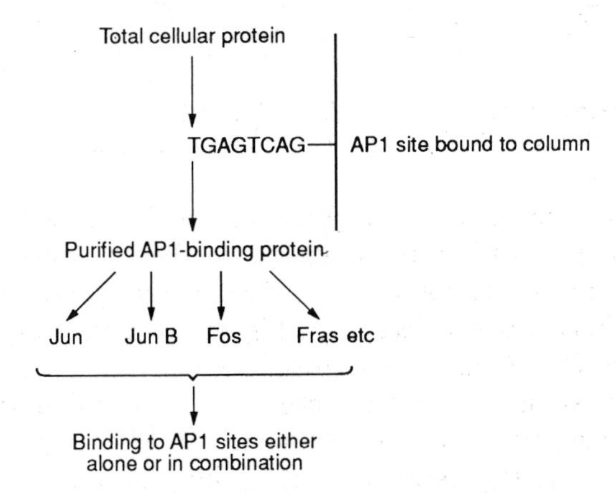

Figure 7.4 Passage of total cellular proteins through a column containing an AP1 site results in the purification of several cellular proteins, including Jun, Jun B, Fos and Fos-related antigens (Fras) which are capable of binding to the AP1 site either alone or in combination.

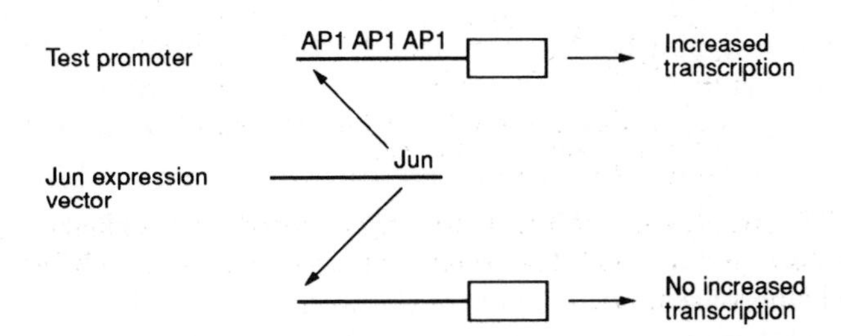

Figure 7.5 Artificial expression of the Jun protein in an expression vector results in the activation of a target promoter containing several AP1-binding sites but has no effect on a similar promoter lacking these sites, indicating that Jun can specifically activate gene expression via AP1-binding sites.

In addition to Jun and Jun-related proteins, purified AP1 preparations also contain the product of another oncogene, c-*fos*, as well as several Fos-related proteins known as the Fras (Fos-related antigens; see Figure 7.4). Unlike Jun, however, Fos cannot bind to the AP1 site alone but can do so only in the presence of another protein p39, which is identical to Jun (Rauscher *et al.*, 1988). Hence in addition to its ability to bind to AP1 sites alone, Jun can also mediate binding to this site by the Fos protein. Such DNA binding by Fos and Jun is dependent on the formation of a dimeric molecule. Although Jun can form a DNA-binding homodimer, Fos cannot do so. Hence DNA binding by Fos is dependent upon the formation of a heterodimer between Fos and Jun which binds to the AP1 site with approximately 30-fold greater affinity than the Jun homodimer (Figure 7.6; Halazonetis *et al.*, 1988).

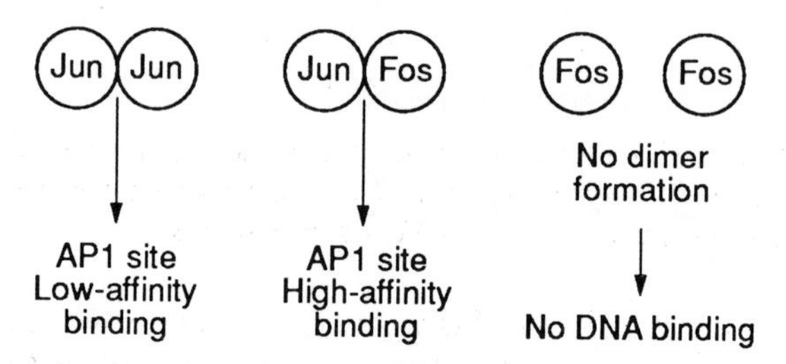

Figure 7.6 Heterodimer formation between Fos and Jun results in a complex capable of binding to an AP1 site with approximately 30-fold greater affinity than a Jun homodimer, whilst a Fos homodimer cannot bind to the AP1 site.

It is clear, therefore, that both Fos and Jun, which were originally isolated in oncogenic retroviruses, are also cellular transcription factors which play an important role in inducing specific cellular genes following phorbol ester treatment. Increased levels of Fos and Jun occur in cells following treatment with phorbol esters (Lamph *et al.*, 1988), indicating that these substances act, at least in part, by increasing the levels of Fos and Jun which in turn bind to the AP1 sites in phorbol ester-responsive genes and activate their expression.

Similar increases in the levels of Fos and Jun as well as Jun B and the Fos-related protein Fra-1 are also observed when quiescent cells are stimulated to grow by treatment with growth factors or serum (Lamph *et al.*, 1988; Rauscher *et al.*, 1988), indicating that these substances act, at least in part, by increasing the levels of Fos and Jun which in turn will switch on genes whose products are

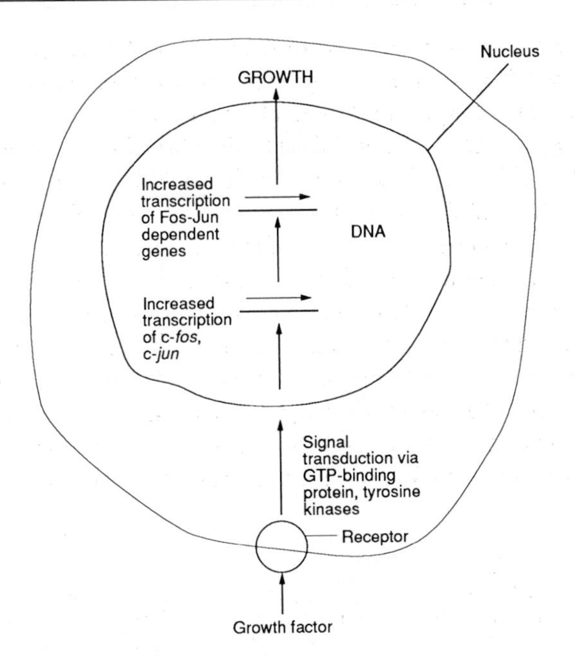

Figure 7.7 Growth factor stimulation of cells results in increased transcription of the c-*fos* and c-*jun* genes, which in turn stimulates transcription of genes which are activated by the Fos–Jun complex.

necessary for growth itself (Figure 7.7). Hence Fos and Jun play a critical role in normal cells, as transcription factors inducing phorbol ester- or growth-dependent genes.

Normally, levels of Fos and Jun increase only transiently following growth factor treatment, resulting in a period of brief controlled growth. Clearly, continuous elevation of these proteins, such as would occur when cells become infected with a retrovirus expressing one of them, would result in cells which exhibited continuous uncontrolled growth and were not subject to normal growth regulatory signals. Since such uncontrolled growth is one of the characteristics of cancer cells, it is relatively easy to link the role of Fos and Jun in inducing genes required for growth with their ability to cause cancer. Normally, however, the transformation of a cell to a transformed cancerous phenotype requires more than simply its conversion to a continuously growing immortal cell (review: Land *et al.*, 1983). Since repeated treatments with phorbol esters can promote tumour formation in immortalized cells, the prolonged induction of phorbol ester-responsive genes by elevated levels of Fos and Jun may result in the conversion of already continuously growing cells into the tumorigenic phenotype characteristic of cancer cells (Figure 7.8).

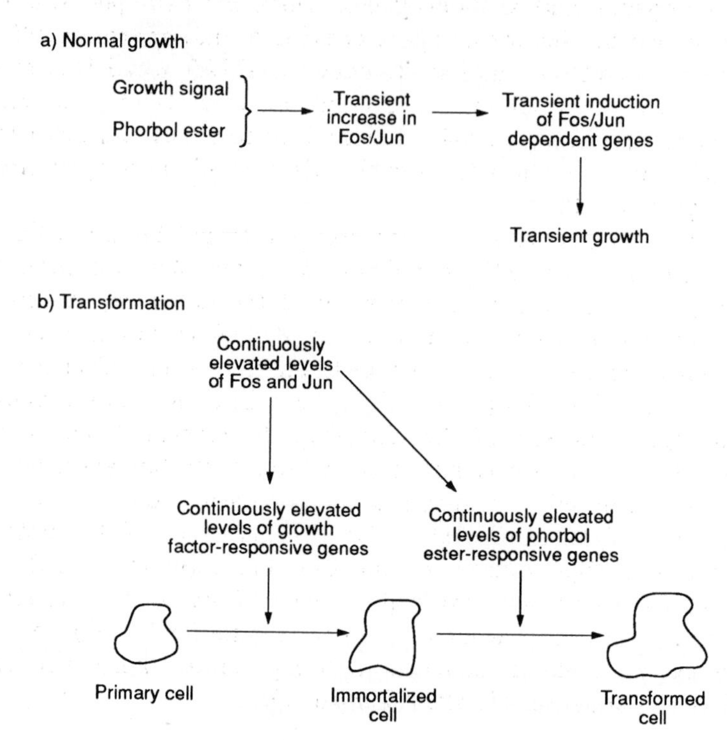

Figure 7.8 Effects of Fos and Jun on cellular growth. (a) In normal cells a brief exposure to a growth signal or phorbol ester will lead to a brief period of growth via the transient induction of Fos and Jun and hence of Fos/Jun-dependent genes. (b) In contrast, the continuous elevation of Fos and Jun produced, for example, by infection with a retrovirus expressing Fos or Jun results in continuous unlimited growth and cellular transformation.

Hence the ability of Fos and Jun to cause cancer represents an aspect of their ability to induce transcription of specific cellular genes. In agreement with this idea, mutations in Fos which abolish its ability to dimerize with Jun and hence prevent it from binding to AP1 sites also abolish its ability to transform cells to a cancerous phenotype (Schuermann *et al.*, 1989). It should be noted, however, that in addition to their over-expression within a retrovirus, there is also some evidence that mutational changes render the viral proteins more potent transcriptional activators than the equivalent cellular proteins. Thus the v-Jun protein appears to activate transcription more efficiently than c-Jun, due to a deletion in a region which normally mediates the interaction of c-Jun and a negatively acting cellular transcription factor (Baichwal and Tjian, 1990).

Interestingly, in addition to its central role in the growth response, the Fos–Jun–AP1 system also appears to represent a target for other oncogenes and anti-oncogenes. Thus, for example, the *ets* oncogene, which, like *fos* and *jun*, encodes a cellular transcription factor, acts via a DNA-binding site known as PEA3, which is located adjacent to the AP1 site in a number of TPA-responsive genes such as collagenase and stromelysin, and the Ets protein co-operates with Fos and Jun to produce high-level activation of these promoters (Wasylyk *et al.*, 1990).

Similarly the *fos* gene itself represents a target for the cellular anti-oncogene protein RB-1 encoded by the retinoblastoma susceptibility gene. This protein is a member of the anti-oncogene class which antagonize the action of oncogenes and whose inactivation by mutation therefore results in tumour formation (review: Cooper and Whyte, 1989). The RB-1 protein is capable of binding to DNA (Lee *et al.*, 1987c) and has recently been shown to be capable of decreasing transcription from the c-*fos* promoter in a co-transfection experiment, acting via a specific retinoblastoma response sequence in the promoter (Robbins *et al.*, 1990). Thus although the RB-1 protein clearly has multiple effects and, for example, interacts directly at the protein–protein level with oncogene proteins, one of its anti-oncogenic effects is mediated by down-regulation of Fos levels, which in turn down-regulates AP1-containing, Fos-dependent promoters (Robbins *et al.*, 1990; Figure 7.9).

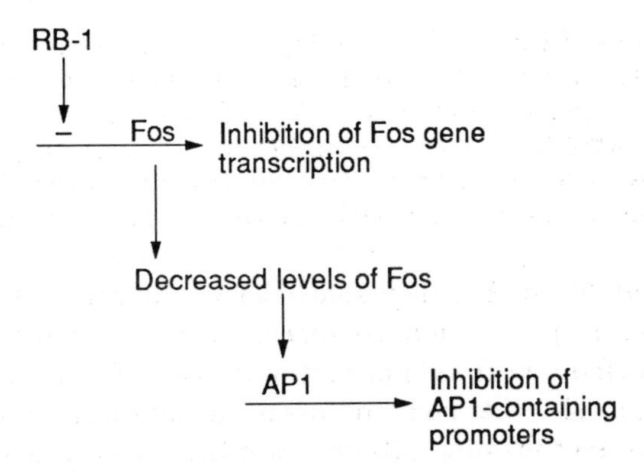

Figure 7.9 The action of the anti-oncogene Rb-1 results in the inhibition of transcription of the c-*fos* gene, leading to a consequent fall in c-Fos levels, resulting in the inhibition of genes which contain AP1 sites and which are therefore normally activated by c-Fos.

Hence the *fos* and *jun* oncogene products play a critical role in the regulation of growth and phorbol ester-inducible genes in normal cells, interacting with the products of other oncogenes and of anti-oncogenes to produce the controlled activity of their target genes necessary for normal controlled growth.

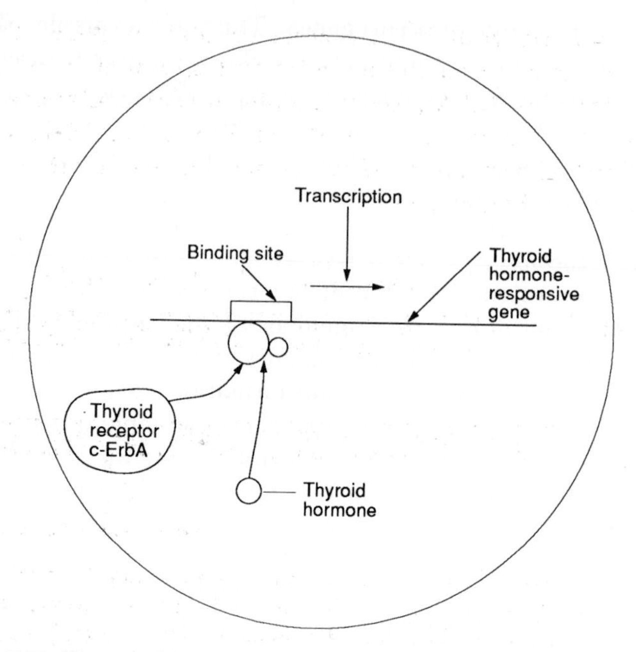

Figure 7.10 The c-*erbA* gene encodes the thyroid hormone receptor and activates transcription in response to thyroid hormone.

7.3 V-*ERBA* AND THE THYROID HORMONE RECEPTOR

The v-*erbA* oncogene is one of two oncogenes carried by avian erythroblastosis virus (AEV) (review: Beug *et al.*, 1985). The cellular equivalent of this oncogene, c-*erbA*, has been shown to encode the cellular receptor for thyroid hormone (Sap *et al.*, 1986; Weinberger *et al.*, 1986), which is a member of the steroid–thyroid hormone receptor superfamily discussed in Section 4.3. Following the binding of thyroid hormone, the receptor–hormone complex binds to its appropriate recognition site in the DNA of thyroid-responsive genes and activates their transcription (Figure 7.10).

Hence the protein encoded by the c-*erbA* gene represents a bona fide cellular transcription factor involved in the activation of thyroid-responsive genes. Unlike the case of the *fos* and *jun* gene products which regulate genes involved in growth, it is not immediately obvious how the form of thyroid hormone receptor encoded by the viral v-*erbA* gene can transform cells to a cancerous phenotype.

The solution to this problem is provided by a comparison of the cellular ErbA protein, which is a functional thyroid hormone receptor, and the viral ErbA protein encoded by AEV. Thus in addition to being fused to the retroviral gag protein at its N-terminus, the viral ErbA protein contains several mutations in the regions of the receptor responsible for binding to DNA and for binding thyroid hormone as well as a small deletion in the hormone-binding domain (Figure 7.11).

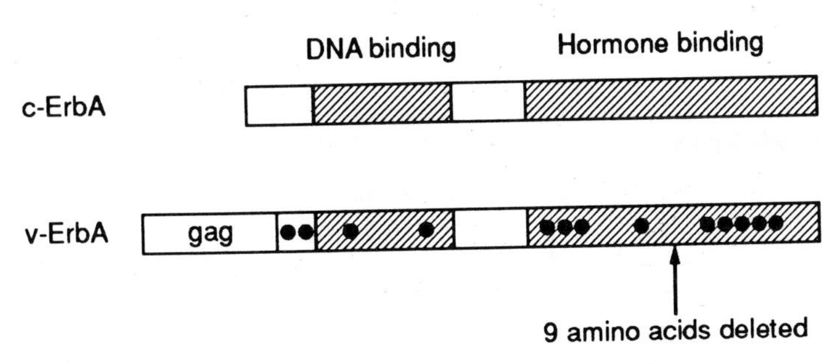

Figure 7.11 Relationship of the cellular ErbA protein and the viral protein. The black dots indicate single amino acid differences between the two proteins while the arrow indicates the region where nine amino acids are deleted in the viral protein.

Interestingly, although these changes do not abolish the ability of the viral ErbA protein to bind to DNA, they do prevent it from binding thyroid hormone and thereby becoming converted to a form which can activate transcription (Sap *et al.*, 1986). Moreover, it has been shown (Sap *et al.*, 1989) that the viral *erbA* gene can inhibit the induction of thyroid hormone-responsive genes when cells are treated with thyroid hormone by binding to the thyroid hormone response elements in their promoters and preventing binding of the activating complex of thyroid hormone and the cellular ErbA protein (Figure 7.12).

Hence the viral ErbA protein acts as a dominant repressor of thyroid-responsive genes, being both incapable of activating transcription itself and able to prevent activation by intact receptor. This mechanism of action is clearly similar to the repression of thyroid hormone-responsive genes by the naturally occurring alternatively spliced form of the thyroid hormone receptor which, as discussed in Section 4.3.4, lacks the hormone-binding domain and therefore cannot bind hormone. Thus the same mechanism of gene repression by a non-hormone-binding receptor is used naturally in the cell and by an oncogenic virus.

One of the targets for repression by the viral ErbA protein is the erythrocyte anion transporter gene (Zenke *et al.*, 1988), which is one of the genes normally induced when avian erythroblasts differentiate

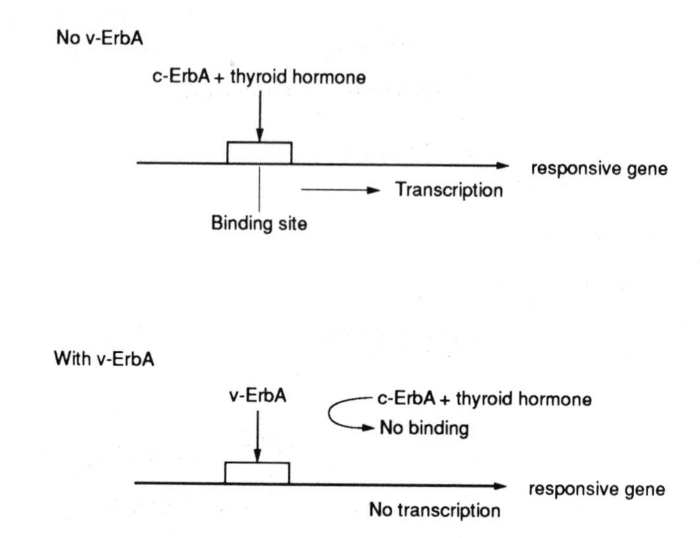

Figure 7.12 Inhibitory effect of the viral ErbA protein on gene activation by the cellular protein, in response to thyroid hormone. Note the similarity to the action of the α2 form of the c-ErbA protein, illustrated in Figure 4.19.

into erythrocytes. This differentiation process has been known for some time to be inhibited by the ErbA protein and it is now clear that it achieves this effect by blocking the induction of the genes needed for differentiation. In turn such inhibition will allow continued proliferation of these cells, rendering them susceptible to transformation into a tumour cell type by the product of the other AEV oncogene, v-*erbB*, which encodes a truncated form of the epidermal growth factor receptor (Downward *et al.*, 1984) and therefore renders cell growth independent of external growth factors (Figure 7.13).

The two cases of Fos/Jun and ErbA therefore represent contrasting examples of the involvement of transcription factors in oncogenesis, in terms of both the mechanism of transformation and the manner in which the cellular form of the oncogene becomes an active transforming gene. Thus in the case of Fos and Jun, transformation is achieved by the continuous activation of genes necessary for growth in normal cell types. Moreover, it occurs, at least in part, via the natural activity of the cellular oncogene in inducing these genes being enhanced by their over-expression such that it occurs at an inappropriate time or place (Figure 7.14a). In contrast, in the ErbA case transformation is achieved by inhibiting the expression of genes whose products are required for the differentiation of a particular cell type, therefore allowing growth to continue. Moreover, this occurs via the activity of a mutated form of the transcription

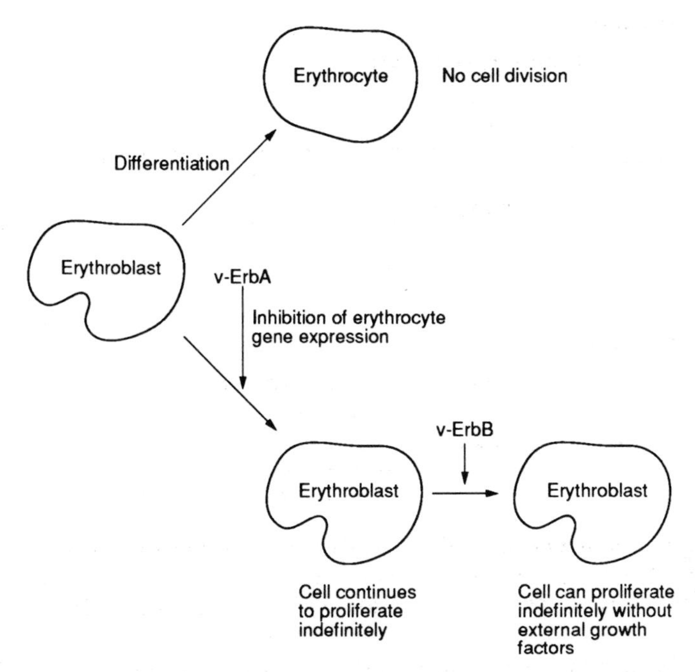

Figure 7.13 Inhibition of erythrocyte-specific gene expression by the v-ErbA protein prevents erythrocyte differentiation and allows transformation by the v-ErbB protein.

factor which, rather than carrying out its normal function more efficiently, actually interferes with the normal role of the thyroid hormone receptor in inducing thyroid hormone-responsive genes required for differentiation (Figure 7.14b).

7.4 THE *MYB* ONCOGENE

In the case of the *fos/jun* and *erbA* oncogenes it was relatively easy to identify the binding sites within DNA and the target genes whose expression is affected by the corresponding cellular transcription factors. This was not the case, however, in the case of the *myb* oncogene originally isolated from two avian retroviruses E26 and AMV (reviews: Cole, 1990; Luscher and Eisenman, 1990b; Shen-Ong, 1990). Thus although both v-Myb and c-Myb proteins are located in the nucleus and bind to DNA, they did not appear to be related to any known transcription factor or DNA-binding activity.

In order to identify the binding site of the Myb protein, therefore, Biedenkapp *et al.* (1988) screened random clones of chicken DNA

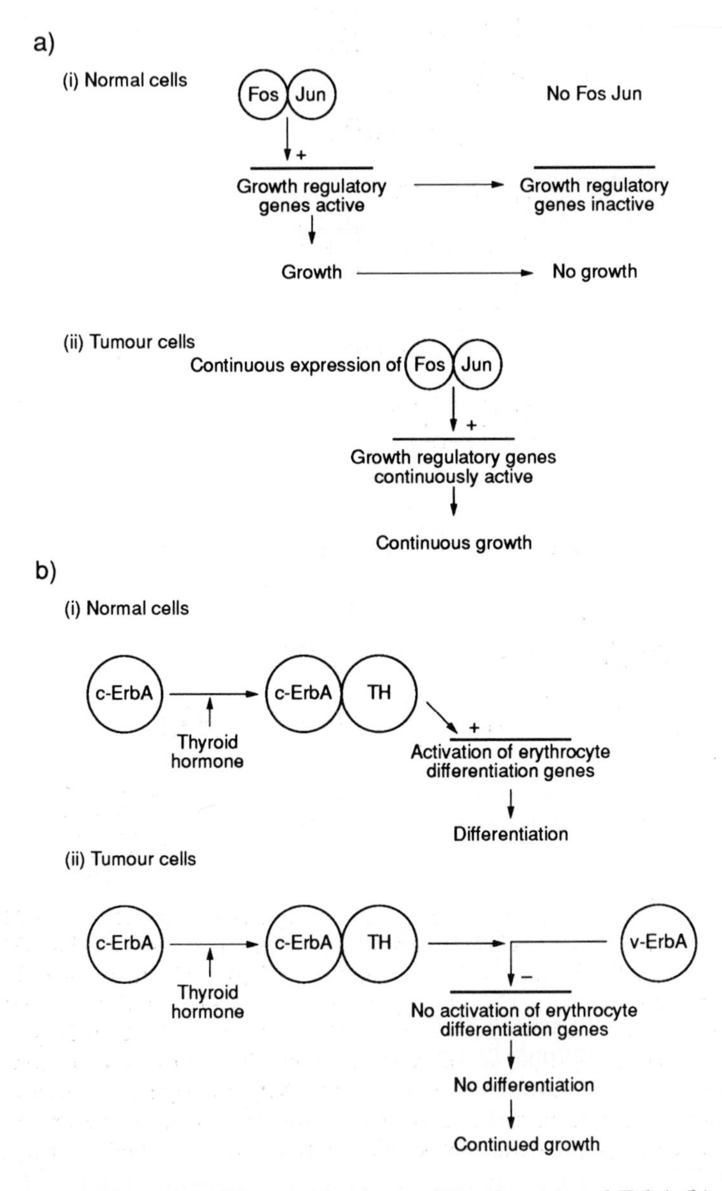

Figure 7.14 Transformation mechanisms of Fos/Jun (a) and ErbA (b). Note that Fos/Jun-induced transformation occurs because the proteins induce the continual activation of growth regulatory genes which are normally expressed only transiently, whilst v-ErbA-induced transformation occurs because the protein interferes with the action of its cellular homologue and hence inhibits the induction of genes involved in erythrocyte differentiation.

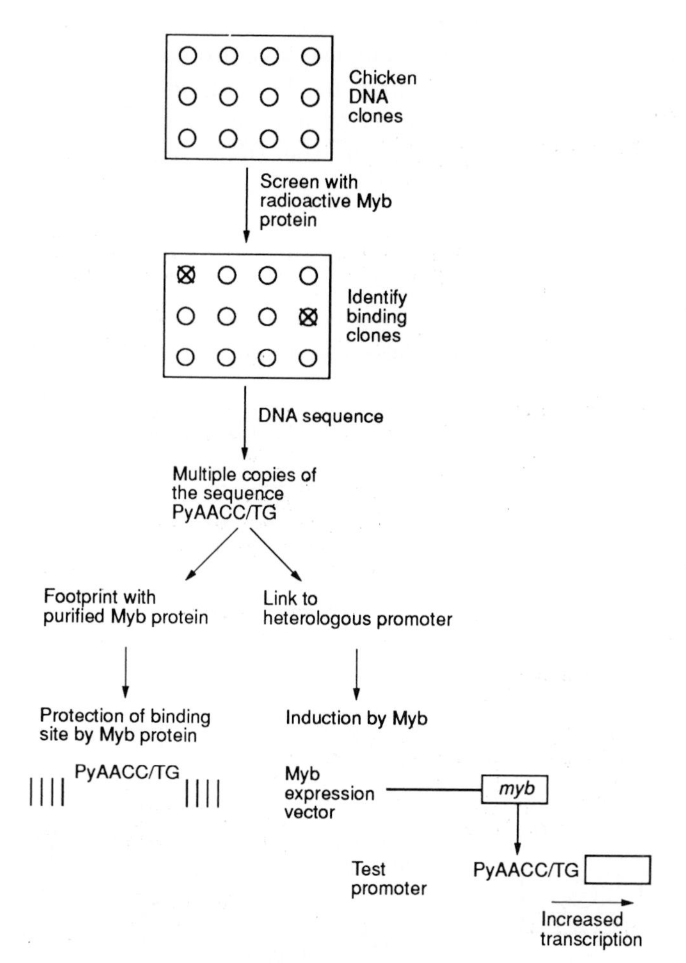

Figure 7.15 Identification of the DNA binding site recognized by the Myb protein. A large number of clones containing random pieces of chicken DNA were screened for the ability to bind radioactive Myb proteins. Clones which bound the protein were subsequently shown to all possess multiple copies of a specific DNA sequence, indicating that this was the Myb-binding site. This was subsequently confirmed by showing that purified Myb protein would produce a DNase I footprint on this DNA (see Figure 2.4) and that Myb protein produced in cultured cells from an expression vector could activate a target gene containing this sequence in its promoter (compare Figure 7.5).

with purified Myb protein in order to identify DNA sequences which could bind Myb protein. They identified six out of 1200 clones which did so and showed that each of these clones contained multiple copies of the sequence PyAACC/TG, over which the Myb protein formed a DNase I footprint (see Section 2.2.2), allowing the Myb-binding site to be identified (Figure 7.15). Subsequently it was shown that a plasmid expressing the Myb protein can trans-activate a target promoter containing its binding site, indicating that the Myb protein is a transcriptional activator which functions by binding to a specific site in the DNA (Weston and Bishop, 1989).

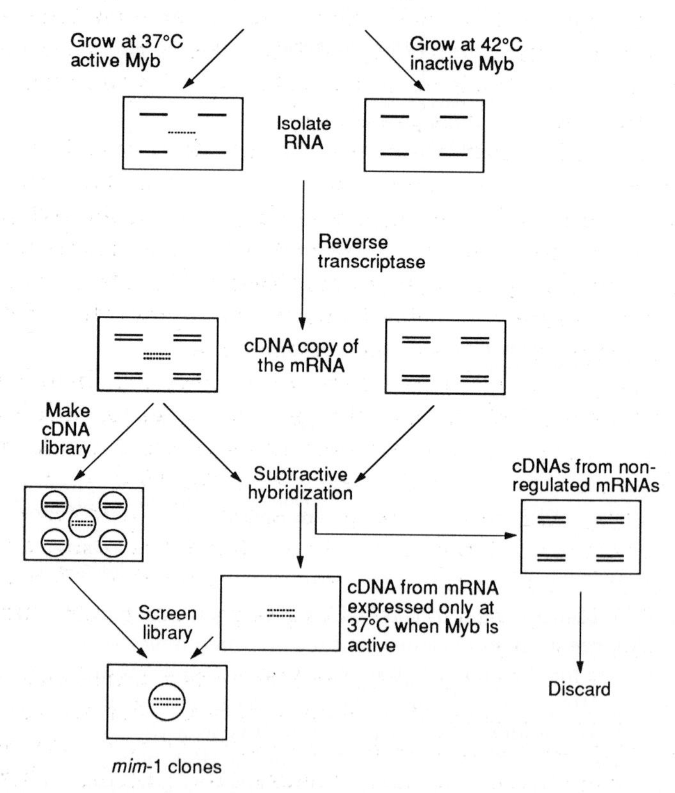

Figure 7.16 Identification of the *mim*-1 gene as a gene whose expression is induced by Myb expression. Cells were infected with a retrovirus expressing a temperature-sensitive form of the Myb protein which was active at 37°C but was inactivated when the temperature was raised to 42°C. Subtractive hybridization techniques were then used to isolate a gene whose corresponding mRNA was present in cells at 37°C which were expressing active Myb but not in cells maintained at 42°C and hence expressing inactive Myb. The single gene isolated in this way was named *mim*-1 (compare the use of a similar technique to isolate MyoD; see Figure 5.8).

Although such studies clearly identify the Myb-binding site and demonstrate its ability to activate transcription, they do not identify the actual cellular genes which are normally activated by Myb. To do this Ness *et al.* (1989) made use of a temperature-sensitive form of the Myb protein which transforms cells at low temperatures but is inactive and fails to transform at higher temperatures. Hence, in cells transformed with a virus encoding this protein, the protein can be inactivated and the transformed state reversed by raising the temperature. By using the technique of differential cDNA library screening (which was also used to isolate the *MyoD* gene — see Section 5.3.1), Ness *et al.* isolated a gene, *mim*-1, which was expressed in the transformed cells only at low temperature (37°C) when the Myb protein was active and not at high temperature (42°C) when it was inactive (Figure 7.16). The *mim*-1 gene was also expressed at high levels in cells transformed with the normal Myb protein but not in untransformed cells.

The conclusion from these experiments that *mim*-1 is a gene activated by Myb was confirmed by the finding that the *mim*-1 promoter contains three consensus binding sites for the Myb protein and that this promoter can be activated by co-transfection with a plasmid expressing the Myb protein (Ness *et al.*, 1989). Hence Myb is clearly a sequence-specific transcription factor which activates the transcription of at least one cellular gene.

Both c-*myb* and the *mim*-1 gene are expressed at high levels in early differentiation stages of the monocyte lineage, with levels of expression declining as these cells differentiate towards mature macrophages (Duprey and Boettiger, 1985; Ness *et al.*, 1989). Interestingly, the v-*myb* oncogene specifically transforms these early monocytic differentiation stages, having no effect on fibroblasts, for example, and can actually reverse the differentiation state of relatively mature macrophages, producing more immature cells (Ness *et al.*, 1987).

Hence, like the ErbA protein, the Myb protein transforms cells by inhibiting or reversing their differentiation, allowing continued proliferation to occur. This is apparently achieved, however, by activating genes required in the undifferentiated cells, rather than, as in the case of ErbA, by repressing genes whose products are required for differentiation (Figure 7.17). Thus, as discussed above, the Myb protein is a potent trans-activator capable of inducing the transcription of genes containing its binding site, including the *mim*-1 gene. Moreover, the ability of various mutant Myb proteins to transform cells correlates precisely with their ability to trans-activate gene expression (Lane *et al.*, 1990), indicating that this ability is essential for transformation.

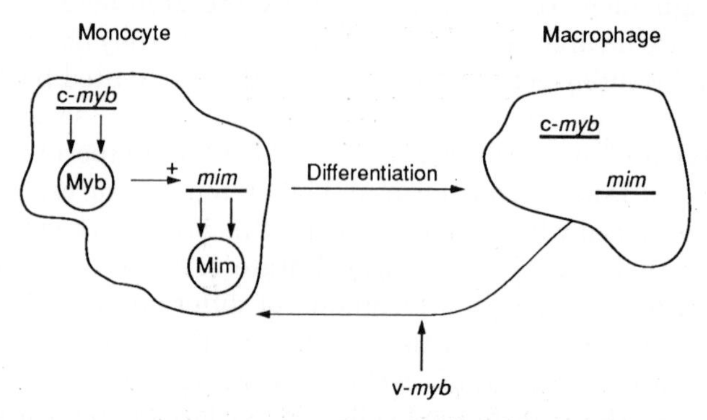

Figure 7.17 Differentiation of cells from immature monocytes to mature macrophages is associated with a switching off of c-*myb* expression and hence of *mim*-1. Artificial expression of v-*myb* in macrophages is capable of reversing their differentiated phenotype and converting them to monocytes, presumably by activating the expression of *mim*-1 and/or other genes expressed in undifferentiated monocytes but not in mature macrophages.

Interestingly, as in the case of c-*erbA* and v-*erbA*, the products of the c-*myb* and v-*myb* genes differ from one another. In the case of the *myb* genes, however, the v-*myb* gene lacks the region encoding the N-terminal region of the protein. This region contains a site for phosphorylation, such phosphorylation inhibiting the DNA-binding activity of the intact cellular protein. Hence, in contrast to the cellular Myb protein, whose DNA-binding activity is regulated by phosphorylation and dephosphorylation, the viral Myb protein has a constitutive DNA-binding activity and is therefore more readily able to activate its target genes (Luscher *et al.*, 1990; Figure 7.18). Hence the *myb* and *erbA* cases represent two contrasting examples of the creation of a transforming protein by mutation. Thus the viral Myb protein has been altered by mutation into a more potent transcriptional activator capable of inducing genes required in undifferentiated cell types, whilst the ErbA protein has been altered into a repressor capable of inhibiting the activating effects of its cellular homologue on genes required in differentiated cells.

As well as presenting a contrasting case to that of the ErbA protein, the studies on the Myb protein discussed in this section represent an excellent example of the power of molecular techniques in characterizing the properties of a factor which was originally identified solely on the basis of its presence in a transforming virus and is now believed to play a critical role in the growth and differentiation of monocytic cells and in their transformation.

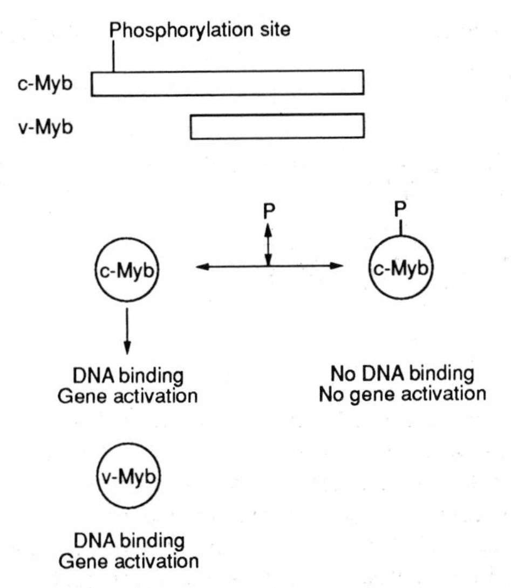

Figure 7.18 Comparison of the c-Myb and v-Myb proteins. Note that v-Myb lacks the N-terminal portion of the protein which contains a phosphorylation site. DNA-binding ability of the c-Myb protein can be inhibited by phosphorylation of this site, allowing its activity to be regulated. In contrast, since v-Myb lacks this site its activity cannot be controlled in this way.

It is noteworthy, however, that these techniques have failed in the case of the c-*myc* oncogene, which was one of the earliest cellular oncogenes to be identified and which is activated in a wide variety of transformed cells (reviews: Cole, 1986, 1990; Luscher and Eisenman, 1990a). Thus the Myc protein has a number of properties suggesting that it is a transcription factor, notably nuclear localization, the possession of several motifs characteristic of transcription factors such as the helix–loop–helix and leucine zipper elements (see Section 8.4), and the ability to activate target promoters in co-transfection assays (Kingston *et al.*, 1984). Moreover, mutations in the leucine zipper motif abolish the transforming ability of the Myc protein (Dang *et al.*, 1989), suggesting that Myc transforms by acting as a transcription factor and affecting the expression of other genes. Despite exhaustive efforts, however, no DNA sequence to which the Myc protein binds has been defined, making it uncertain whether its target genes are activated directly by Myc or indirectly via another gene which is itself directly activated by Myc.

One possible solution to this problem is provided by the presence in the Myc protein of the helix–loop–helix and leucine zipper motifs which mediate the dimerization of proteins containing them both with themselves and with other proteins containing these motifs (see Section 8.3). It may be, therefore, that, as with Fos, specific DNA binding by the Myc protein is dependent on heterodimerization with another factor, preventing identification of DNA binding sites using Myc protein alone. In agreement with this idea Blackwood and Eisenman (1991) have recently identified a novel factor, Max, which associates with Myc via the helix–loop–helix and leucine zipper motifs to form a heterodimer capable of high affinity sequence specific DNA binding.

7.5 CONCLUSIONS

The ability to affect cellular transcriptional regulatory processes is crucial to the ability of many different viruses to transform cells. Thus, for example, the large T oncogenes of the small DNA tumour viruses SV40 and polyoma and the Ela protein of adenovirus can all affect cellular gene expression, and this ability is essential for the transforming ability of these viruses (review: Kingston *et al.*, 1985).

In this chapter we have seen that several RNA viruses also have this ability, containing transcription factors which can act as oncogenes either by promoting the expression of genes required for growth or for the maintenance of the proliferating undifferentiated state or by inhibiting the expression of genes required for the production of non-proliferating differentiated cells.

Although the oncogenes of both DNA and some RNA tumour viruses can therefore affect transcription, their origins are completely different. Thus whilst the oncogenes of the DNA viruses do not have equivalents in cellular DNA and appear to have evolved within the viral genome, the oncogenes of retroviruses have, as we have seen, been picked up from the cellular genome. The fact that, despite their diverse origins, both types of oncogenes can affect transcription indicates that the modulation of transcription represents an effective mechanism for the transformation of cells.

In addition, however, the origin of retroviral oncogenes from the cellular genome allows several other features of transcription to be studied. Thus, for example, the conversion of a normal cellular transcription factor into a cancer-causing viral oncogene allows insights to be obtained into the processes whereby oncogenes become activated.

In general, such oncogenes, whether they encode growth factors, growth factor receptors or other types of protein, can be activated within a virus either by over-expression driven by a strong retroviral promoter or by mutation. The transcription factors we have discussed in this chapter illustrate both these processes. Thus the *fos* and *jun* oncogenes, for example, become cancer-causing, both by continuous expression of proteins which are normally made only transiently, leading to constitutive stimulation of genes required for growth, and by mutations in the viral forms of the protein which render them more potent transcriptional activators. Similarly, the *erbA* and *myb* oncogenes are activated by deletion of a part of the protein-coding regions leading to a protein with different or enhanced properties. Although such effects of mutation or over-expression have initially been defined in tumorogenic retroviruses, it is likely that such changes can also occur within the cellular genome, over-expression of the c-*myc* oncogene, for example, being characteristic of many different human tumours (review: Cole, 1986).

In addition, since cellular oncogenes clearly also play an important role in the regulation of normal cellular growth and differentiation, their identification via tumorogenic retroviruses has, paradoxically, greatly aided the study of normal cellular growth regulatory processes. Thus, for example, the prior isolation of the c-*fos* and c-*jun* genes greatly aided the characterization of the AP1-binding activity and of its role in stimulating genes involved in cellular growth. Similarly, the c-*myb* gene whose product appears to play a critical role in gene regulation during monocyte differentiation was initially identified by the isolation of the v-*myb* gene from a retrovirus.

It is clear, therefore, that as with other oncogenes, the study of the oncogenes which encode transcription factors can provide considerable information on both the processes regulating normal growth and differentiation and on how these processes are altered in cancer.

REFERENCES

Angel, P., Imagawa, M., Chiu, R., Stein, B., Imbra, R. J., Rahmsdirf, H. J., Jonat, C., Herrlich, P. and Karin, M. (1987). Phorbol ester-inducible genes contain a common cis element recognized by a TPA-modulated trans-acting factor. *Cell* **49**, 729–739.

Angel, P., Allegretto, E. A., Okino, S. T., Hattori, K., Boyle, W. J., Hunter, T. and Karin, M. (1988). Oncogene *jun* encodes a sequence-specific trans-activator similar to AP-1. *Nature* **332**, 166–170.

Baichwal, U. R. and Tjian, R. (1990). Control of c-Jun activity by interaction of a cell-specific inhibitor with regulatory domain delta: Differences between v- and c-Jun. *Cell* **63**, 815–825.

Beug, H., Kahn, P., Vennstrom, B., Hayman, M. J. and Graf, T. (1985). How do retroviral oncogenes induce transformation in mammalian cells. *Proceedings of the Royal Society of London, Series B* **226**, 121–126.

Biedenkapp, H., Borgmeyer, U., Sippel, A. E. and Klempnauer, K.-H. (1988). Viral *myb* encodes a sequence-specific DNA-binding activity. *Nature* **335**, 835–837.

Bishop, J. M. (1987). The molecular genetics of cancer. *Science* **235**, 305–311.

Blackwood, E. M. and Eisenman, R. N. (1991) Max: A helix–loop–helix zipper protein that forms a sequence-specific DNA binding complex with Myc. *Science* **251**, 1211–1217.

Bos, T. J., Bohmann, D., Tsuchie, H., Tjian, R. and Vogt, P. K. (1988). v-*jun* encodes a nuclear protein with enhancer binding properties of AP-1. *Cell* **52**, 705–712.

Cole, M. D. (1986). The *myc* oncogene: its role in transformation and differentiation. *Annual Review of Genetics* **20**, 361–384.

Cole, M. D. (1990). The *myb* and *myc* nuclear oncogenes as transcriptional activators. *Current Opinion in Cell Biology* **2**, 502–508.

Cooper, J. A. and Whyte, P. (1989). RB and the cell cycle: entrance or exit? *Cell* **58**, 1009–1011.

Curran, T. and Franza, B. R. (1988). Fos and Jun: the AP-1 connection. *Cell* **55**, 315–397.

Dang, C. V., McGuire, M., Blackmore, M. and Lee, W. M. F. (1989). Involvement of the leucine zipper region in oligomerization and transforming activity of human c-*myc* protein. *Nature* **337**, 664–666.

Downward, J., Yarden, Y., Mayes, E., Scrace, G., Totty, N., Stockwell, P., Ullrich, A., Schlessinger, J. and Watefield, M. D. (1984). Close similarity of epidermal growth factor receptor and v-*erb-B* oncogene protein sequences. *Nature* **307**, 521–527.

Duprey, S. P. and Boettiger, D. (1985). Developmental regulation of c-*myb* in normal myeloid progenitor cells. *Proceedings of the National Academy of Sciences, USA* **82**, 6937–6941.

Gilmore, T. D. (1990). NF-kappaB, KBF1, *dorsal* and *rel*ated matters. *Cell* **62**, 841–843.

Goebl, M. G. (1990). The PU-1 transcription factor is the product of the putative oncogene Spi-1. *Cell* **61**, 1165–1166.

Gutman, A. and Wasylyk, B. (1991). Nuclear targets for transcription regulation by oncogenes. *Trends in Genetics* **7**, 49–54.

Halazonetis, T. D., Georgopoulos, K., Greenberg, M. E. and Leder, P. (1988). C-*jun* dimerizes with itself and with c-*fos* forming complexes of different DNA binding abilities. *Cell* **55**, 917–924.

Hill, D. E., Hope, I. A., Mackie, J. P. and Struhl, K. (1986). Saturation mutagenesis of the yeast *his 3* regulatory site: requirements for transcriptional induction and for binding by GCN4 activator protein. *Science* **234**, 451–457.

Kingston, R. E., Baldwin, A. S. and Sharp, P. A. (1984). Regulation of heat shock protein 70 gene expression by c-*myc*. *Nature* **312**, 280–282.

Kingston, R. E., Baldwin, A. S. and Sharp, P. A. (1985). Transcription control by oncogenes. *Cell* **41**, 3–5.

Lamph, W. W., Wamsley, P., Sassone-Corsi, P. and Verma, I. M. (1988). Induction of proto-oncogene Jun/AP-1 by serum and TPA. *Nature* **334**, 629–631.

Land, H., Parada, L. F. and Weinberg, R. (1983). Cellular oncogenes and multistep carcinogenesis. *Science* **222**, 771–778.

Lane, T., Ibanez, C., Garcia, A., Graf, T. and Lipsick, J. C. (1990). Transformation by v-*myb* correlates with trans-activation of gene expression. *Molecular and Cellular Biology* **10**, 2491–2598.

Lee, W., Haslinger, A., Karin, M. and Tjian, R. (1987a). Activation of transcription by two factors that bind promoter and enhancer sequences of the human metallothionein gene and SV40. *Nature* **325**, 369–372.

Lee, W., Mitchell, P. and Tjian, R. (1987b). Purified transcription factor AP-1 interacts with TPA-inducible enhancer elements. *Cell* **49**, 741–752.

Lee, W.-H., Shew, J.-Y., Hong, F. D., Sery, T. W., Donso, L. A., Bookstein, A. and Lee, E.Y.-H.P. (1987c). The retinoblastoma susceptibility gene encodes a nuclear phosphoprotein associated with DNA binding activity. *Nature* **329**, 642–645.

Luscher, B. and Eisenman, R. N. (1990a). New light on Myc and Myb. Part I, Myc. *Genes and Development* **4**, 2025–2035.

Luscher, B. and Eisenman, R. N. (1990b). New light on Myc and Myb. Part II, Myb. *Genes and Development* **4**, 2235–2241.

Luscher, B., Christenson, E., Litchfield, D. W., Krebs, E. G. and Eisenman, R. N. (1990). *Myb* DNA binding inhibited by phosphorylation at a site deleted during oncogenic activation. *Nature* **344**, 517–522.

Ness, S. A., Beug, H. and Graf, T. (1987). v-*myb* dominance over v-*myc* in doubly transformed myelomonocytic cells. *Cell* **51**, 41–50.

Ness, S. A., Marknell, A. and Graf, T. (1989). The v-*myb* oncogene product binds to and activates the promyelocyte-specific *mim*-1 gene. *Cell* **59**, 1115–1125.

Rauscher, F. J., Cohen, D. R., Curran, T., Bos, T. J., Bogt, P. K., Bohmann, D., Tjian, R. and Franza, B. R. (1988). Fos-associated protein p39 is the product of the c-*jun* oncogene. *Science* **240**, 1010–1016.

Reith, W., Satola, S., Sanchey, C. H., Amaldi, I., Lisowska-Grospiere, B., Griscelli, C., Hadam, M. R. and Mach, B. (1988). Congenital immunodeficiency with a regulatory defect in MHC class II gene expression lacks a specific HLA-DR promoter binding protein RF-X. *Cell* **53**, 897–906.

Robbins, P. D., Horowitz, J. M. and Mulligan, R. C. (1990). Negative regulation of human c-*fos* expression by the retinoblastoma gene product. *Nature* **346**, 668–671.

Sap, J., Munoz, A., Damm, K., Goldberg, Y., Ghysdael, J., Leutz, A., Beug, H. and Vennstrom, B. (1986). The c-*erb-A* protein is a high-affinity receptor for thyroid hormone. *Nature* **324**, 635–640.

Sap, J., Munoz, A., Schmitt, A., Stunnenberg, H. and Vennstrom, B. (1989). Repression of transcription at a thyroid hormone response element by the v-*erbA* oncogene product. *Nature* **340**, 242–244.

Schuermann, M., Neuberg, M., Hunter, J. B., Henuwein, T., Ryseck, R.-P., Bravo, R. and Muller, R. (1989). The leucine repeat motif in *fos* protein mediates complex formation with Jun/AP1 and is required for transformation. *Cell* **56**, 507–516.

Shen-Ong, G. L. C. (1990). The *myb* oncogene. *Biochimica et Biophysica Acta* **1032**, 39–52.

Vogt, P. K., Bos, T. J. and Doolittle, R. F. (1987). Homology between the DNA-binding domain of the GCN4 regulatory protein of yeast and the carboxyl terminal region of a protein coded for by the oncogene *jun*. *Proceedings of the National Academy of Science USA* **84**, 3316–3319.

Wasylyk, B., Wasylyk, C., Flores, P., Begue, A., Leprince, D. and Stehelin, D. (1990). The c-*ets* proto-oncogenes encode transcription factors that co-operate with c-Fos and c-Jun for transcriptional activation. *Nature* **346**, 191–193.

Weinberg, R. A. (1985). The action of oncogenes in the cytoplasm and nucleus. *Science* **230**, 770–776.

Weinberger, C., Thompson, C. C., Ong, E. S., Lebo, R., Gruol, D. J. and Evans, R. M. (1986). The c-*erb-A* gene encodes a thyroid hormone receptor. *Nature* **324**, 641–646.

Weston, K. and Bishop, J. M. (1989). Transcriptional activation by the v-*myb* oncogene and its cellular progenitor c-*myb*. *Cell* **58**, 85–93.

Zenke, M., Kahn, D., Disela, C., Vennstrom, B., Leutz, A., Keegan, K., Hayman, M. J., Choe, H.-R., Yew, N., Engel, J. D. and Beug, H. (1988). v-*erb A* specifically suppresses transcription of the avian erythrocyte anion transporter (Band 3) gene. *Cell* **52**, 107–119.

CHAPTER EIGHT

DNA binding by transcription factors

8.1 INTRODUCTION

In previous chapters we have considered the role of transcription factors in various processes such as constitutive, inducible and tissue-specific gene expression. In order to fulfil these roles, transcription factors must possess certain features allowing them to modulate gene expression. These features will be discussed in the remaining chapters of this book. Clearly, the first essential feature that these factors require is the ability to bind to DNA in a sequence-specific manner and this is discussed in this chapter. Following binding, the factor must interact with other factors or with the RNA polymerase itself in order to influence transcription either positively or negatively and this aspect is discussed in Chapter 9. Finally, in the case of factors modulating inducible, tissue-specific or developmentally regulated gene expression, some means must exist to regulate the synthesis or activity of the factor so that it is active only in a particular situation. This regulation of factor synthesis or activity is discussed in Chapter 10.

Following the cloning of many different eukaryotic transcription factors, the domain-mapping experiments described in Section 2.3.3 have led to the identification of several distinct structural elements in different factors which can mediate DNA binding. These motifs will be discussed in turn, using transcription factors which contain them to illustrate their properties (reviews: Schleif, 1988; Struhl, 1989; Latchman, 1990).

8.2 THE HELIX–TURN–HELIX MOTIF IN THE HOMEOBOX

As discussed in Section 6.2.2, the DNA-binding abilities of the homeobox-containing transcription factors are mediated by the homeobox region of the protein. Thus this region, when synthesized without the remainder of the protein, either by expression in bacteria (Muller *et al.*, 1988) or by chemical synthesis (Mihara and Kaiser, 1988), can bind to DNA in the identical sequence-specific manner exhibited by the intact protein.

This ability to define the 60-amino-acid homeodomain as the region binding to DNA has led to intensive study of its structure in the hope of elucidating how the protein binds to DNA in a sequence-specific manner (review: Affolter *et al.*, 1990). In particular, the crystal structure of the Antennapedia (Antp) homeodomain has been determined by nuclear magnetic resonance spectroscopy (NMR) of the purified protein fragment (Quian *et al.*, 1989).

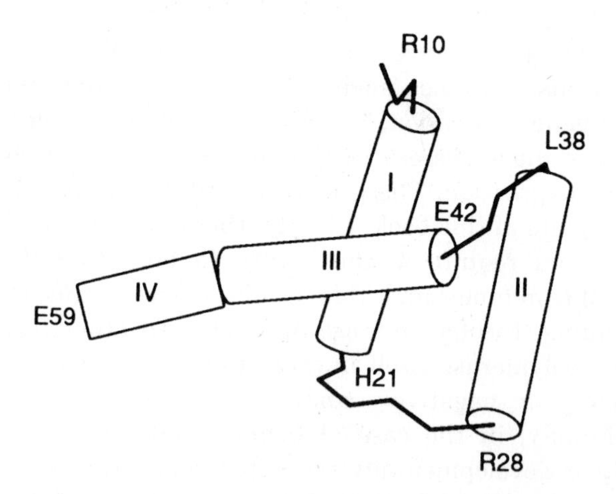

Figure 8.1 Structure of the Antennapedia homeodomain as determined by nuclear magnetic resonance spectroscopy. Note the four α-helical regions (I–IV) represented as cylinders with the amino acids at their ends indicated by numbers and the one-letter amino acid code.

In this study the Antp homeodomain was shown to contain four α-helical regions (Figure 8.1). The first two of these are virtually anti-parallel to each other, with the other two helices arranged at right angles to the first. Most interestingly, helices II and III are separated by a β-turn, forming a helix–turn–helix motif (Figure 8.2). This helix–turn–helix structure is very similar to the DNA-binding motif of several bacteriophage regulatory proteins such as

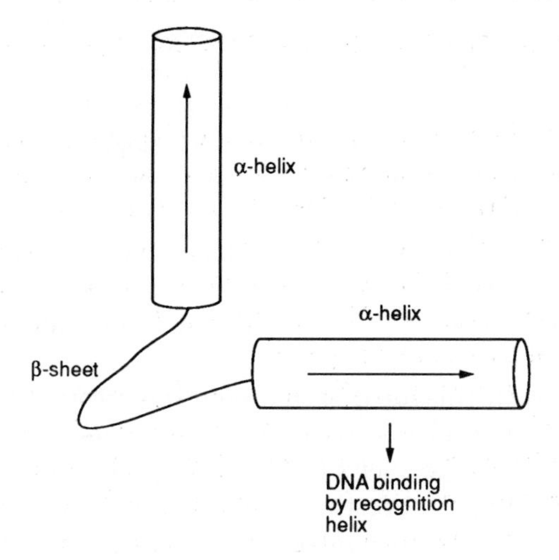

Figure 8.2 The helix–turn–helix motif.

the lambda cro protein or the phage 434 repressor, which have been crystallized and subjected to intensive structural study.

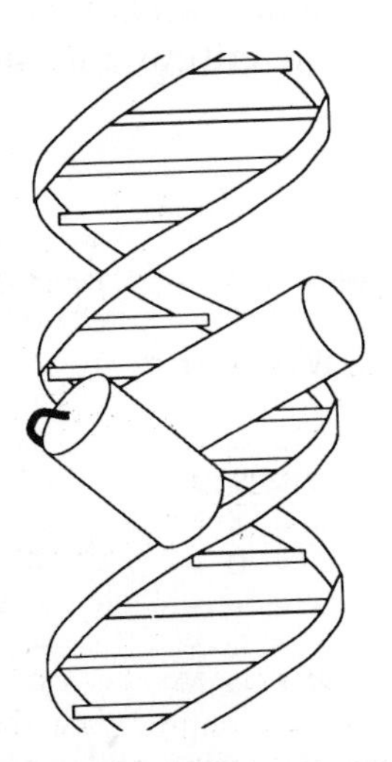

Figure 8.3 Binding of the helix–turn–helix motif to DNA with the recognition helix in the major groove of the DNA.

In these bacteriophage proteins X-ray crystallographic studies have shown that the helix–turn–helix motif does indeed contact DNA. One of the two helices lies across the major groove of the DNA, whilst the other lies partly within the major groove, where it can make sequence-specific contacts with the bases of DNA. It is this second helix (known as the recognition helix) that therefore controls

the sequence-specific DNA-binding activity of these proteins (Figure 8.3; for further details see Schleif (1988)).

The similarity in structure of helices II and III in the Antp homeodomain to the two helices of the bacteriophage proteins has led to the suggestion that these two helices in the homeodomain are similarly aligned relative to the DNA, with helix III constituting the recognition helix responsible for sequence-specific DNA binding. Hence the precise amino acid sequence in the recognition helix in different homeodomain proteins would determine the DNA sequence which they bound.

In agreement with this idea, exchanging the recognition helix in the Bicoid (Bcd) homeodomain for that of Antp resulted in a protein with the DNA-binding specificity of Antp and not that of Bcd (Hanes and Brent, 1989). Most interestingly, a Bcd protein with the DNA-binding specificity of Antp could also be obtained by exchanging only the ninth amino acid in the recognition helix, replacing the lysine residue in Bcd with the glutamine residue found in the Antp protein (Figure 8.4), whereas the exchange of other residues which differ between the two proteins has no effect on the DNA-binding specificity. Hence the ninth amino acid within the recognition helix of the homeodomain plays a critical role in determining DNA-binding specificity.

	Recognition helix										Binding to Bicoid site TCTAATCCC	Binding to Antp site TCAATTAAAT
Bicoid	T	A	Q	V	K	I	W	F	K	N	+	–
	A	–	–	–	–	–	–	–	–	–	+	–
	–	–	–	–	A	–	–	–	–	–	+	–
	–	–	–	–	–	–	–	–	A	–	–	–
	A	–	–	–	A	A	–	–	A	–	–	–
	E	R	–	–	–	–	–	–	Q	–	–	+
	E	R	–	–	–	–	–	–	–	–	+	–
	–	–	–	–	–	–	–	–	Q	–	–	+
Antp	E	R	Q	I	K	I	W	F	Q	N	–	+

Figure 8.4 Effect of changing the amino acid sequence in the recognition helix of the Bicoid protein on its binding to its normal recognition site and that of the Antennapedia (Antp) protein. Note the critical effect of changing the ninth amino acid in the helix, which completely changes the specificity of the Bicoid protein.

It is likely that the amino group of lysine found at the ninth position in the Bcd protein makes hydrogen bonds with the O6 and N7 positions of a guanine residue in the Bcd-specific DNA binding site, whereas the amide group of glutamine found at the corresponding position in the Antp recognition helix forms hydrogen bonds with the N6 and N7 positions of an adenine residue at the equivalent position within the Antp-specific DNA binding site. Hence the replacement of lysine with glutamine results in the loss of two potential hydrogen bonds to a Bcd site and the gain of two potential hydrogen bonds to an Antp site, explaining the observed change in DNA-binding specificity.

A similar critical role for the ninth amino acid in determining the precise DNA sequence which is recognized is also seen in other homeobox-containing proteins, replacement of the serine found at this position in the paired protein with the lysine found in Bcd or the glutamine found in Antp allowing the paired protein to recognize respectively Bcd- or Antp-specific DNA sequences (Treisman *et al.*, 1990). Hence the DNA sequence recognized by a homeobox-containing protein appears to be primarily determined by the ninth amino acid in the recognition helix, proteins with different amino acids at this position recognizing different DNA sequences, whereas proteins such as Antp and Fushi-tarazu, which have the same amino acid at this position, recognize the same DNA sequence.

This critical role of the ninth amino acid is in contrast to the situation in the bacteriophage proteins, in which the helix–turn–helix motif was originally defined. In these proteins, the most N-terminal residues (1–3) in the recognition helix play a critical role in determining DNA-binding specificity (reviews: Pabo and Sauer, 1984; Ptashne, 1986). As shown in Figure 8.4, however, these amino acids appear to play little or no role in determining the DNA-binding specificity of eukaryotic helix–turn–helix proteins, suggesting, therefore, that the recognition helix of these proteins is oriented differently in the major groove of the DNA.

This idea has recently been confirmed by structural studies of the Antp and engrailed homeodomains actually bound to DNA (Otting *et al.*, 1990; Kissinger *et al.*, 1990). These studies have shown that, as in the bacteriophage proteins, the recognition helix directly contacts the bases of DNA but that this helix is oriented within the major groove somewhat differently in the homeobox proteins, such that the critical base-specific contacts are, as predicted, made by the C-terminal end of the helix, which contains residue nine.

It is clear, therefore, that the helix–turn–helix motif in the homeobox both mediates the DNA binding of the protein and also, via the recognition helix, controls the precise DNA sequence which is recognized.

This critical role for the helix–turn–helix motif in the homeodomain is also seen in the POU proteins which, as discussed in Section 6.3.2, contain the homeobox as part of a much larger conserved domain which also contains a POU-specific region. As noted in Chapter 6, however, whilst the isolated homeodomain can bind to DNA in a sequence-specific manner, both the binding affinity and sequence specificity are greatly increased in the presence of the POU-specific domain, which directly contacts specific bases within the DNA recognition sequence (Theill *et al.*, 1989; Kristie and Sharp, 1990). Unlike the homeodomain, the POU-specific domain cannot form a helix–turn–helix motif, although it can form numerous α-helices. The mechanism by which it binds to DNA is therefore uncertain and may involve a previously uncharacterized DNA-binding structure.

8.3 THE ZINC FINGER MOTIF

8.3.1 The two-cysteine two-histidine finger

Transcription factor TFIIIA plays a critical role in regulating the transcription of the 5*S* ribosomal RNA genes by RNA polymerase III (review: Cilberto *et al.*, 1983). When this transcription factor was purified, it was found to have a repeated structure and to be associated with between 7 and 11 atoms of zinc per molecule of purified protein (Miller *et al.*, 1985). When the gene encoding TFIIIA was cloned, it was shown that this repeated structure consisted of the unit Tyr/Phe-X-Cys-X-Cys-$X_{2,4}$-Cys-X_3-Phe-X_5-Leu-X_2-His-$X_{3,4}$-His-X_5, which is repeated nine times within the TFIIIA molecule. This repeated structure therefore contains two invariant cysteine and two invariant histidine residues which were predicted to bind a single zinc atom, accounting for the multiple zinc atoms bound by the intact molecule.

This motif is referred to as a zinc finger on the basis of its proposed structure in which a loop of 12 amino acids containing the conserved leucine and phenylalanine residues as well as several basic amino acids projects from the surface of the molecule, being anchored at its base by the cysteine and histidine residues which tetrahedrally coordinate an atom of zinc (Figure 8.5). The proposed interaction of zinc with the conserved cysteine and histidine residues in this structure was subsequently confirmed by X-ray adsorption spectroscopy of the purified TFIIIA protein (Diakun *et al.*, 1986).

Following its identification in the RNA polymerase III transcription factor TFIIIA, similar Cys_2-His_2-containing zinc finger motifs were identified in a number of RNA polymerase II transcription factors such as Sp1, which contains three contiguous zinc fingers (Kadonga *et al.*, 1987), and the *Drosophila* Kruppel protein, which

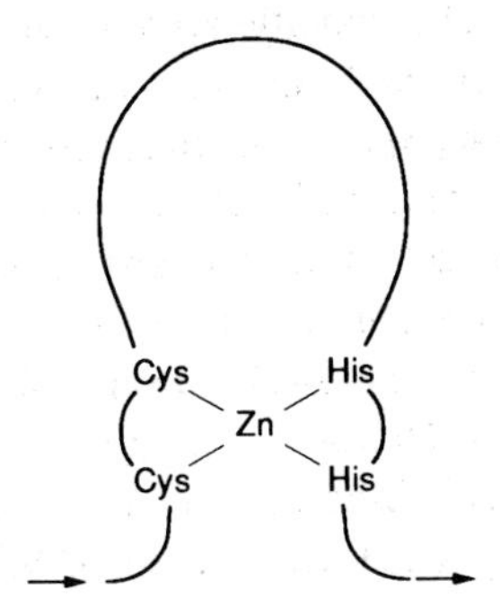

Figure 8.5 Schematic representation of the zinc finger motif. The finger is anchored at its base by the conserved cysteine and histidine residues which tetrahedrally coordinate an atom of zinc.

contains four finger motifs (Rosenberg *et al.*, 1986; see Section 6.1). A list of zinc finger-containing transcription factors is given in Table 8.1 (reviews: Evans and Hollenberg, 1988; Struhl, 1989).

Table 8.1 Transcriptional regulatory proteins containing Cys_2-His_2 zinc fingers

Organism	Gene	Number of fingers
Drosophila	Kruppel	4
	Hunchback	6
	Snail	4
	Glass	5
Yeast	ADR1	2
	SWI5	3
Xenopus	TFIIIA	9
	Xfin	37
Rat	NGF-1A	3
Mouse	MK1	7
	MK2	9
	Egr1	3
	Evi1	10
Human	Sp1	3
	TDF	13

In all cases studied, the zinc finger motifs have been shown to constitute the DNA-binding domain of the protein, with DNA binding being dependent upon their activity. Thus, in the case of TFIIIA, DNA binding is dependent on the presence of zinc, allowing the finger structures to form (Hanas *et al.*, 1983), whilst progressive deletion of more and more zinc finger repeats in the molecule results in a parallel loss of DNA-binding activity (Vrana *et al.*, 1988). Similarly, in the case of Sp1, DNA binding is dependent on the presence of zinc, and most importantly the sequence-specific binding activity of the intact protein can be reproduced by a protein fragment containing only the zinc finger region (Kadonga *et al.*, 1987).

A similar dependence of DNA binding on the zinc finger motif is also seen in the *Drosophila* Kruppel protein which, as discussed in Chapter 6, is essential for correct thoracic and abdominal development. In this case a single mutation in one of the conserved cysteine residues in the finger, replacing it with a serine which cannot bind zinc, results in the production of a mutant fly indistinguishable from that produced by a complete deletion of the gene (Redemann *et al.*, 1988), indicating the vital importance of the zinc finger (Figure 8.6).

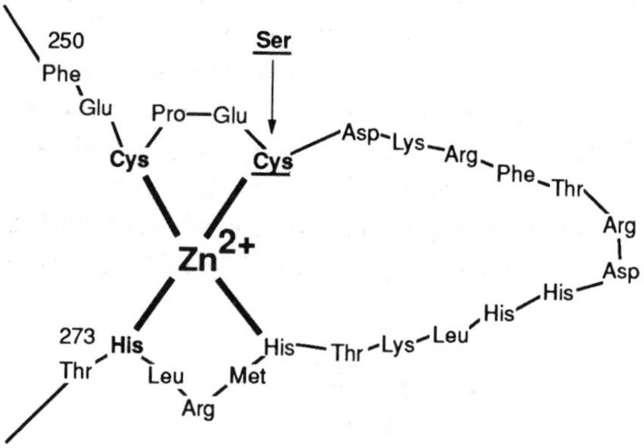

Figure 8.6 Structure of a zinc finger in the *Drosophila* Kruppel protein indicating the cysteine to serine change which abolishes the ability to bind zinc and results in a mutant fly indistinguishable from that obtained when the entire gene is deleted.

As with the helix–turn–helix motif of the homeobox, therefore, the zinc finger motif forms the DNA-binding element of the transcription factors which contain it. Interestingly, however, a single zinc finger taken from the yeast ADRI protein is unable to mediate sequence-specific DNA binding in isolation whereas a protein fragment containing both the two fingers present in the intact protein can do so (Parraga *et al.*, 1988). This suggests, therefore, that DNA binding by the zinc finger is dependent upon interactions with adjacent fingers and explains why zinc finger-containing transcription factors always contain multiple copies of the zinc finger motif (see Table 8.1).

The initial structure proposed for the zinc finger, in which the zinc-coordinating cysteines and histidines form a base from which a loop of basic amino acids projects, provides an obvious basis for the

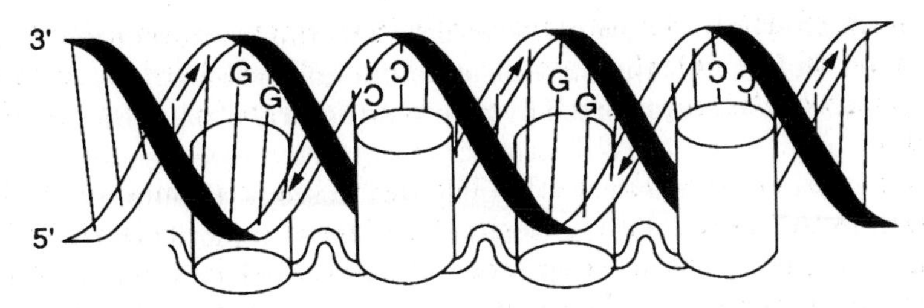

Figure 8.7 Model of the binding of the zinc fingers in TFIIIA to the 5*S* DNA. Note that adjacent fingers make contact with the DNA from opposite sides of the helix.

DNA-binding ability of the zinc finger. Thus, in the case of TFIIIA, it was proposed that the tips of the fingers, which were rich in basic amino acids such as arginine and lysine, would directly contact the acidic DNA in the major groove, with successive fingers binding on opposite sides of the helix (Figure 8.7) (review: Klug and Rhodes, 1987). This structure, in which each finger makes contact with adjacent regions of DNA, would allow the relatively small TFIIIA molecule to make contact, via its multiple fingers, with the relatively large 50-base-pair regulatory region of the 5*S* molecule.

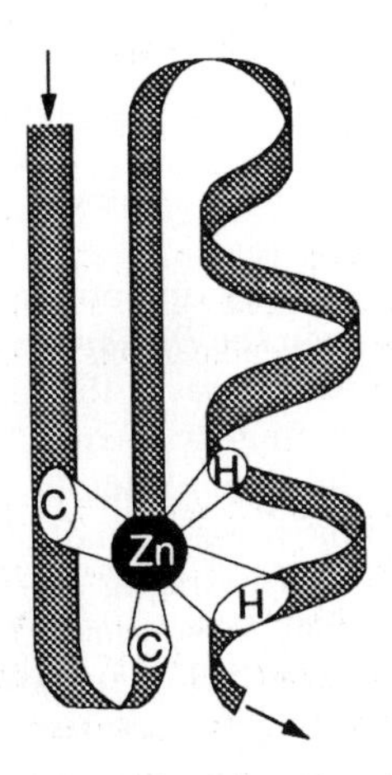

Figure 8.8 Alternative structure for the zinc finger in which two anti-parallel β-sheets (straight lines) are packed against an adjacent α-helix (wavy line).

Such a structure for the zinc finger, in which the zinc coordination via cysteine and histidine serves as a scaffold for the intervening region which makes direct contact with the DNA, is now generally accepted. However, it is unclear whether the intervening amino acids form a simple loop structure as proposed in the original model.

Thus an alternative view of the structure of the zinc finger has been inferred from the structures of other metalloproteins (Figure 8.8; Berg, 1988) and is supported by recent structural studies on the zinc fingers in the yeast SWI5 transcription factor (see Section 5.4.2) using nuclear magnetic resonance spectroscopy (Neuhaus *et al.*, 1990). In this structure, the finger region forms a motif consisting of two anti-parallel β-sheets with an adjacent α-helix packed against one face of the β-sheet. Upon contact with DNA, the α-helix would lie in the major groove of the DNA and make sequence-specific contacts with the bases of DNA, whilst the β-sheets lie further away from the helical axis of the DNA and contact the DNA backbone.

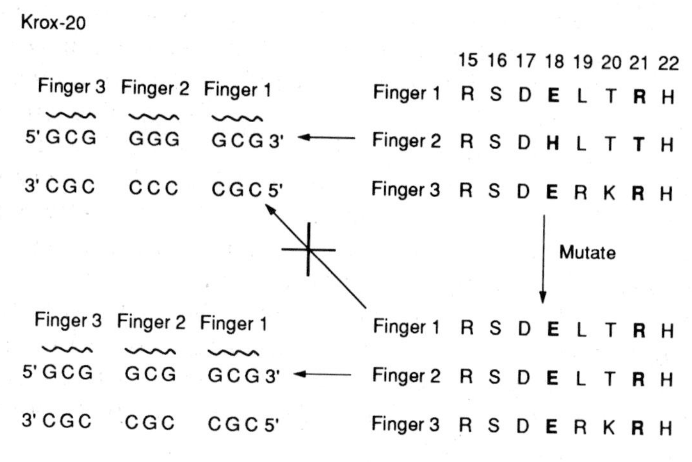

Figure 8.9 DNA-binding specificity and amino acid sequence of the three cysteine–histidine zinc fingers in the *Drosophila* Krox-20 protein. Note that each finger binds to three specific bases in the recognition sequence and that finger 2, which differs from fingers 1 and 3 in the DNA sequence it recognizes, also differs in the amino acids at positions 18 and 21 in the finger (bold letters). Mutating these amino acids to their equivalents in fingers 1 and 3 changes the DNA-binding specificity of finger 2 to that of fingers 1 and 3, indicating that these amino acids play a critical role in determining the DNA sequence which is recognized.

Most interestingly, this proposed structure suggests that a critical role in sequence-specific DNA binding will be played by amino acids at the N-terminus of the α-helix, most notably the amino acids immediately preceding the first histidine residue. In agreement with this idea, a recent study (Nardelli *et al.*, 1991) has revealed a critical role for two amino acids in this region in determining the DNA-binding specificity of the *Drosophila* Krox-20 transcription factor. Thus this factor contains three zinc fingers and interacts with the DNA sequence 5′-GCGGGGGCG-3′. If each finger contacts three bases within this sequence, then the central finger must recognize the sequence GGG whereas the two outer fingers will each recognize the sequence GCG (Figure 8.9). When the amino acid sequence of each of the Krox-20 fingers was compared, it was found that the two outer fingers contain a glutamine residue at position 18 of the finger and an arginine at position 21, whereas the central finger differs in that it has histidine and threonine residues at these positions. As expected if these two amino acid differences are critical in determining the DNA sequence which is recognized, altering these two residues in the central finger to their equivalents in the outer two fingers resulted in a factor which failed to bind to the normal Krox-20-binding site but instead bound to the sequence 5′-GCGGCGGCG-3′, in which each finger binds the sequence GCG. This experiment therefore indicates the critical role of two amino acids at the N-terminus of the α-helix in producing the DNA-binding specificity of zinc fingers of this type and also shows that, at least in the case of Krox-20, each successive finger interacts with three bases of DNA within the recognition sequence. Clearly, such an important role for the amino acids at the N-terminus of an α-helix parallels the similar critical role for the equivalent amino acids in the recognition helix of the bacteriophage DNA recognition proteins (see Section 8.2).

Hence, like the helix–turn–helix motif, the cysteine–histidine zinc finger plays a critical role in mediating the DNA-binding abilities of transcription factors which contain it, with sequence-specific recognition of DNA being determined in both cases by amino acids within an α-helix.

8.3.2 The multi-cysteine zinc finger

A similar zinc-binding domain to that discussed above has also been identified in the DNA-binding domains of the members of the steroid–thyroid hormone receptor family (see Section 4.3 and Evans (1988) for a discussion of these receptors). As with the cysteine–histidine fingers, this motif has been shown by X-ray adsorption spectroscopy to bind zinc in a tetrahedral configuration (Freedman *et al.*, 1988). However, in this case, coordination is achieved by four cysteine residues rather than the two-cysteine two-histidine structure discussed above. Similar multi-cysteine motifs have been identified in several other DNA-binding transcription factors, such as the yeast proteins GAL4, PPRI and LAC9, as well as in the adenovirus transcription factor E1A (Table 8.2) (review: Evans and Hollenberg, 1988), indicating that this type of motif is not confined to the steroid–thyroid hormone receptors.

Table 8.2 Transcriptional regulatory proteins with multiple cysteine fingers

Finger type	Factor	Species
Cys_4-Cys_5	Steroid, thyroid receptors	Mammals
Cys_4	E1A	Adenovirus
Cys_6	GAL4, PPRI, LAC9	Yeast

In the case of the steroid–thyroid hormone receptors the DNA-binding domain has the consensus sequence Cys-X_2-Cys-X_{13}-Cys-X_2-Cys-$X_{15,17}$-Cys-X_5-Cys-X_9-Cys-X_2-Cys-X_4-Cys. This motif is therefore capable of forming a pair of fingers with each four cysteines coordinating a single zinc atom (Figure 8.10), and as with the cysteine–histidine finger proteins, DNA binding of the receptors is dependent on the presence of zinc (Sabbah *et al.*, 1987).

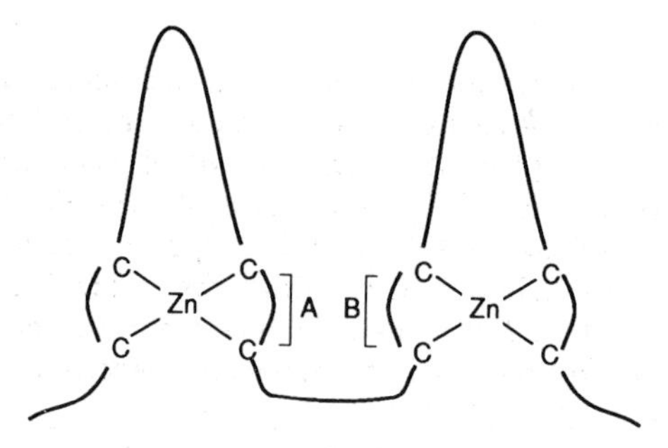

Figure 8.10 Schematic representation of the four-cysteine zinc finger. Regions labelled A and B are of critical importance in determining respectively the DNA sequence which is bound by the finger and the optimal spacing between the two halves of the palindromic sequence which is recognized.

However, the multi-cysteine finger cannot be converted into a functional cysteine–histidine finger by substituting two of its cysteine residues with histidines, suggesting that the two types of finger are functionally distinct (Green and Chambon, 1987). Moreover, unlike the cysteine–histidine zinc finger, which is present in multiple copies within the proteins which contain it, the unit of two multi-cysteine fingers present in the steroid receptors is found only once in each receptor. Interestingly, recent structural studies of the two multi-cysteine fingers in the glucocorticoid (Hard *et al.*, 1990) and oestrogen (Schwabe *et al.*, 1990) receptors have indicated that the two fingers form one single structural motif consisting of two α-helices perpendicular to one another, with the cysteine–zinc linkage holding the base of a loop at the N-terminus of each helix (Figure

8.11). This is quite distinct from the modular structure of the two-cysteine two-histidine finger, where each finger constitutes an independent structural element whose configuration is unaffected by the presence or absence of adjacent fingers (Neuhaus *et al.*, 1990).

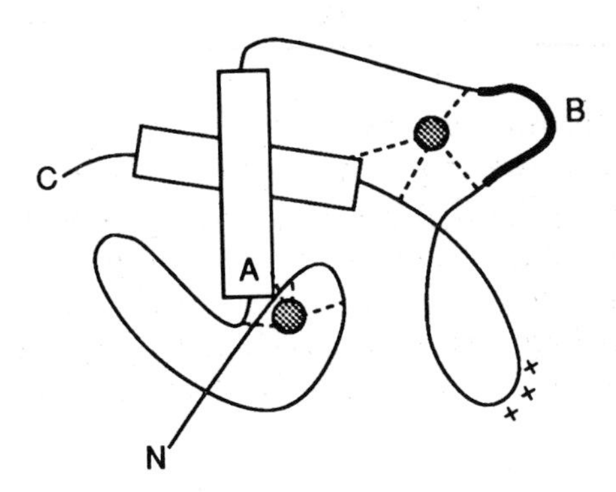

Figure 8.11 Schematic model of a pair of zinc fingers in the oestrogen receptor. Note the helical regions (indicated as cylinders) with the critical residues for determining the DNA sequence which is bound located at the terminus of the recognition helix (indicated as A), the zinc atoms (shaded), conserved basic residues (+++) and the region which interacts with another receptor molecule and determines the optimal spacing between the two halves of the palindromic sequence which is recognized (B) are indicated. Note that A and B indicate the same regions as in Figure 8.10.

Thus, although these two DNA-binding motifs are similar in their coordination of zinc, they differ in the lack of histidines and of the conserved phenylalanine and leucine residues in the multi-cysteine finger as well as structurally. It is likely, therefore, that they represent distinct functional elements which may not be evolutionarily related (for further discussion see Frankel and Pabo (1988)).

Whatever the present relationship between these motifs, it is clear that the multi-cysteine finger mediates the DNA binding of the steroid–thyroid receptors. Thus mutations which eliminate or alter critical amino acids in this motif interfere with DNA binding by the receptor (Giguere *et al.*, 1986; Hollenberg *et al.*, 1987; Figure 8.12). The role of the cysteine fingers in mediating DNA binding by the steroid–thyroid hormone receptors can also be demonstrated by taking advantage of the observation (discussed in Section 4.3.1) that the different receptors bind to distinct but related sequences in the DNA of hormone-responsive genes (see Beato (1989) for review and Table 4.2 for a comparison of these binding sites). Thus if the cysteine-rich region of the oestrogen receptor is replaced by that of the glucocorticoid receptor, the resulting chimaeric receptor has the

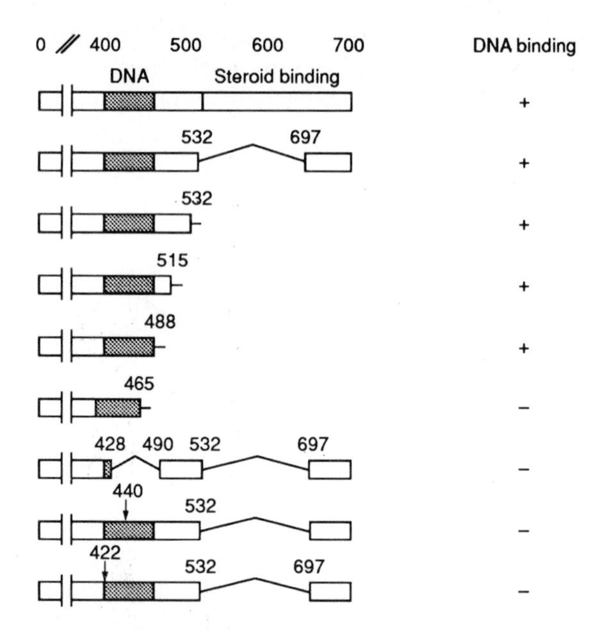

Figure 8.12 Effect of various deletions or mutations on the DNA binding of the glucocorticoid receptor. Note that DNA binding is only prevented by deletions which include part of the DNA-binding domain (shaded) or by mutations within it (arrows) but not by deletions in other regions such as the steroid-binding domain. Numbers indicate amino acid residues.

DNA-binding specificity of the glucocorticoid receptor but continues to bind oestrogen, since all the other regions of the molecule are derived from the oestrogen receptor (Green and Chambon, 1987; Figure 8.13). Hence the DNA-binding specificity of the hybrid receptor is determined by its cysteine-rich region, resulting in the hybrid receptor inducing the expression of glucocorticoid-responsive genes (which carry its DNA-binding site) in response to oestrogen (to which it binds).

These so called 'finger-swop' experiments therefore provide further evidence in favour of the critical role for the multi-cysteine fingers in DNA binding, exchanging the fingers of two receptors leading to an exchange of the DNA-binding specificity. In addition, however, because of the existence of short distinct DNA-binding regions of this type in receptors which bind to distinct but related DNA sequences, they provide a unique opportunity to dissect the elements in a DNA-binding structure which mediate binding to specific sequences (review: Berg, 1989).

Thus by exchanging one or more amino acids between two different receptors it is possible to investigate the effects of these changes on DNA-binding specificity and hence elucidate the role of

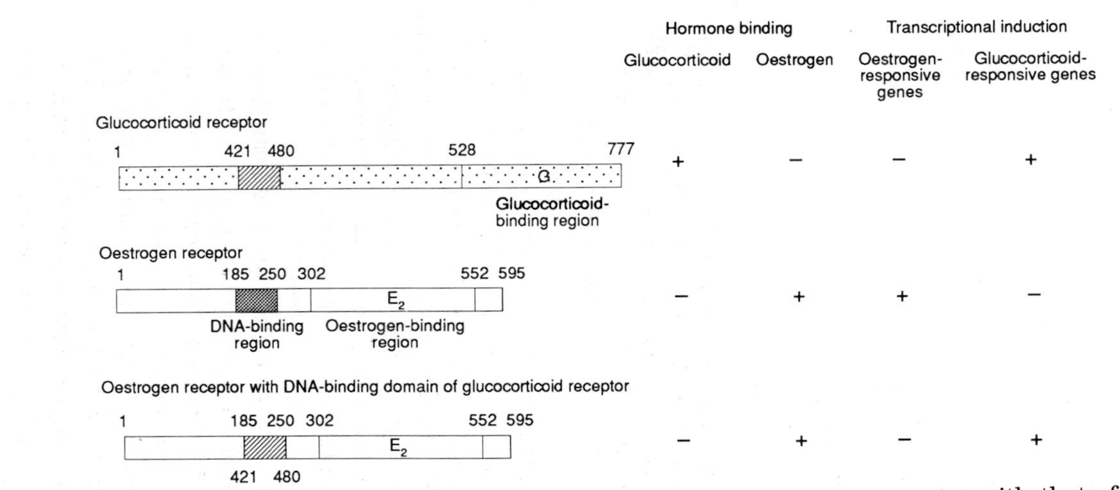

Figure 8.13 Effect of exchanging the DNA-binding domain (shaded) of the oestrogen receptor with that of the glucocorticoid receptor on the binding of hormone and gene induction by the hybrid receptor.

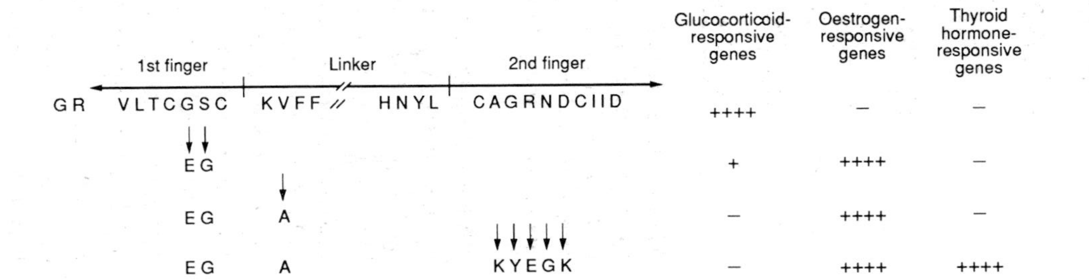

Figure 8.14 Effect of amino acid substitutions in the zinc finger region of the glucocorticoid receptor on the ability to bind to and activate genes which are normally responsive to different steroid hormones.

individual amino acid differences in producing the different patterns of sequence-specific binding. For example, the replacement of the two amino acids between the third and fourth cysteines of the N-terminal finger in the glucocorticoid receptor with their equivalents in the oestrogen receptor changes the DNA-binding specificity of the chimaeric receptor to that of the oestrogen receptor (Danielsen *et al.*, 1989; Umesono and Evans, 1989; Figure 8.14). Hence the exchange of two amino acids in a critical region of a protein of 777 amino acids (indicated as A in Figure 8.10) can completely change the DNA-binding specificity of the glucocorticoid receptor, resulting in it binding to and activating genes which are normally oestrogen-responsive. The specificity of this hybrid receptor for such oestrogen-responsive genes can be further enhanced by exchanging another amino acid located between the two fingers (Figure 8.14), indicating that this region also plays a role in controlling the specificity of DNA binding.

As discussed in Section 4.3.1, the steroid receptors bind to palindromic recognition sequences within DNA, with the receptor binding to DNA as a dimer in which each receptor molecule interacts with one half of the palindrome. In addition to differences in the actual sequence recognized, steroid–thyroid hormone receptors can also differ in the optimal spacing between the two separate halves of the palindromic DNA sequence which is recognized (Table 4.2). Thus the oestrogen receptor and the thyroid hormone receptor both recognize the identical palindromic sequence in the DNA but differ in that in the thyroid receptor-binding sites the two halves of the palindrome are adjacent, whereas in the oestrogen receptor-binding sites they are separated by three extra bases. The further alteration of the chimaeric receptor illustrated in Figure 8.14, the change of five amino acids in the second finger to their thyroid hormone receptor equivalents, is sufficient to allow the receptor to recognize thyroid hormone receptor-binding sites (Umesono and Evans, 1989; Figure 8.14). These amino acids in the second finger (indicated as B in Figure 8.10) appear to play a critical role, therefore, in determining the optimal spacing of the palindromic sequence which is recognized.

As discussed above, structural studies of the two zinc fingers in the oestrogen and glucocorticoid receptors (Hard *et al.*, 1990; Schwabe *et al.*, 1990) suggest that they form a single structural motif with two perpendicular α-helices (Figure 8.11). In this proposed structure, the critical amino acids for determining the spacing in the palindromic sequence recognized are located on the surface of the molecule, allowing them to interact with equivalent residues on another receptor monomer during dimerization (indicated as B in

Figure 8.15). Hence differences in the interaction of these regions in the different receptors determine the spacing of the two monomers within the receptor dimer and thus the optimal spacing in the palindromic DNA sequence which is recognized.

Interestingly, within this structure, the critical residues for determining the precise DNA sequence which is recognized are located at the N-terminus of the first α-helix (indicated as A in Figure 8.15), further supporting the critical role of such helices in DNA binding. Moreover, in the proposed structure of the oestrogen receptor dimer (Schwabe *et al.*, 1990) the DNA-binding helices in each monomer will be separated by 34 Å, allowing each of these recognition helices to make sequence-specific contacts in adjacent major grooves of the DNA molecule.

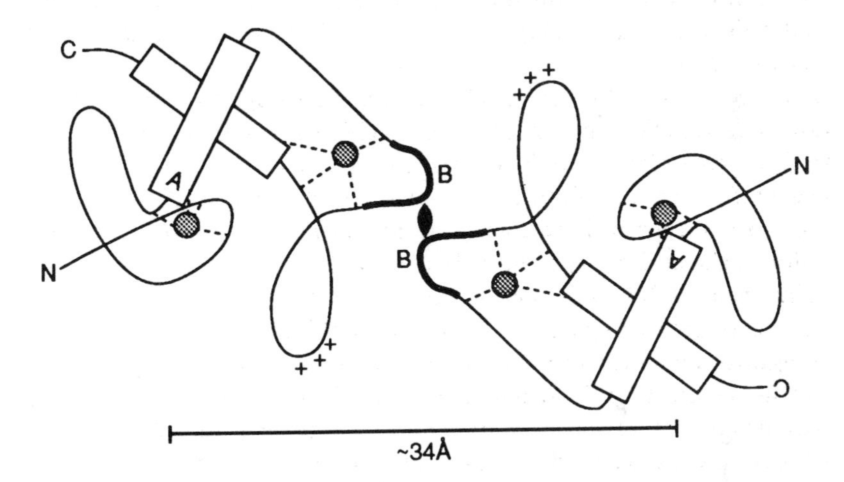

Figure 8.15 Interaction of two oestrogen receptor molecules to form a DNA-binding dimer. Compare with Figure 8.11 and note the interaction of the B regions on each molecule. The resulting dimer has a spacing of 34 Å between the two DNA-binding regions, allowing binding in successive major grooves of the DNA molecule.

The definition of the DNA-binding domain of the steroid–thyroid hormone receptors as a short sequence containing two multi-cysteine fingers has therefore allowed the elucidation of the features in this motif which mediate the different sequence specificities of the different receptors and their relationship to the structure of the motif. In particular, a helical region of the first finger plays a critical role in determining the precise DNA sequence which is recognized by binding in the major groove of the DNA, whilst a distinct region in the second zinc finger controls the spacing of adjacent sequences which is optimal for the binding of each receptor by interacting with the equivalent region in another receptor monomer and hence affecting the structure of the receptor dimer which forms.

8.4 THE LEUCINE ZIPPER AND THE BASIC DNA-BINDING DOMAIN

As discussed in the preceding sections of this chapter, the study of motifs common to several different transcription factors has led to the identification of the role of these motifs in DNA binding. A similar approach has led to the identification of the leucine zipper motif (review: Abel and Maniatis, 1989). Thus this structure has been detected in several different transcription factors such as the CAAT box-binding protein C/EBP (see Section 3.3.2), the yeast factor GCN4 and the proto-oncogene products Myc, Fos and Jun (see Sections 7.2 and 7.4). It consists of a leucine-rich region in which successive leucine residues occur every seventh amino acid (Figure 8.16).

```
C/EBP   L TSDNDR L RKRVEQ L SRELDT L RGIFRQ L
Jun B   L EDKVKT L KAENAG L SSAAGL L REQVAQ L
Jun     L EEKVKT L KAQNSE L ASTANM L REQVAQ L
GCN 4   L EDKVEE L LSKNYH L EHEVAR L KKLVGE R
Fos     L QAETDQ L EDEKSA L QTEIAN L LKEKEK L
Fra 1   L QAETDK L EDEKSG L QREIIE L QKQKER L
c-Myc   V QAEEQK L ISEEDL L RKRREQ L KHKLEQ L
n-Myc   L QAEEHQ L LLEKEK L QARQQQ L LKKIEH A
l-Myc   L VGAEKK M ATEKRQ L RCRQQQ L QKRIAY L
```

Figure 8.16 Alignment of the leucine-rich region in several cellular transcription factors. Note the conserved leucine residues (L) which occur every seven amino acids.

In all these cases, the leucine-rich region can be drawn as an α-helical structure in which adjacent leucine residues occur every two turns on the same side of the helix. Moreover, these leucine residues appear to play a critical role in the functioning of the protein. Thus with one exception (a single methionine in the Myc protein), the central leucine residues of the motif are conserved in all the factors which contain it (Figure 8.16). It was therefore proposed (Landshulz *et al.*, 1988) that the long side chains of the leucine residues extending from one polypeptide would interdigitate with those of the analogous helix of a second polypeptide, forming a motif known as the leucine zipper which would result in the dimerization of the factor (Figure 8.17). This effect could also be achieved by a methionine residue which, like leucine, has a long side chain with

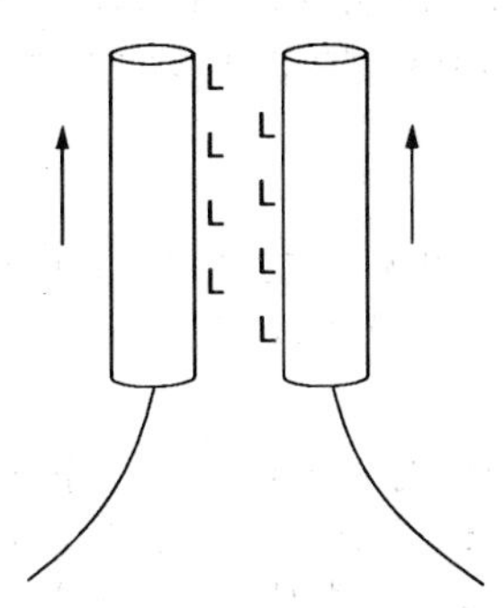

Figure 8.17 Model of the leucine zipper and its role in the dimerization of two molecules of a transcription factor.

no lateral methyl groups, but not by other hydrophobic amino acids such as valine or isoleucine which have methyl groups extending laterally from the β-carbon atom.

In agreement with this idea, substitutions of individual leucine residues in C/EBP or other leucine zipper-containing proteins such as Myc, Fos and Jun with isoleucine or valine abolish the ability of the intact protein to form a dimer, indicating the critical role of this region in dimerization (Landshulz *et al.*, 1989); for review see Johnson and McKnight (1989). A comparison of the effects of various mutations of this type on the ability of the mutant protein to dimerize suggested that the two leucine-rich regions associate in a parallel manner, with both helices oriented in the same direction (as illustrated in Figure 8.17) rather than in an anti-parallel configuration as originally suggested (Landshulz *et al.*, 1989). This idea was confirmed by structural studies of the leucine zipper region of GCN4 (O'Shea *et al.*, 1989). These studies indicated that each zipper motif forms a right-handed α-helix, with dimerization occurring via the association of two parallel helices that coil around each other to form a coiled coil motif similar to that found in fibrous proteins such as the keratins and myosins (Figure 8.18).

In addition to its role in dimerization, the leucine zipper is also essential for DNA binding by the intact molecule. Thus mutations in the zipper which prevent dimerization also prevent DNA binding from occurring (Landshulz *et al.*, 1989). Unlike the zinc finger or helix–turn–helix motifs, however, the zipper is not itself the DNA-binding domain of the molecule and does not directly contact the DNA. Rather it facilitates DNA binding by an adjacent region of the molecule which in C/EBP, Fos and Jun is rich in basic amino acids and can therefore interact directly with the acidic DNA. The leucine zipper is believed, therefore, to serve an indirect structural role in

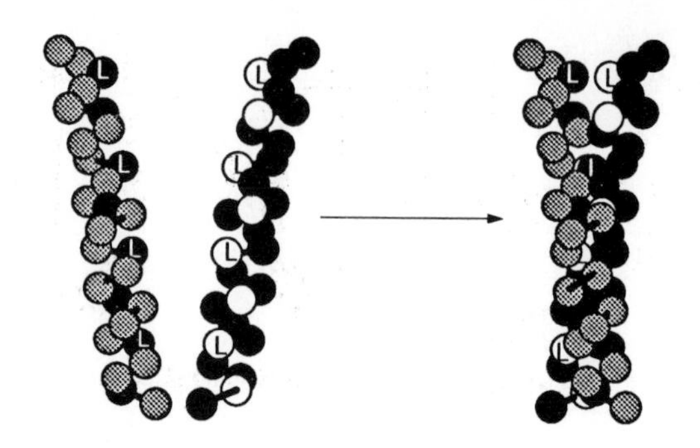

Figure 8.18 Coiled coil structure of the leucine zipper formed by two helical coils wrapping around each other. L indicates a leucine residue.

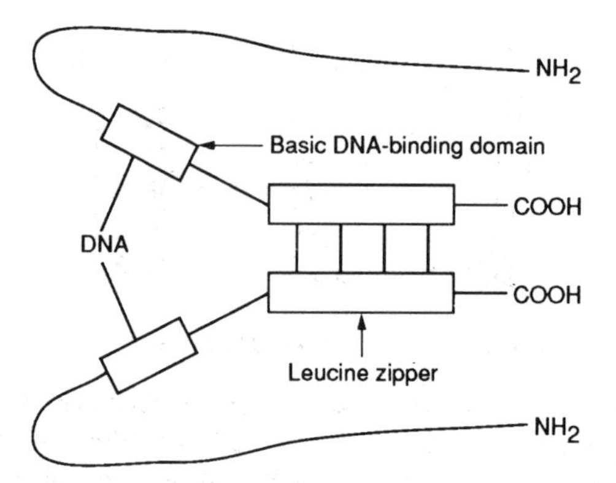

Figure 8.19 Model for the structure of the leucine zipper and the adjacent DNA-binding domain following dimerization of the transcription factor C/EBP.

DNA binding, facilitating dimerization which in turn results in the correct positioning of the two basic DNA-binding domains in the dimeric molecule for DNA binding to occur (Figure 8.19).

In agreement with this idea, mutations in the basic domain abolish the ability to bind to DNA without affecting the ability of the protein to dimerize, as expected for mutations which directly affect the DNA-binding domain (Landshulz *et al.*, 1989). Similarly, exchange of the basic region of GCN4 for that of C/EBP results in a hybrid protein with the DNA-binding specificity of C/EBP, whilst exchange of the leucine zipper region has no effect on the DNA-binding specificity of the hybrid molecule (Agre *et al.*, 1989; Figure 8.20).

Hence the DNA-binding specificity of leucine zipper-containing transcription factors is determined by the sequence of their basic

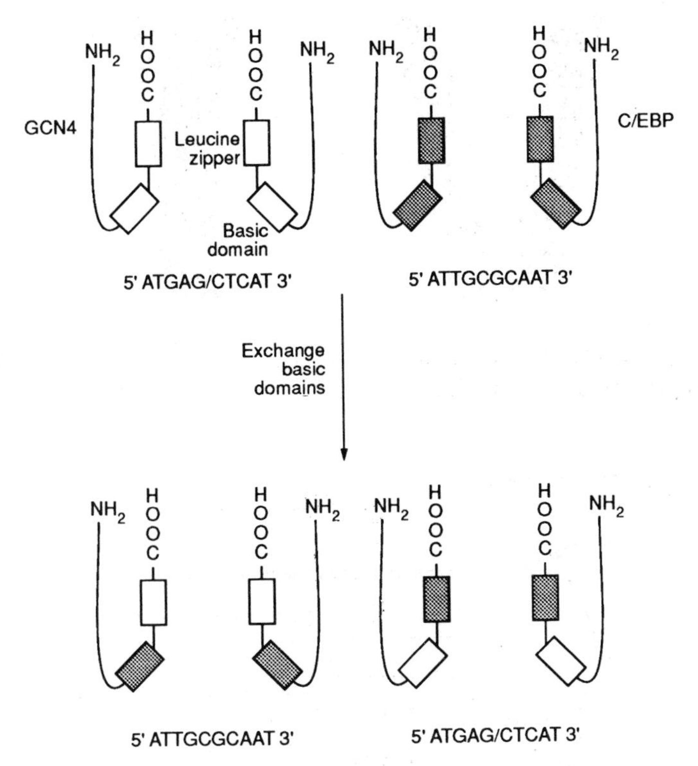

Figure 8.20 Effect of exchanging the basic domains of GCN4 and C/EBP on the DNA-binding specificity. Note that the DNA-binding specificity is determined by the origin of the basic domain and not that of the leucine zipper.

domain, with the leucine zipper allowing dimerization to occur and hence facilitating DNA binding by the basic domain. As expected from this idea, the basic DNA-binding domain can interact with DNA in a sequence-specific manner in the absence of the leucine zipper if it is first dimerized via an intermolecular disulphide bond (Talanin *et al.*, 1990; Figure 8.21).

Following dimerization via the leucine zipper, the intact transcription factor will form a rotationally symmetrical dimer which contacts the DNA via the bifurcating basic regions (Figure 8.19) which form α-helical structures. These two helices then track along the DNA in opposite directions corresponding to the dyad symmetrical structure of the DNA recognition site and form a clamp or scissors grip around the DNA, similar to the grip of a wrestler on his opponent, resulting in very tight binding of the protein to DNA (Vinson and McKnight, 1989). Most interestingly, recent structural studies have suggested that the basic region does not assume a fully

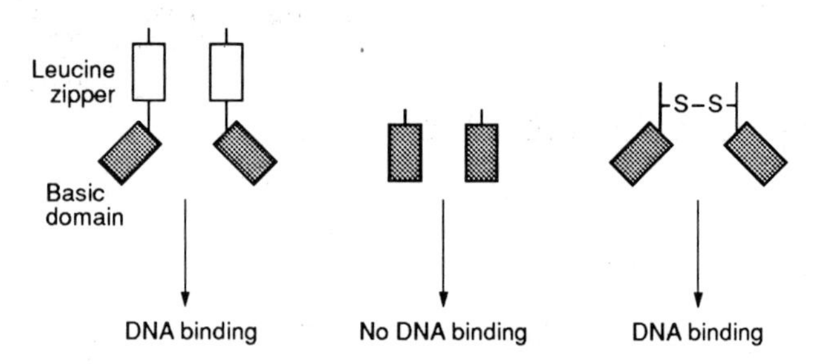

Figure 8.21 DNA binding of molecules containing basic DNA-binding domains can occur following dimerization mediated by leucine zippers or by a disulphide bridge (S–S) but cannot be achieved by unlinked monomeric molecules.

α-helical structure until it contacts the DNA, when it undergoes a configurational change to a fully α-helical form. Hence the association of the transcription factor with the appropriate DNA sequence results in a conformational change in the factor leading to a tight association with that sequence (for discussion see Sauer (1990)).

Although originally identified in the leucine zipper-containing proteins, the basic DNA-binding domain has also been identified by a homology comparison in a number of other transcription factors which do not contain a leucine zipper (Prendergast and Ziff, 1989). These factors include the immunoglobulin enhancer-binding proteins E12 and E47 discussed in Sections 5.2 and 5.3.1, the muscle-determining gene *MyoD* (Section 5.3) and the *Drosophila* daughterless protein.

In all these cases, the basic DNA-binding domain is juxtaposed to a region which can form a helix–loop–helix motif. This helix–loop–helix motif is distinct from the helix–turn–helix motif in the homeobox (Section 8.2) in that it can form two amphipathic helices, containing all the charged amino acids on one side of the helix, which are separated by a non-helical loop (Murre *et al.*, 1989a). This helix–loop–helix motif is thought to play a similar role to the leucine zipper, allowing dimerization of the transcription factor molecule and thereby facilitating DNA binding by the basic motif (Murre *et al.*, 1989b); for discussion see Jones (1990).

In agreement with this, deletions or mutations in the basic domain of the MyoD protein do not abolish dimerization but do prevent DNA binding, paralleling the effect of similar mutations in C/EBP (Figure 8.22). Similarly, mutations or deletions in the helix–loop–helix region abolish both dimerization and DNA binding, parallelling the effects of similar mutations in leucine zipper-containing proteins. Moreover, the DNA-binding ability of MyoD

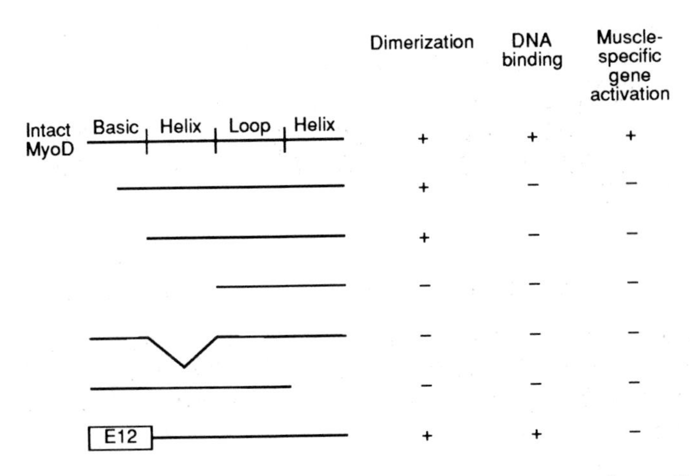

Figure 8.22 Effect of deleting the basic domain or the adjacent helix–loop–helix motif on dimerization, DNA binding and activation of muscle-specific gene expression by the MyoD transcription factor. Note that deletion of any part of the helix–loop–helix motif abolishes dimerization and consequent DNA binding and gene activation, whilst deletion of the basic domain directly abolishes DNA binding and consequent gene activation. Substitution of the basic domain of the constitutive factor E12 for that of MyoD restores DNA binding but not the ability to activate muscle-specific gene expression.

from which the basic DNA-binding domain has been deleted can be restored by substituting the basic domain of the E12 protein (Davis *et al.*, 1990). Most interestingly, however, such substitution does not allow the hybrid protein to activate muscle-specific gene expression, suggesting that, in addition to mediating DNA binding, the basic region of MyoD also contains elements involved in the activation of muscle-specific genes (Davis *et al.*, 1990; Figure 8.22).

Hence both the leucine zipper and the helix–loop–helix motif act by causing dimerization, allowing DNA binding by the adjacent basic motif. Interestingly, the Myc oncoproteins contain both a helix–loop–helix motif and a leucine zipper region adjacent to the basic DNA-binding region (Landshulz *et al.*, 1988; Murre *et al.*, 1989a). It is possible, therefore, that proteins containing the basic DNA-binding region may form an evolutionarily related family comprising three subfamilies having either a leucine zipper, a helix–loop–helix motif or both (for discussion see Prendergast and Ziff (1989)).

This essential role of dimerization (mediated by the leucine zipper or the helix–loop–helix motifs) in allowing DNA binding by the basic DNA-binding domain-containing proteins provides an additional aspect to the regulation of these factors (for discussion see Jones

(1990)). Thus in addition to the formation of homodimers, it is possible to hypothesize that heterodimers will also form between two different leucine zipper- or two different helix–loop–helix-containing factors, allowing the production of dimeric factors with novel DNA-binding specificities or affinities for different sites. One example of this process is seen in the proto-oncogene products Fos and Jun. Thus, as discussed in Section 7.2, the Fos protein cannot bind to AP1 sites in DNA when present alone but can form a heterodimer with the Jun protein which is capable of binding to such sites with 30-fold greater affinity than a Jun homodimer (Figure 8.23). The formation of Jun homodimers and Jun–Fos heterodimers is dependent upon the leucine zipper regions of the proteins (Kouzarides and Ziff, 1988; Turner and Tjian, 1989). Moreover, the failure of Fos to form homodimers is similarly dependent on its leucine zipper region. Thus if the leucine zipper domain of Fos is replaced by that of Jun, the resulting protein can dimerize (Neuberg *et al.*, 1989) and the chimaeric protein can bind to DNA through the basic DNA-binding region of Fos, which is therefore a fully functional DNA-binding domain. Hence the ability of leucine zipper proteins to bind to DNA is determined both by the nature of the leucine zipper which facilitates homodimerization and/or heterodimerization and by the basic DNA-binding motif which allows DNA binding following dimerization.

In addition to its positive role in allowing DNA binding by factors which cannot do so as homodimers, heterodimerization between two related factors can also have an inhibitory role. Thus, as discussed

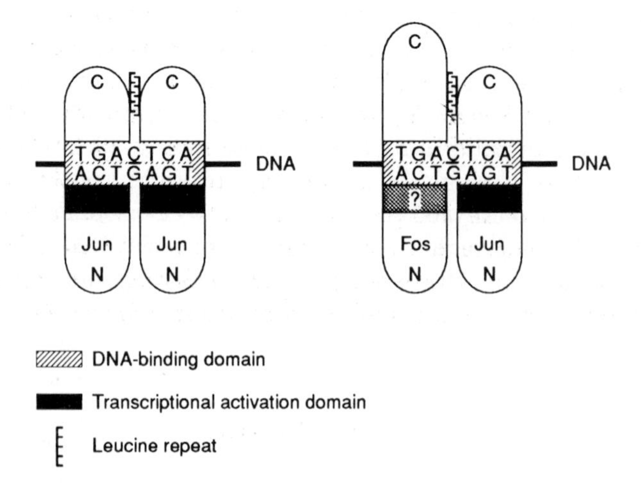

Figure 8.23 Model for DNA binding by the Jun homodimer and the Fos–Jun heterodimer.

in Section 5.3.2, the DNA-binding ability of functional helix–loop–helix proteins which contain a basic DNA-binding domain can be inhibited by association with the Id protein. This protein contains a helix–loop–helix motif, allowing it to associate with other members of this family, but lacks the basic DNA-binding domain. The heterodimer of Id and a functional protein therefore lacks the dimeric basic regions necessary for DNA binding and the activity of the functional transcription factor is thereby inhibited by Id.

Hence the role of the leucine zipper and helix–loop–helix motifs in dimerization can be put to use in gene regulation in both positive and negative ways, either allowing DNA binding by factors which could not do so in isolation, or inhibiting the binding of fully functional factors.

8.5 OTHER DNA-BINDING MOTIFS

Although the majority of DNA-binding domains which have been identified in known transcription factors fall into the three classes we have discussed in the preceding sections, not all do so. Thus, for example, the DNA-binding domains of transcription factors such as AP2, and the CAAT box-binding factors CTF/NF1 and HAP2 and 3 (see Section 3.3.2), are distinct from the known motifs and from each other. As more and more factors are cloned, it is likely that other factors with DNA-binding motifs similar to those of these proteins will be identified and that they will become founder members of new families of DNA-binding motifs. Indeed, this process is already under way, the relatedness of the DNA-binding domains of the *Drosophila* transcription factor fork head and the mammalian liver transcription factor HNF-3 having led to the identification of a new family of proteins containing the so-called fork head DNA-binding motif (Weigel and Jackle, 1990). Similarly, another recently identified domain common to the *ets*-1 proto-oncogene, the *Drosophila* E74 gene, the mouse PU-1 gene and several other transcription factors has been proposed as a new DNA-binding motif mediating binding to the purine-rich sequences recognized by these factors (Karim *et al.*, 1990).

8.6 CONCLUSIONS

In this chapter we have discussed a number of different DNA-binding motifs common to several different transcription factors

Table 8.3 DNA-binding motifs

Motif	Structure	Factors containing domain	Comments
Homeobox	Helix–turn–helix	Numerous *Drosophila* homeotic genes, related genes in other organisms	Structurally related to similar motif in bacteriophage proteins
POU	Helix–turn–helix and adjacent helical region	Mammalian Oct-1, Oct-2, Pit-1, nematode *unc86*	Related to homeodomain
Cysteine–histidine zinc finger	Multiple fingers, each coordinating a zinc atom	TFIIIA, Kruppel, Sp1 etc.	May form β-sheet and adjacent α-helical structure
Cysteine–cysteine zinc finger	Single pair of fingers, each coordinating a zinc atom	Steroid–thyroid hormone receptor family	Related motifs in EIA, GAL4 etc.
Basic domain	α-helical	C/EBP c-Fos, c-Jun, c-Myc, MyoD etc.	Associated with leucine zipper and/or helix–loop–helix dimerization motifs
Fork head	Unknown	Fork head, HNF 3A	—
ETS	Unknown	c-Ets, c-Erg, *Drosophila* E74, PU.1	Binds purine-rich sequences

which can mediate DNA binding. These motifs are listed in Table 8.3.

As we have seen, it has proved possible in many cases to define the precise amino acids in a particular motif which mediate binding to a particular DNA sequence. It should be noted, however, that many transcription factors have the ability to bind to several dissimilar sequences using the same DNA-binding domain. For example, as discussed in Chapter 4, the glucocorticoid receptor binds to a specific DNA sequence in genes which are induced by glucocorticoid and to a distinct DNA sequence in genes which are repressed by this hormone, the same DNA-binding domain of the protein being used in each case (Sakai *et al.*, 1988). Similarly the yeast *CYC1* and *CYC7* genes contain entirely distinct sequences, both of which bind the HAP1 transcription factor, allowing gene activation to occur (Pfeifer *et al.*, 1987).

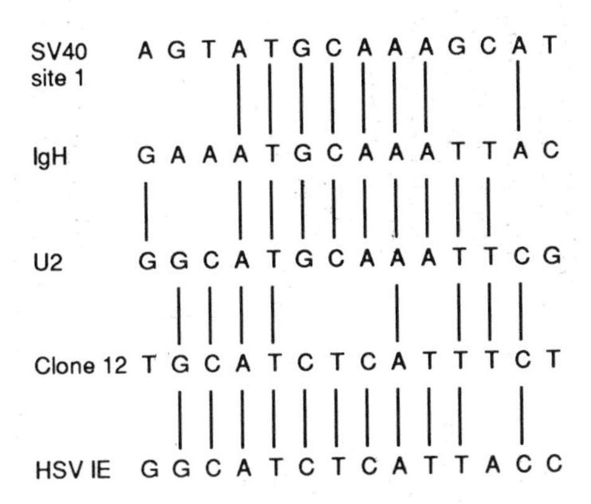

Figure 8.24 Relationship between the various diverse sequences bound by the Oct-1 transcription factor in the simian virus 40 enhancer, the immunoglobulin IgH chain gene enhancer (IgH), the U2 snRNA gene, clone 12 (a mutated version of a site in the SV40 enhancer which binds Oct-1) and the herpes simplex virus immediate-early genes (HSV IE).

This phenomenon of a single factor binding to highly divergent sequences has been most extensively analysed in the case of the octamer-binding protein Oct-1. Thus this factor binds to a sequence in the SV40 enhancer which shares less than 30% homology (4 out of 14 bases) or little more than a random match with another Oct-1-binding sequence in the herpes simplex virus immediate-early gene promoters (Figure 8.24). By analysing a series of other Oct-1-binding elements however, Baumruker *et al.* (1988) were able to show that the two apparently unrelated Oct-1–binding sites could be linked by a smooth progression via a series of other binding sites which were related to one another (Figure 8.24). This suggests, therefore, that Oct-1 can bind to very dissimilar sequences because there are few, if any, obligatory contacts with specific bases in potential binding sites. Rather, specific binding to a particular sequence can occur via many possible independent interactions with DNA, only some of which will occur with any particular binding site. Hence the binding to apparently unrelated sequences does not reflect two distinct binding specificities but indicates that the protein can make many different contacts with DNA, the sequences which can specifically bind the protein being those with which it can make a certain proportion of these possible contacts.

In addition to the ability of one factor to bind to different sequences, it is also possible for the same sequence to be bound by more than one factor. Thus, as well as the Oct-1 factor, several other proteins binding to the octamer motif exist including the B cell-specific factor Oct-2 (see Section 5.2.2) and several others expressed specifically in the brain (Scholer *et al.*, 1989). In all these factors, DNA binding is mediated by the POU domain discussed above (Section 8.2).

In contrast, however, whilst the transcription factors CTF/NF1 and C/EBP both bind to the CAAT box sequence (see Section 3.3.2), they do so via completely different DNA-binding domains, with C/EBP having a basic DNA-binding domain (Section 8.4) whilst CTF/NF1 has a DNA-binding domain distinct from that of any other factor (Section 8.5).

It is unlikely, therefore, that the existence of several distinct DNA-binding domains reflects the need of the factors which contain them to bind to distinct types of DNA sequences. Rather, it seems perfectly possible that one DNA-binding motif could be present in all factors, with variations of it in different factors producing the observed binding to different DNA sequences. This is particularly so in view of the fact that in diverse DNA-binding motifs such as the homeobox, the basic DNA-binding domain and the two types of zinc fingers, the amino acids which determine sequence-specific binding to DNA are all located within similar α-helical structures. This idea evidently begs the question of why different DNA-binding motifs exist.

It is possible that this situation has arisen simply by different motifs which could produce DNA binding having arisen in particular factors during evolution and having been retained since they efficiently fulfilled their function. Alternatively it may be that the existence of different motifs reflects differences in the factors containing them other than the specific DNA sequence which is recognized. For example, the highly repeated zinc finger motif may be of particular use where, as in the case of transcription factor TFIIIA, the factor must contact a large regulatory region in the DNA. Similarly, a motif such as the basic domain, which can only bind to DNA following dimerization, will be of particular use where the activity of the factor must be regulated, whether positively or negatively, via dimerization with another factor.

Whatever the case, it is clear that DNA binding by transcription factors is dependent upon specific domains of defined structure within the molecule. Following such DNA binding, the bound factor must influence the rate of transcription either positively or negatively. The manner in which this occurs and the regions of the factors which achieve this effect are discussed in Chapter 9.

REFERENCES

Abel, T. and Maniatis, T. (1989). Action of leucine zippers. *Nature* **341**, 24–25.

Affolter, M., Schier, A. and Gehring, W. J. (1990). Homeodomain proteins and the regulation of gene expression. *Current Opinion in Cell Biology* **2**, 485–495.

Agre, P., Johnson, P. F. and McKnight, S. L. (1989). Cognate DNA binding specificity retained after leucine zipper exchange between GCN4 and C/EBP. *Science* **246**, 922–926.

Baumruker, T., Sturm, R. and Herr, W. (1988). OBP 100 binds remarkably degenerate octamer motifs through specific interaction with flanking sequences. *Genes and Development* **2**, 1400–1413.

Beato, M. (1989). Gene regulation by steroid hormones. *Cell* **56**, 335–344.

Berg, J. M. (1988). Proposed structure for the zinc-binding domains from transcription factor III A and related proteins. *Proceedings of the National Academy of Sciences, USA* **85**, 99–102.

Berg, J. M. (1989). DNA binding specificity of steroid receptors. *Cell* **57**, 1065–1068.

Cilberto, G., Castagnoli, L. and Cortese, R. (1983). Transcription by RNA polymerase III. *Current Topics in Developmental Biology* **18**, 59–88.

Danielsen, M., Hinck, C. and Ringold, G. M. (1989). Two amino acids within the knuckle of the first zinc finger specify DNA response element activation by the glucocorticoid receptor. *Cell* **57**, 1131–1138.

Davis, R. L., Cheng, P.-F., Lassar, A. B. and Weintraub, H. (1990). The MyoD DNA binding domain contains a recognition code for muscle-specific gene activation. *Cell* **60**, 733–746.

Diakun, G. P., Gairall, L. and Klug, A. (1986). EXAFS study of the zinc-binding sites in the protein transcription factor III A. *Nature* **324**, 698–699.

Evans, R. M. (1988). The steroid and thyroid hormone receptor gene superfamily. *Science* **240**, 889–895.

Evans, R. M. and Hollenberg, S. M. (1988). Zinc fingers: guilt by association. *Cell* **52**, 1–3.

Frankel, A. D. and Pabo, C. O. (1988). Fingering too many proteins. *Cell* **53**, 675.

Freedman, L. P., Luisi, B. F., Korszin, Z. R., Basavappa, R., Sigler, P. B. and Yamamoto, K. R. (1988). The function and structure of the metal coordination sites within the glucocorticoid receptor binding domain. *Nature* **334**, 543–546.

Giguere, V., Hollenberg, S. M., Rosenfeld, M. G. and Evans, R. M. (1986). Functional domains of the human glucocorticoid receptor. *Cell* **46**, 645–652.

Green, S. and Chambon, P. (1987). Oestradiol induction of a glucocorticoid-response gene by a chimaeric receptor. *Nature* **325**, 75–78.

Hanas, J. S., Bogenhagen, D. F. and Wu, C.-W. (1983). Co-operative model for the binding of *Xenopus* transcription factor A to the 5S RNA gene. *Proceedings of the National Academy of Sciences, USA* **80**, 2142–2145.

Hanes, S. D. and Brent, R. (1989). DNA specificity of the bicoid activator protein is determined by homeodomain recognition helix 9. *Cell* **57**, 1275–1283.

Hard, T., Kellenbach, E., Boelens, R., Maler, B. A., Dahlman, K., Freedman, L. F., Carlstedt-Duke, J., Yamamoto, K. R., Gustafsson, J. A. and Kapstein, R. (1990). Structure of glucocorticoid receptor domain. *Science* **249**, 157–160.

Hollenberg, S. M., Giguere, V., Segui, P. and Evans, R. M. (1987). Colocalization of DNA-binding and transcriptional activation functions in the human glucocorticoid receptor. *Cell* **49**, 39–46.

Johnson, P. F. and McKnight, S. L. (1989). Eukaryotic transcriptional regulatory proteins. *Annual Review of Biochemistry* **58**, 799–839.

Jones, N. (1990). Transcriptional regulation by dimerization: two sides to an incestuous relationship. *Cell* **61**, 9–11.

Kadonga, J. T., Carner, K. R., Masiarz, F. R. and Tjian, R. (1987). Isolation of cDNA encoding the transcription factor Sp1 and functional analysis of the DNA binding domain. *Cell* **51**, 1079–1090.

Karim, F. D., Urness, L. D., Thummel, C. S., Klemsz, M. J., McKercher, S. R., Celada, A., Van Beveren, C., Maki, R. A., Gunther, C. V., Nye, J. A. and Graves, B. J. (1990). The Ets-domain: a new DNA-binding motif that recognizes a purine-rich core DNA sequence. *Genes and Development* **4**, 1451–1453.

Klug, A. and Rhodes, D. (1987). Zinc fingers: a novel protein motif for nucleic acid recognition. *Trends in Biochemical Sciences* **12**, 464–469.

Kissinger, C. R., Liu, B., Martin-Bianco, E., Kornberg, T. B. and Pabo, C. O. (1990). Crystal structure of an engrailed homeodomain–DNA complex at 2.8 Å resolution: a framework for understanding homeodomain–DNA interactions. *Cell* **63**, 579–590.

Kouzarides, T. and Ziff, E. (1988). The role of the leucine-zipper in the Fos-Jun interaction. *Nature* **336**, 646–651.

Kristie, T. M. and Sharp, P. A. (1990). Interactions of the Oct-1 POU subdomains with specific DNA sequences and with the HSV alpha-transactivator protein. *Genes and Development* **4**, 2383–2396.

Landschulz, W. H., Johnson, P. F. and McKnight, S. L. (1988). The leucine zipper: a hypothetical structure common to a new class of DNA binding proteins. *Science* **240**, 1759–1764.

Landschulz, W. H., Johnson, P. F. and McKnight, S. L. (1989). The DNA binding domain of the rat liver nuclear protein C/EBP is bipartite. *Science* **243**, 1681–1688.

Latchman, D. S. (1990). Eukaryotic transcription factors. *Biochemical Journal* **270**, 281–289.

Mihara, J. and Kaiser, E. T. (1988). A chemically synthesized Antennapedia homeo domain binds to a specific DNA sequence. *Science* **242**, 925–927.

Miller, J., McLachlan, A. D. and Klug, A. (1985). Repetitive zinc-binding domains in the protein transcription factor III A from *Xenopus* oocytes. *EMBO Journal* **4**, 1609–1614.

Muller, M., Affolter, M., Leupin, W., Offing, G., Winthrich, K. and Gehring, U. S. (1988). Isolation and sequence specific DNA binding of the Antennapedia homeodomain. *EMBO Journal* **7**, 4299–4304.

Murre, C., McCaw, P. S. and Baltimore, D. (1989a). A new DNA binding and dimerization motif in immunoglobulin enhancer binding, daughterless, MyoD and myc proteins. *Cell* **56**, 777–783.

Murre, C., McCaw, P. S., Vaessin, H., Caudy, M., Jan, L.Y., Jan, Y. N., Cabera, C. V., Buskin, J. N., Hauschka, S. D., Lassar, A. B., Weintraub, H. and Baltimore, D. (1989b). Interactions between heterologous helix–loop–helix proteins generate complexes that bind specifically to a common DNA sequence. *Cell* **58**, 537–544.

Nardelli, J., Gibson, T. J., Vesque, C. and Charnay, P. (1991). Base sequence discrimination by zinc-finger DNA binding domains. *Nature* **349**, 175–178.

Neuberg, M., Adamkiewicz, J., Hunter, J. B. and Muller, R. (1989). Fos protein containing the Jun leucine zipper forms a homodimer which binds to the AP1 binding site. *Nature* **341**, 243–245.

Neuhaus, D., NaKaseko, Y., Nagai, A. and Klug, A. (1990). Sequence specific [^{1}H] NMR resonance assignments and secondary structure identification for 1- and 2-zinc finger constructs from SWI5. *FEBS Letters* **262**, 179–184.

O'Shea, E. K., Rutkowski, R. and Kim, P. S. (1989). Evidence that the leucine zipper is a coiled coil. *Science* **243**, 538–542.

Otting, G., Quian, Y. Q., Billeter, M., Muller, M., Affolter, M., Gehring, M. and Wuthrich, K. (1990). Protein–DNA contacts in the structure of a homeodomain DNA complex determined by nuclear magnetic resonance spectroscopy in solution. *EMBO Journal* **9**, 3085–3092.

Pabo, C. and Sauer, R. T. (1984). Protein–DNA recognition. *Annual Review of Biochemistry* **53**, 293–326.

Parraga, G., Horvath, S. J., Eisen, A., Taylor, W. E., Hood, L., Young, E. T. and Klevitt, R. E. (1988). Zinc dependent structure of a single-finger domain of yeast ADRI. *Science* **241**, 1489–1492.

Pfeifer, K., Prezant, T. and Guarente, L. (1987). Yeast HAPI activator binds to two upstream sites of different sequence. *Cell* **49**, 19–27.

Prendergast, G. C. and Ziff, E. B. (1989). DNA-binding motif. *Nature* **341**, 392.

Ptashne, M. (1986). *A Genetic Switch*. Cambridge and Palo Alto: Cell Press and Blackwell Scientific Publications.

Quian, Y. Q., Billeter, M., Otting, G., Muller, M., Gehring, W. J. and Wuthrich, K. (1989). The structure of the Antennapedia homeodomain determined by NMR spectroscopy in solution: comparison with prokaryotic repressors. *Cell* **59**, 573–580.

Redemann, N., Gaul, U. and Jackle, H. (1988). Disruption of a putative Cys–zinc interaction eliminates the biological activity of the Kruppel finger protein. *Nature* **332**, 90–92.

Rosenberg, U. B., Schroder, C., Preiss, A., Kienlin, A., Cote, S., Riede, I. and Jackle, H. (1986). Structural homology of the *Drosophila Kruppel* gene with *Xenopus* transcription factor III A. *Nature* **319**, 336–339.

Sabbah, M., Redeuilh, G., Secco, C. and Baulieu, E.-E. (1987). The binding activity of estrogen receptor DNA and heat shock protein (Mr 90,000) is dependent on receptor bound metal. *Journal of Biological Chemistry* **262**, 8631–8635.

Sakai, D. D., Helms, S., Carlstedt-Duke, J., Gustafsson, J.-A., Rottman, F. M. and Yamamoto, K. R. (1988). Hormone-mediated repression: a negative glucocorticoid response element from the bovine prolactin gene. *Genes and Development* **2**, 1144–1154.

Sauer, R. T. (1990). Scissors and helical forks. *Nature* **347**, 514–515.

Schleif, R. (1988). DNA binding by proteins. *Science* **241**, 1182–1187.

Scholer, H. R., Hatzpoulos, A. K., Balling, A. R., Suzuki, N. and Gruss, P. (1989). A family of octamer-specific proteins during mouse embryogenesis: evidence for germline-specific expression of an Oct factor. *EMBO Journal* **8**, 2543–2550.

Schwabe, J. W. R., Neuhaus, D. and Rhodes, D. (1990). Solution structure of the DNA binding domain of the oestrogen receptor. *Nature* **348**, 458–461.

Struhl, K. (1989). Helix–turn–helix, zinc finger, and leucine zipper motifs for eukaryotic transcriptional regulatory proteins. *Trends in Biochemical Sciences* **14**, 137–140.

Talanin, R. V., McKnight, J. C. and Kim, P. S. (1990). Sequence-specific DNA binding by a short peptide dimer. *Science* **249**, 769–771.

Theill, L. E., Castrillo, J.-L., Wu, D. and Karin, M. (1989). Dissection of functional domains of the pituitary-specific transcription factor GHF-1. *Nature* **342**, 945–948.

Treisman, J., Gonczy, P., Vashishtha, M., Harris, E. and Desplan, C. (1990). A single amino acid can determine the DNA binding specificity of homeodomain proteins. *Cell* **59**, 553–562.

Turner, R. and Tjian, R. (1989). Leucine repeats and an adjacent DNA binding domain mediate the formation of functional cfos-cjun heterdimers. *Science* **243**, 1689–1694.

Umesono, K. and Evans, R. M. (1989). Determinants of target gene specificity for steroid/thyroid hormone receptors. *Cell* **57**, 1139–1146.

Vinson, C. R., Sigler, P. B. and McKnight, S. L. (1989). Scissors-grip model for DNA recognition by a family of leucine zipper proteins. *Science* **246**, 911–916.

Vrana, K. E., Churchill, M. A. E., Tullius, T. D. and Brown, D. D. (1988). Mapping functional regions of transcription factor TF III A. *Molecular and Cellular Biology* **8**, 1684–1696.

Weigel, D. and Jackle, H. (1990). The fork head domain: a novel DNA binding motif of eukaryotic transcription factors. *Cell* **63**, 455–456.

CHAPTER NINE

Activation and repression of gene expression by transcription factors

9.1 INTRODUCTION

Although binding to DNA is evidently a necessary prerequisite for a factor to affect transcription, it is not in itself sufficient. Thus following binding to DNA, the bound factor must interact with other transcription factors or with the RNA polymerase itself in order to affect the rate of transcription. Although such an interaction very often results in the activation of transcription, this does not always occur, and a number of cases have now been defined in which the binding of a factor can result in the repression of transcription. Activation and repression of transcription will therefore be discussed in turn.

9.2 ACTIVATION OF TRANSCRIPTION

9.2.1 Identification and significance of activation domains

Extensive studies on a variety of transcription factors have shown that they have a modular structure in which distinct regions of the protein mediate particular functions such as DNA binding (see Chapter 8) or interaction with specific effector molecules such as steroid hormones (see Section 4.3.1). It is likely, therefore, that a specific region of each individual transcription factor will be involved in its ability to activate transcription following DNA binding. As described in Section 2.3.3, such activation domains have been identified by so called 'domain-swap' experiments in which various regions of one factor are linked to the DNA-binding domain of another factor and the ability to activate transcription assessed.

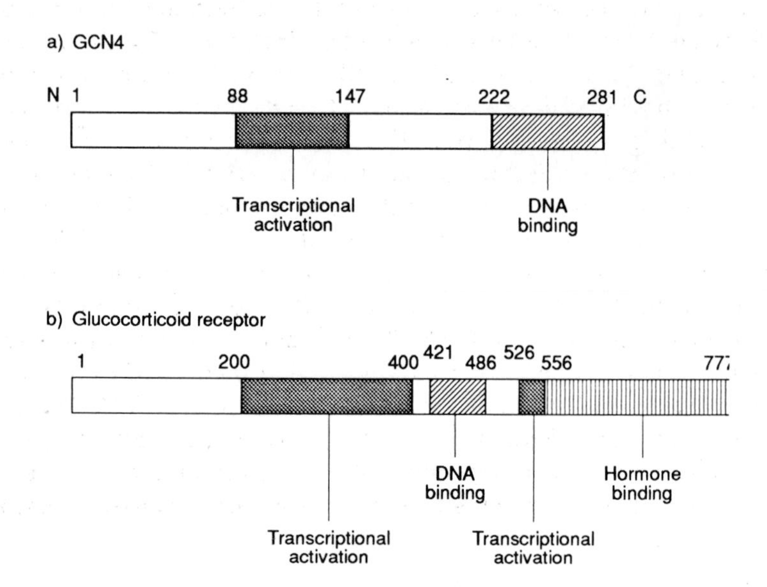

Figure 9.1 Domain structure of the yeast GCN4 transcription factor (a) and the mammalian glucocorticoid receptor (b). Note the distinct domains which are active in DNA binding or transcriptional activation.

In general, these experiments have confirmed the modular nature of transcription factors with distinct domains mediating DNA binding and transcriptional activation. Thus, in the case of the yeast factor GCN4 two distinct regions, each of 60 amino acids, have been identified which mediate, respectively, DNA binding and transcriptional activation (Figure 9.1a; Hope and Struhl, 1986). Similarly, domain-swop experiments have identified two regions of the glucocorticoid receptor, one at the N-terminus of the molecule and the other near the C-terminus which can independently mediate gene activation when linked to the DNA-binding domain of another transcription factor (Godowski *et al.*, 1988; Hollenberg and Evans, 1988) and both of these are distinct from the DNA-binding domain of the molecule. Interestingly, the C-terminal activation domain is located close to the hormone-binding domain of the receptor (Figure 9.1b), and can mediate the activation of transcription only following hormone addition. It, therefore, plays an important role in the steroid-dependent activation of transcription following hormone addition (see Section 10.3.3).

Studies on a variety of transcription factors have therefore strongly indicated their modular nature, with distinct regions of the molecule mediating DNA binding and transcriptional activation. An extreme example of this modularity is provided by the interaction of the cellular transcription factor Oct-1 (see Sections 8.2 and 8.6) and the herpes simplex virus trans-activating protein VP16 (review: Goding and O'Hare, 1989). Thus, although VP16 contains a very strong activating region which can strongly induce transcription when artificially fused to the DNA binding domain of the yeast GAL4 transcription factor (Sadowski *et al.*, 1988), it contains no DNA-binding domain and cannot therefore bind to DNA itself. Transcriptional activation by VP16 following viral infection is therefore dependent upon its ability to form a protein–protein complex with the cellular Oct-1 protein. This complex then binds to the octamer-related TAATGARAT (R = purine) motif in the viral immediate-early genes via the DNA-binding domain of Oct-1, and transcription is activated by the activation domain of VP16 (O'Hare and Goding, 1988; Preston *et al.*, 1988). Hence, in this case, the DNA-binding and transcriptional activation domains are actually located on different proteins in the DNA-binding complex (Figure 9.2).

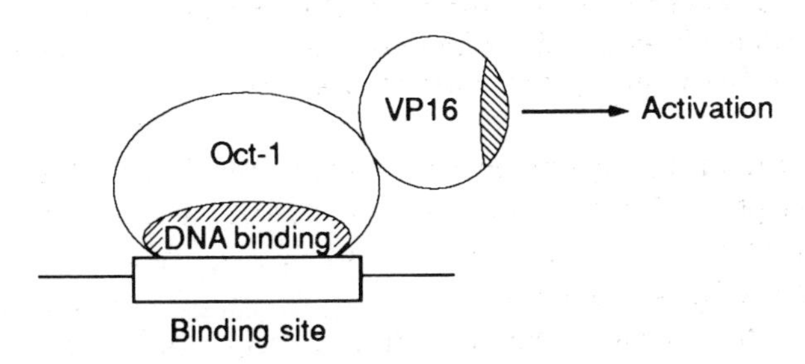

Figure 9.2 Activation of gene transcription by interaction of the cellular factor Oct-1, which contains a DNA-binding domain, and the herpes simplex virus VP16 protein, which contains an activation domain but cannot bind to DNA.

This requirement for the Oct-1 factor to be provided with a strong activating domain from VP16 in order to activate herpes simplex virus transcription also has implications for the regulation of cellular gene transcription by octamer-binding proteins (reviews: Robertson, 1988; Schaffner, 1989). Thus, as discussed in Section 5.2.2, the activation of the immunoglobulin gene promoters is dependent upon their binding of the B cell-specific octamer-binding protein Oct-2 resulting in their observed B cell-specific pattern of expression. Yet, paradoxically, the DNA-binding specificity of the

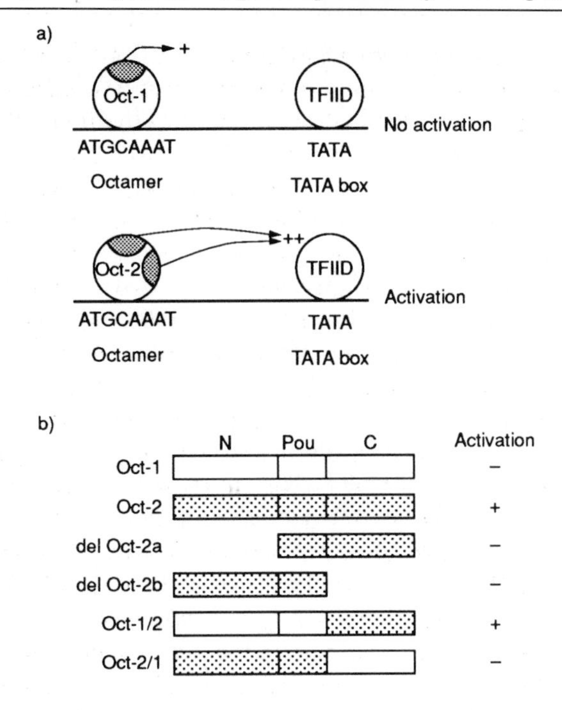

Figure 9.3 Transcriptional activation by Oct-1 and Oct-2. (a) Oct-2 but not Oct-1 is capable of activating a simple promoter containing only a TATA box and an octamer motif. This difference in activity is due to the presence of only one activation domain (shaded) in Oct-1 whereas Oct-2 has two activation domains. (b) Mapping of the activation domains in Oct-2. Note that removal of either the N-terminal (del Oct-2a) or C-terminal (del Oct-2b) activation domains abolishes the ability of Oct-2 to activate a simple octamer-containing promoter. Activity can be restored, however, by substituting the N-terminal and POU domains from Oct-1 for those of Oct-2 (Oct-1/2) but not by substituting the C-terminal domain of Oct-1 for that of Oct-2 (Oct-2/1) Hence the difference in activity between Oct-1 and Oct-2 is due to differences in the C-terminal region, where Oct-2 has an additional activation domain.

ubiquitous Oct-1 protein is identical to that of Oct-2 (Sive and Roeder, 1986; Staudt *et al.*, 1986) and hence it can bind equally well to the immunoglobulin promoters. This paradox is explained by the finding that whilst both Oct-1 and Oct-2 contain an N-terminal transcriptional activation domain, only Oct-2 has an additional activation domain at its C-terminus (Tanaka and Herr, 1990). The activation of a simple promoter which contains only an octamer and a TATA box requires a molecule with both N- and C-terminal activation domains and is hence produced only by Oct-2 and not by Oct-1 (Figure 9.3). Interestingly, a hybrid molecule with the N-terminal activation domain and DNA-binding domain derived from

Oct-1 and the C-terminal activation domain of Oct-2 can activate transcription of such a promoter, indicating that the difference in the C-terminal regions of Oct-1 and Oct-2 is critical for this effect (Figure 9.3). Hence the presence of two co-operating activation domains in Oct-2 allows it, but not Oct-1, to activate the immunoglobulin gene promoters which contain only a TATA box and the octamer motif and which are, therefore, active in a B cell-specific manner dependent on the B cell-specific expression of Oct-2.

Despite its inability to activate transcription of the immunoglobulin genes, Oct-1 can play a role in activating transcription of some cellular genes. Thus, for example, the genes encoding the small nuclear RNAs such as U1 and U2 contain octamer motifs in their promoters and the constitutive expression of these genes is dependent upon the binding of Oct-1 to the octamer motif (Mattaj *et al.*, 1985). However, these genes do not contain a TATA box but instead have a specialized promoter element known as a proximal sequence element (PSE) (review: Dahlberg and Lund, 1988). Oct-1 can activate a promoter containing only an octamer and a PSE even though, as discussed above, it cannot activate a promoter containing an octamer and a TATA box (Tanaka *et al.*, 1988). This suggests, therefore, that the ability of Oct-1 to activate the SnRNA promoters is dependent on their possession of a unique promoter element which binds a protein distinct from the TATA box-binding protein TFIID and with which Oct-1 can interact (Figure 9.4a). In agreement with this distinction between the PSE and the TATA box, conventional activating factors such as GAL4 which activate transcription from a TATA box cannot do so when their specific binding sites are placed upstream of a PSE (Tanaka *et al.*, 1988).

Although this explanation is likely to be correct for the SnRNA genes it cannot apply to the histone H2B, gene which is also expressed in an Oct-1-dependent manner but contains a conventional TATA box. Moreover, unlike the SnRNA genes, the histone H2B gene is expressed in a cell cycle-dependent manner, with transcription being highest in S phase (Sittman *et al.*, 1983). Paradoxically, this cell cycle-dependent expression pattern is dependent upon binding of the constitutively expressed Oct-1 protein to the octamer motif in the histone H2B promoter, deletion of this motif abolishing the cell cycle-specific pattern of expression (La Bella *et al.*, 1988). Since Oct-1 is expressed throughout the cell cycle it seems likely that its role in H2B gene expression is dependent upon its association with a cellular factor which is expressed only at certain phases of the cell cycle and which supplies a strong activation domain in the same manner as the viral VP16 protein, allowing Oct-1 to activate a TATA box-containing promoter and producing

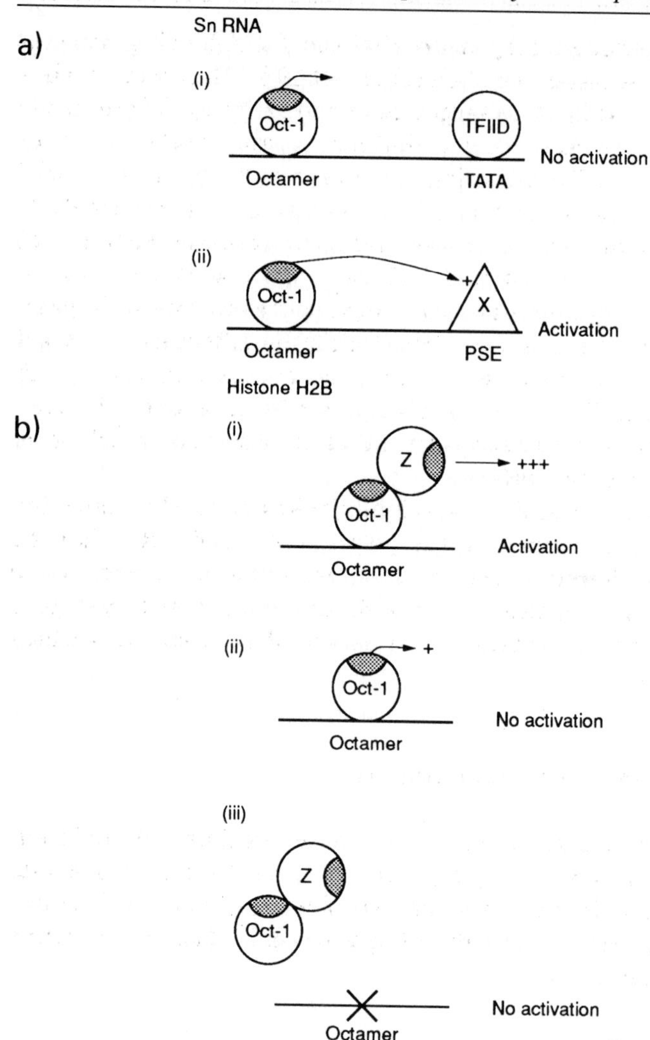

Figure 9.4 Gene activation by Oct-1. (a) Although Oct-1 cannot activate transcription via TFIID bound to a TATA box, it can do so via the factor (x) which binds to the proximal sequence element (PSE) in the snRNA genes. (b) Activation of the histone H2B gene by Oct-1 is likely to involve binding of another factor (Z) which has a strong activation domain. Activation of histone H2B is therefore dependent both on binding of Oct-1 to the octamer motif and on the presence of Z, which may be expressed in a cell cycle-specific manner, accounting for the cell cycle-specific regulation of H2B gene expression.

the observed cell cycle-dependent pattern of gene expression (Figure 9.4b).

Thus the differences in the activation domains of Oct-1 and Oct-2 result in important differences in the target genes which they regulate even though they have identical DNA-binding domains (for further discussion see Schaffner (1989)). Moreover, such a mechanism for modulating target gene specificity by differences in the activation domains of two different factors rather than by differences in their DNA-binding domains is unlikely to be confined to the octamer-binding proteins. Thus, as discussed in Section 6.2.3, several *Drosophila* homeobox-containing transcription factors such as eve, Ftz or zen can bind to the identical DNA sequence but have very different effects on embryonic development (review: Hayashi and Scott, 1990). As with Oct-1 and Oct-2, the differences in target gene specificity of these factors may be determined not by differences in their DNA-binding ability but by differences in their abilities to activate transcription either in isolation or following association with other factors.

Whether this mechanism is actually relevant to the homeobox factors or not, it is clear from the case of Oct-1 and Oct-2 that the nature of the different activation domains in different transcription factors can play as critical a role in determining their target gene specificity as can differences in the nature of the sequences which they recognize.

9.2.2 Nature of activation domains

Following the identification of activation domains in different transcription factors, it rapidly became clear that they fell into several distinct families with common features which will be discussed in turn (for a typical example of each class of activation domain see Figure 9.5).

Acidic domains

Comparison of several different activation domains including those of the yeast factors GCN4 and GAL4 and the activation domain at the N-terminus of the glucocorticoid receptor and that of VP16 which were discussed above (Section 9.2.1), indicated that, although they do not show any strong amino acid sequence homology to each other, they all have a large proportion of acidic amino acids, producing a strong net negative charge (Figure 9.5). Thus the 82-amino-acid activating region of the glucocorticoid receptor contains 17 acidic residues (Hollenberg and Evans, 1988) whilst the same number of negatively charged amino acids is found within the 60-amino-acid activating region of GCN4 (Hope and Struhl, 1986). These findings indicated, therefore, that these activation regions consist of so-called 'acid blobs' or 'negative noodles' with a high

GAL4 acidic domain

D S A A A H H **D** N S T I P L **D** F M P R **D**
A L H G F **D** W S **E E D D** M S **D** G L P F L
K T **D** P N N N G F

Sp1 glutamine-rich domain B

Q G Q T P **Q** R V S G L **Q** G S D A L N I **Q**
Q N **Q** T S G G S L **Q** A G **Q Q** K E G E **Q** N
Q Q T **Q Q Q Q** I L I **Q** P **Q** L V **Q** G G **Q** A
L **Q** A L **Q** A A P L S G **Q** T F T T **Q** A I S
Q E T L **Q** N L **Q** L **Q** A V P N S G P I I I
R T P T V G P N G **Q** V S W **Q** T L **Q** L **Q** N
L **Q** V **Q** N P **Q** A **Q** T I T L A P M **Q** G V S
L G **Q**

CTF/NFI Proline-rich domain

P P H L N **P** Q D **P** L K D L V S L A C D **P**
A S Q Q **P** G R L N G S G Q L K M **P** S H C
L S A Q M L A **P P P P** G L **P** R L A L **P P**
A T K **P** A T T S E G G A T S **P** S Y S **P P**
D T S **P**

Figure 9.5 Structure of typical members of each of the three classes of activation domains. Acidic, glutamine or proline residues are highlighted in the appropriate case.

proportion of negatively charged amino acids which are involved in the activation of transcription (reviews: Ptashne, 1988; Sigler, 1988).

In agreement with this idea, mutations in the activation domain of GAL4 which increase its net negative charge increase its ability to activate transcription, whereas those which decrease its negative charge result in a reduction of this activating ability (Gill and Ptashne, 1987). Similarly, if recombination is used to create a GAL4 protein with several more negative charges, the effect on gene activation is additive, a mutant with four more negative charges than the parental wild type activating transcription nine-fold more efficiently than the wild type (Gill and Ptashne, 1988).

Although the acidic nature of the activation domain is clearly important for its function, it is not the only feature required, since it is possible to decrease the activity of the GAL4 activation domain without reducing the number of negatively charged residues (Gill and Ptashne, 1987). Moreover, studies of a number of acidic activation domains have shown that the negatively charged residues are not randomly distributed but are displayed along one side of an α-helical structure, allowing them to contact another protein, whereas the other surface of the helix contains only hydrophobic residues (Figure 9.6).

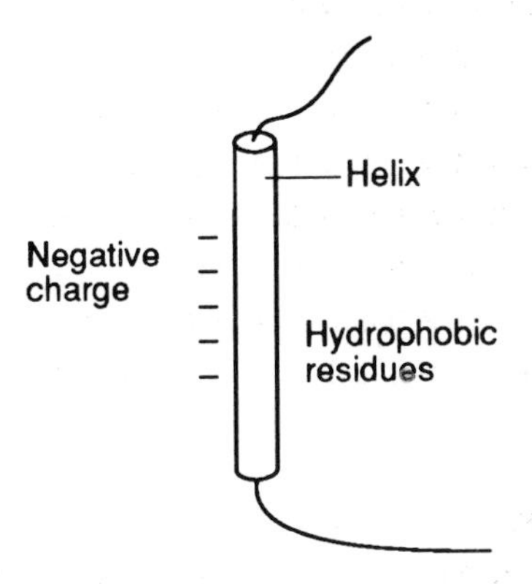

Figure 9.6 Amphipathic helix, in which all the negatively charged amino acids are displayed on one side of the helix.

The importance of such an amphipathic helix in which all the charged residues are located on one side of the helix was confirmed by synthesizing two 15-amino-acid peptides containing the same amino acids but in a different order such that only one of the two peptides could form an amphipathic helix. When each of these was linked to the DNA-binding domain of GAL4 only the amphipathic helix-forming peptide could activate transcription, whereas the other peptide could not do so even though it contained the same amino acids, indicating that a simple preponderance of acidic residues is insufficient to create an activating region (Giniger and Ptashne, 1987; Figure 9.7). Similarly, in an experiment where random portions of the *E. coli* genome were linked to a DNA-binding domain and tested for their ability to activate transcription, the only fragments which could do so were those which could encode peptides containing amphipathic α-helices (Ma and Ptashne, 1987; Figure 9.8).

It is clear, therefore, that a number of activation domains consist of an acidic region in which the negatively charged amino acids are arranged along one side of an amphipathic α-helix. Although activation domains of this type form the majority of the activation domains so far identified in eukaryotic transcription factors from yeast to mammals, other types of activation domains have been identified in higher eukaryotes and these will be discussed in turn.

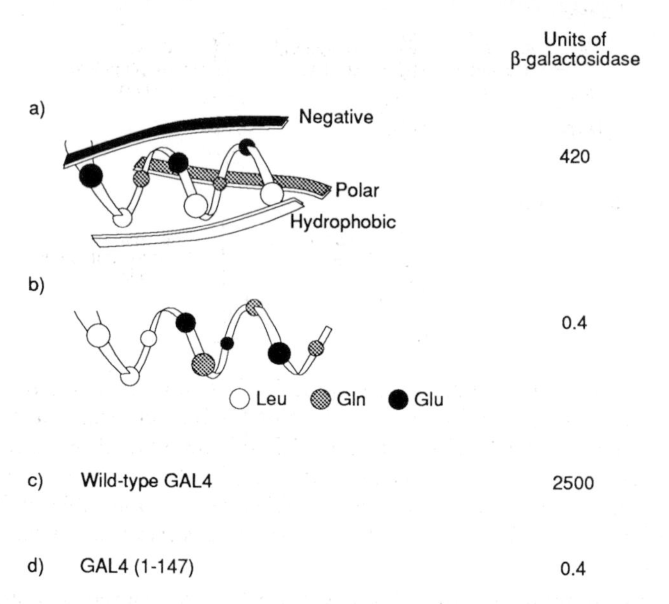

Figure 9.7 Effects on gene expression of a sequence forming an amphipathic helix (a) and a sequence of identical amino acid composition which cannot form an amphipathic helix (b). Gene expression is measured as the amount of β-galactosidase produced by a hybrid β-galactosidase gene containing a binding site for GAL4 in its promoter. Note that the intact GAL4 protein (c) and the truncated GAL4 protein linked to an amphipathic helix (a) can induce gene expression, whereas truncated GAL4 linked to a non-amphipathic helix (b) or alone (d) cannot do so.

Glutamine-rich domains

Analysis of the constitutive transcription factor Sp1 (see Section 3.3.2) revealed that the two most potent activation domains contained approximately 25% glutamine residues and very few negatively charged residues (Courey and Tjian, 1988; Figure 9.5). These glutamine-rich motifs are essential for the activation of transcription mediated by these domains since their deletion abolishes the ability to activate transcription. Most interestingly, however, transcriptional activation can be restored by substituting the glutamine-rich regions of Sp1 with a glutamine-rich region from the *Drosophila* homeobox transcription factor Antennapedia, which has no obvious sequence homology to the Sp1 sequence. Hence, as with the acidic activation domains, the activating ability of a glutamine-rich domain is not defined by its primary sequence but rather by its overall nature in being acidic or glutamine-rich respectively.

Similar glutamine-rich regions have now been defined in

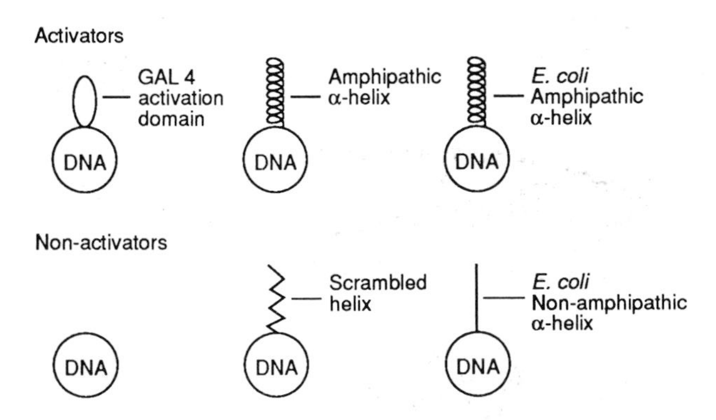

Figure 9.8 Gene activation by the GAL4 DNA-binding domain (circled DNA) can be achieved by linkage to the GAL4 activation domain, a synthetic amphipathic helix or a region of *E. coli* DNA which can encode an amphipathic helix but is not observed with the DNA-binding domain alone or following linkage to a synthetic scrambled non-amphipathic helix or a region of *E. coli* DNA which does not encode an amphipathic helix.

transcription factors other than Sp1 and Antennapedia, including the N-terminal activation domains of the octamer-binding proteins Oct-1 and Oct-2 (see Section 9.2.1), the *Drosophila* homeobox proteins ultrabithorax and zeste, and the yeast HAP1 and HAP2 transcription factors, indicating that this motif may be quite widespread, being found in different transcription factors in different species (review: Mitchell and Tjian, 1989).

Proline-rich domains

Studies on the constitutive factor CTF/NF1 which binds to the CCAAT box motif (Section 3.3.3) defined a third type of activation domain distinct from those previously discussed. Thus the activation domain located at the C-terminus of CTF/NF1 is not rich in acidic or glutamine residues but instead contains numerous proline residues forming approximately one-quarter of the amino acids in this region (Mermod *et al.*, 1989; Figure 9.5). As with the other classes of activation domains, this region is capable of activating transcription when linked to the DNA-binding domains of other transcription factors. Similar proline-rich domains have now been identified in several other transcription factors such as the proto-oncogene product Jun, AP2 and the C-terminal activation domain of Oct-2 (see Section 9.2.1) (review: Mitchell and Tjian, 1989). Thus, as with the glutamine-rich domains, proline-rich domains are not confined to a single factor whilst a single factor such as Oct-2 can contain two activation domains of different types.

In summary, therefore, it is clear that as with DNA binding, several distinct protein motifs can activate transcription (Figure 9.5).

9.2.3 Mechanism of action of activation domains

The widespread interchangeability of acidic activation domains from yeast, *Drosophila* and mammalian transcription factors discussed above strongly suggests that a single common mechanism may mediate transcriptional activation in a wide range of organisms, at least in cases involving acidic activation domains, which have been most intensively studied. This idea is supported by the finding that mammalian transcription factors, such as the glucocorticoid receptor, can activate a gene carrying their appropriate DNA-binding site in yeast cells, whilst the yeast GAL4 factor can do so in cells of *Drosophila*, tobacco plants and mammals (reviews: Guarente, 1988; Ptashne, 1988).

Most interestingly, yeast and mammalian factors can co-operate together in gene activation, the rat glucocorticoid receptor and the yeast activator GAL4, for example, synergistically activating the transcription in mammalian cells of a gene containing binding sites for both factors (Kakidani and Ptashne, 1988). Such a synergistic interaction, in which activation by the yeast and mammalian factors together is far greater than the sum of that produced by either alone, is unlikely to be due to interactions between the two factors, which would never normally come into contact. Rather it is likely to reflect their ability to simultaneously contact and activate another molecule which is part of the general transcription machinery common to yeast and mammalian cells. In agreement with this idea, synergistic activation by GAL4 and the mammalian factor ATF (Lin *et al.*, 1990) or by two bound molecules of GAL4 (Carey *et al.*, 1990) can be observed under conditions where binding sites on the DNA for these factors are saturated (Figure 9.9a). Hence such synergistic activation cannot be dependent upon co-operative binding of the two factors to DNA and must involve the ability of two different transcriptional activators or two molecules of the same factor to interact simultaneously with the same target factor (Figure 9.9b,c).

Such considerations clearly focus attention on the nature of the target factor with which these activators interact. A number of experiments have indicated that this target factor is likely to be a component of the basal transcription complex required for the transcription of a number of different genes and not solely for that of the activated gene. Thus the over-expression of the yeast GAL4

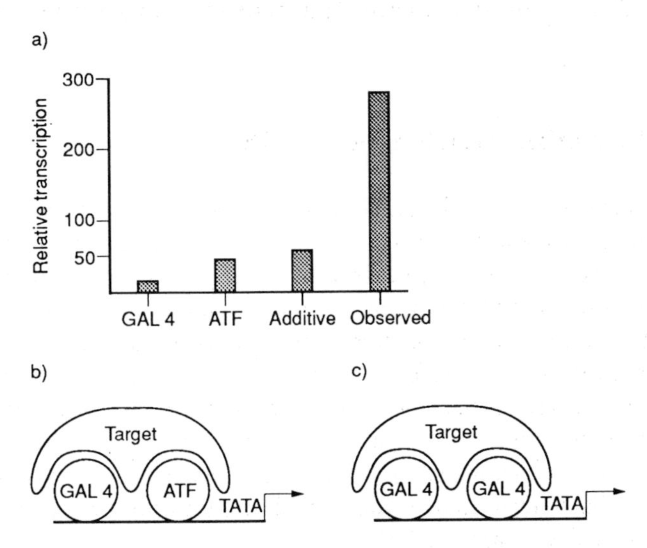

Figure 9.9 (a) Activation of a target promoter containing binding sites for the yeast transcription factor GAL4 and the mammalian transcription factor ATF. Note that the observed activation in the presence of both factors is greater than the additive effect of each factor individually, indicating that the two factors can activate gene expression synergistically. (b, c) Proposed mechanism for synergistic activation by GAL4 and ATF or by two molecules of GAL4 in which each molecule makes simultaneous contact with the target for activation.

protein, which contains a strong activation domain, results in the down-regulation of genes which lack GAL4-binding sites, such as the *CYC1* gene, as well as activating genes which do contain GAL4–binding sites (Gill and Ptashne, 1988). This phenomenon, which has been noted for a number of transcription factors with strong activation domains, is known as squelching (review: Ptashne, 1988). Although the degree of squelching by any given factor is proportional to the strength of its activation domain, squelching differs from activation in that it does not require DNA binding and can be achieved with truncated factors containing only the activation domain and lacking the DNA-binding domain. This phenomenon can, therefore, be explained on the basis that a transcriptional activator, when present in high concentration, can interact with its target factor in solution as well as on the DNA. If this target factor is present at limiting concentrations it will be sequestered away from other genes which require it for transcription, resulting in their inhibition (Figure 9.10).

The existence of squelching indicates, therefore, that the target factor for activation domains is likely to be a component of the basal

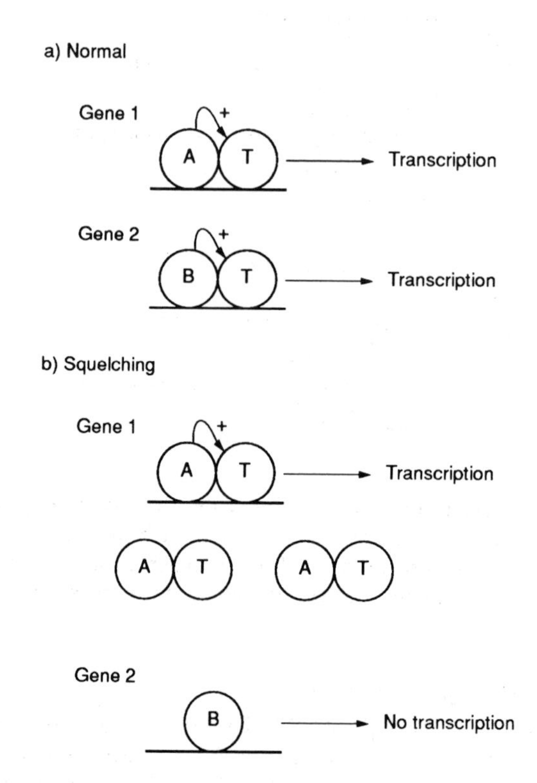

Figure 9.10 The process of squelching. In the normal case illustrated in (a), two distinct activator molecules, A and B, involved in the activation of genes 1 and 2 respectively, both act by interacting with the general transcription factor T and both genes are transcribed. In squelching, illustrated in (b), factor A is present at high concentration and hence interacts with T both on gene 1 and in solution. Hence factor T is not available for transcription of gene 2 and therefore only gene 1 is transcribed whilst transcription of gene 2 is squelched.

transcriptional apparatus which is required for the transcription of a wide range of genes and which is conserved from yeast to mammals, allowing yeast activators to work in mammalian cells and vice versa. Two such possible targets for activation have been envisaged (Figure 9.11). Thus activation might occur directly by interaction of the activator with RNA polymerase itself (Figure 9.11a) or, alternatively, activating molecules could function via another transcription factor such as the TATA box-binding factor TFIID, facilitating its binding into a stable transcriptional complex or allowing it to interact more efficiently with the RNA polymerase (Figure 9.11b). These two possibilities will be discussed in turn.

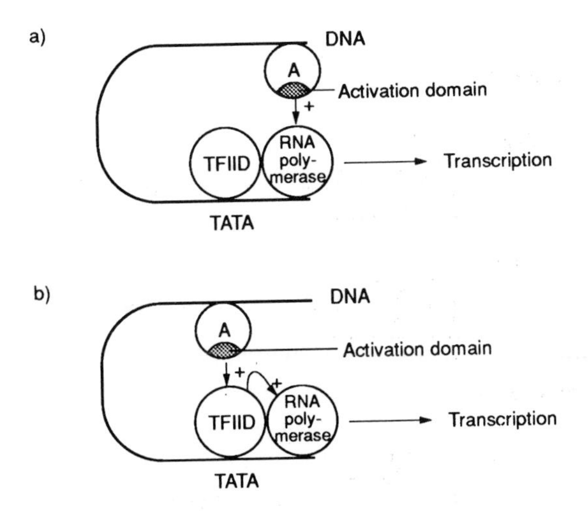

Figure 9.11 Two possible mechanisms by which an activating factor (A) could act. This could occur via direct interaction with the RNA polymerase itself (a) or by interaction with another transcription factor such as the TATA-box binding factor TFIID, which in turn interacts with the polymerase (b).

Activation of RNA polymerase

An obvious mechanism for activation would be for activating domains to interact directly with the RNA polymerase itself to increase its activity. As discussed in Section 3.1, the largest subunit of RNA polymerase II contains multiple copies of a sequence whose consensus is Tyr-Ser-Pro-Thr-Ser-Pro-Ser which is highly conserved in evolution and is essential for its function. This motif is very rich in hydroxyl groups and lacks negatively charged acidic residues.

It has therefore been suggested (Sigler, 1988) that this motif could interact directly either with a negatively charged acidic domain or via hydrogen bonding with amide groups in a glutamine-rich activation domain. This would provide a mechanism for direct interaction between activating domains and RNA polymerase itself, whilst the evolutionary conservation of the target region within the polymerase would explain why yeast activators work in mammalian cells and vice versa. In agreement with this idea it has recently been shown that yeast mutants containing a reduced number of copies of the heptapeptide repeat in RNA polymerase II are defective in their response to activators such as GAL4 (Scafe *et al.*, 1990).

Although these results are consistent with a direct interaction between transcriptional activators and the RNA polymerase, they are equally consistent with an indirect interaction in which the activator contacts another component of the transcriptional machinery which then interacts with the repeated motif in the polymerase. Therefore, despite the attractiveness of a model involving direct interaction between activating factors and the polymerase itself, no direct evidence in its favour has thus far been obtained.

Interaction with other transcription factors

Clearly, transcriptional activation could be achieved by interaction of the activating factor with any of the components of the basal transcriptional machinery discussed in Section 3.2.4. One obvious candidate for the component with which activating factors interact, however, is TFIID since this factor is both required for the transcription of a wide variety of genes both with and without TATA boxes and is highly conserved in evolution, with the yeast factor being able to promote transcription in mammalian cell extracts and vice versa.

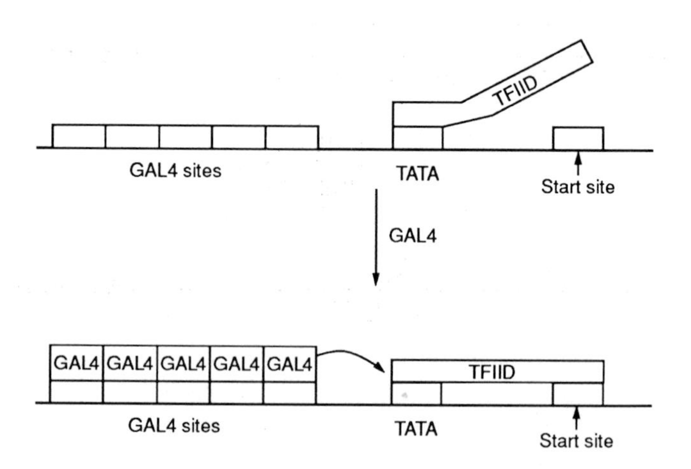

Figure 9.12 Effect of GAL4 binding on the binding of TFIID.

Evidence for an effect of activating factors on TFIID has been obtained in the case of the yeast activating factor GAL4 (Horikoshi *et al.*, 1988a). Thus in the absence of GAL4, TFIID was shown to be bound only at the TATA box of a promoter containing both a TATA box and GAL4-binding sites. In contrast, in the presence of GAL4 bound to its upstream binding sites in the promoter, the binding of TFIID was altered such that it now covered both the TATA box and the start site for transcription (Figure 9.12). In contrast, no change in TFIID binding was observed in the presence of a truncated GAL4 molecule which can bind to DNA but lacks the acidic activation domain. Hence, an acidic activator can produce a change in TFIID binding, resulting in its binding to the start site for transcription and this effect correlates with the ability of GAL4 to activate transcription rather than being a consequence of its binding to DNA. A similar change in the binding of TFIID has also been noted following the binding of the mammalian activating factor ATF to sites upstream of the TATA box (Horikoshi *et al.*, 1988b), suggesting that TFIID is the common target for GAL4 and ATF detected in the experiments of Lin *et al.* (1990) discussed above.

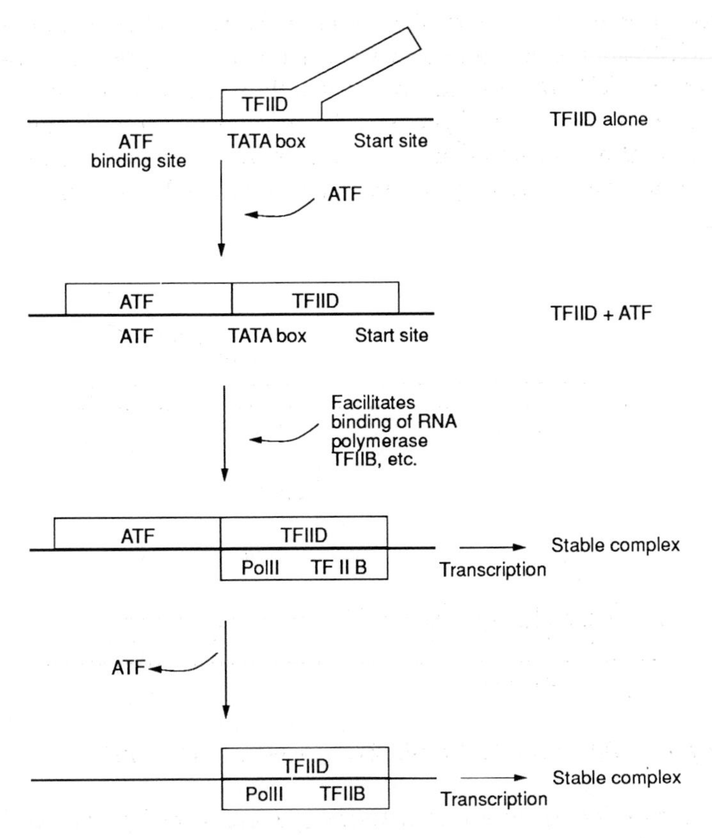

Figure 9.13 Binding of ATF to its binding site alters the binding of TFIID and facilitates the assembly of a stable transcription complex.

Most interestingly, the observed change in TFIID binding produced by ATF has been shown to promote assembly of the basal transcriptional complex by facilitating the binding of its other components such as the RNA polymerase itself and the transcription factors TFIIB and TFIIE. Thus these factors will complex with the DNA only after ATF and TFIID have bound, whilst, once the complex has formed, ATF can be removed without affecting transcription (Hai *et al.*, 1988; Figure 9.13). Hence, like several of the constitutive factors discussed in Section 3.2, ATF appears to be an assembly factor which activates transcription by promoting the assembly of the basal transcription complex by stabilizing the binding of TFIID to the start site of transcription.

On the basis of these studies, it is likely that activation occurs via interaction with TFIID. Recently, however, controversy has arisen over whether activating factors interact directly with TFIID itself or whether they affect TFIID binding by interacting with so-called 'adaptor' molecules which then interact with TFIID (Figure 9.14), and at present it is unclear which of these possibilities is correct (review: Lewin, 1990; Ptashne and Gann, 1990).

i) Direct interaction

+
Act
TFIID
DNA
TATA

ii) Indirect interaction

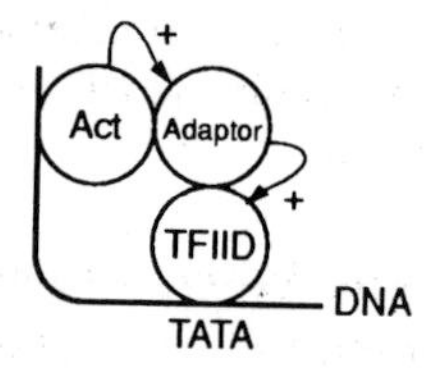

Figure 9.14 Interaction of an activator molecule with TFIID can occur either directly (i) or indirectly (ii) via an intermediate adaptor molecule.

It should be noted, however, that whilst most studies have been carried out on acidic activation domains it is not clear that different types of activation domain function in the same way. Thus in a recent study (Tasset *et al.*, 1990), the acidic activation domain of VP16 was not capable of squelching gene activation by the non-acidic activation domain of the oestrogen receptor, whereas the oestrogen receptor activation domain was capable of squelching gene activation mediated both by its own activation domain and by the acidic domain of VP16. These results are consistent with a model in which a series of adaptor molecules mediate activation, with the acidic activation domain of VP16 contacting a factor which is located earlier in the series than that contacted by the non-acidic activation domain of the oestrogen receptor (Figure 9.15). Hence the factor contacted by the activation domain of the oestrogen receptor would also be essential for activation by VP16 (factor 4 in Figure 9.15), whereas the factor contacted by the acidic activation domain of VP16 (factor 1 in Figure 9.15) would not be required for activation by the oestrogen receptor.

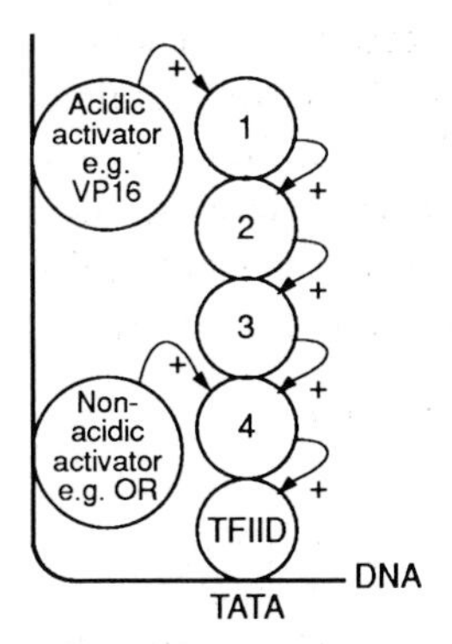

Figure 9.15 Interaction of different activator molecules with different adaptor molecules (1–4) which each activate each other and ultimately activate TFIID. Note that the ability of the non-acidic activation domain of the oestrogen receptor to squelch activation by the acidic activation domain of VP16 but not vice versa can be explained if the oestrogen receptor interacts with an adaptor molecule (4) closer to TFIID in the series than that with which VP16 interacts (1).

It is clear, therefore, that further studies are required to identify the precise molecules which are contacted by acidic and non-acidic activation domains. Moreover, recent data (for discussion see Sharp, 1991) has indicated that in addition to interacting with TFIID, activation domains may also interact directly with TFIIB, another component of the basal transcriptional complex, in order to facilitate its binding to DNA which occurs following TFIID binding (see Section 3.2.4). Hence activating factors may act on the basal transcription complex at several points either directly or via adaptor molecules to facilitate its assembly and thereby activate transcription.

9.3 REPRESSION OF TRANSCRIPTION

Although the majority of transcription factors which have so far been described act in a positive manner, a number of cases have now been reported in which a transcription factor exerts an inhibitory

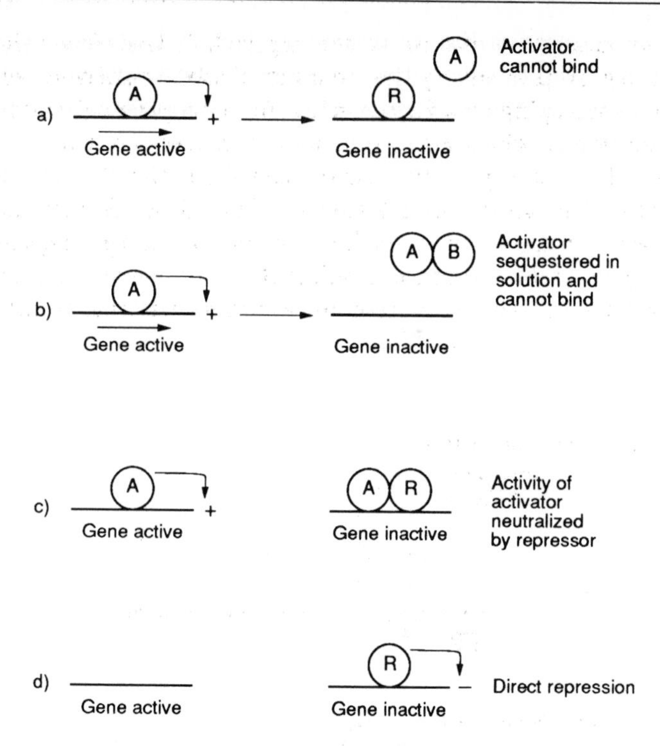

Figure 9.16 Potential mechanisms by which a transcription factor can repress gene expression. This can occur by the repressor (R) binding to DNA and preventing an activator (A) from binding and activating gene expression (a), by the repressor interacting with the activator in solution and preventing its DNA binding (b), by the repressor binding to DNA with the activator and neutralizing its ability to activate gene expression (c) or by direct repression by an inhibitory transcription factor (d).

effect on transcription, and there are several possible mechanisms by which this can occur (reviews: Levine and Manley, 1989; Goodbourn, 1990) (Figure 9.16).

The simplest means of achieving repression of transcription is for a negatively acting factor to bind to the site in DNA normally occupied by a positively acting factor necessary for gene activation (Figure 9.16a). By preventing the binding of the positively acting factor, the negatively acting factor effectively inhibits gene activation. Thus, as discussed in Section 4.3.4, the glucocorticoid receptor can inhibit the transcription of several genes such as the human glycoprotein hormone α-subunit gene in response to glucocorticoid treatment. This is achieved by the receptor binding to a sequence known as the nGRE which is distinct from that to which it binds when mediating gene activation (Figure 4.16). Following binding to such sites, the receptor cannot activate transcription, presumably because its activation domain is not exposed. It therefore inhibits gene activity by preventing the binding of other positively acting factors to overlapping or adjacent sites. As with gene activation by steroid hormones, glucocorticoid-mediated repression will involve activation of the receptor by dissociation from the 90-kDa heat-shock protein following steroid treatment (hsp90; see Section 4.3.2). The released receptor will then bind to its site in the negatively regulated gene and prevent the binding of a positively acting factor, thereby inhibiting gene expression in response to steroid treatment (Figure 9.17).

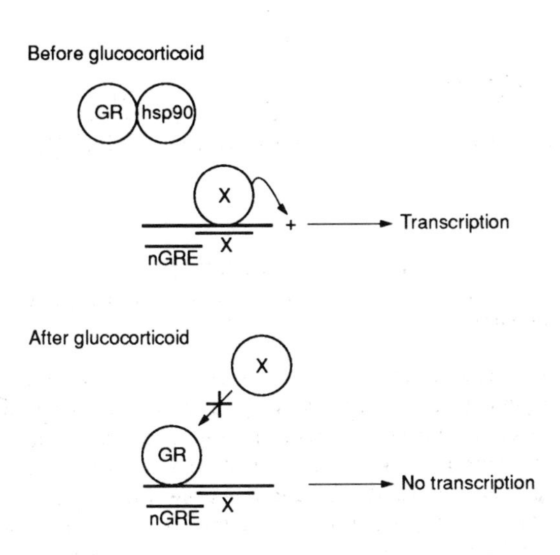

Figure 9.17 Glucocorticoid-mediated inhibition of a gene containing an nGRE binding site for the glucocorticoid receptor. Glucocorticoid treatment releases the receptor from hsp90, allowing it to bind to the nGRE and prevent transcription by inhibiting the binding of an activating transcription factor (x) to its binding site which overlaps the nGRE.

Interestingly, the glucocorticoid receptor is also capable of producing repression by an entirely different mechanism which does not involve DNA binding and is of the type illustrated in Figure 9.16b. Thus, glucocortocoid hormones have been known for some time to be potent inhibitors of the induction of the collagenase gene by phorbol esters, resulting in their having an anti-inflammatory effect. This inhibition is mediated by the glucocorticoid receptor, which abolishes the binding of the Jun and Fos proteins to the AP1 sites in the collagenase gene. Since such binding is essential for gene induction by phorbol esters (for discussion of Fos, Jun and AP1 see Section 7.2), this effectively inhibits gene activation. Unlike the examples of repression by the glucocorticoid receptor discussed above, however, the collagenase promoter does not contain any binding sites for the receptor adjacent to the AP1 sites nor does the receptor apparently bind to the collagenase promoter (Jonat *et al.*, 1990; Yang-Yen *et al.*, 1990). Rather, repression is mediated via the formation of a protein–protein complex between Fos or Jun and the glucocorticoid receptor which prevents binding of Fos and Jun to the AP1 site and hence inhibits transcription (Figure 9.18). Clearly, such an association of Fos or Jun and the glucocorticoid receptor will also prevent the glucocorticoid receptor from binding to DNA and thereby activating glucocorticoid-dependent genes in response to hormone, and this is indeed observed at high concentrations of Fos and Jun (Jonat *et al.*, 1990; Lucibello *et al.*, 1990).

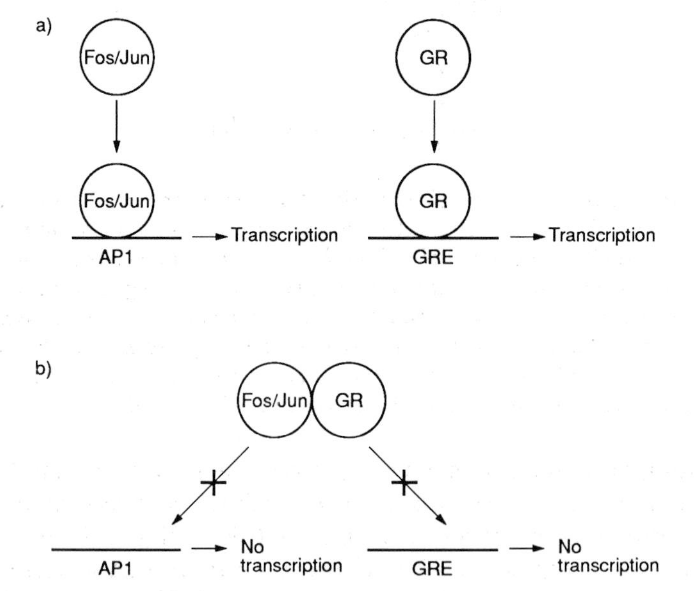

Figure 9.18 Mutual transrepression by Fos/Jun and the glucocorticoid receptor. The formation of a protein–protein complex between Fos/Jun and the glucocorticoid receptor inhibits the expression of genes containing binding sites for either Fos/Jun (AP1 sites) or for the glucocorticoid receptor (GRE) by preventing the proteins from binding to these sites and activating transcription.

Hence mutual transrepression of two different activating proteins can be achieved by their forming a complex which inhibits the DNA binding of both. Such a mechanism is clearly related to the process of squelching discussed earlier in Section 9.3.2, in which a strong

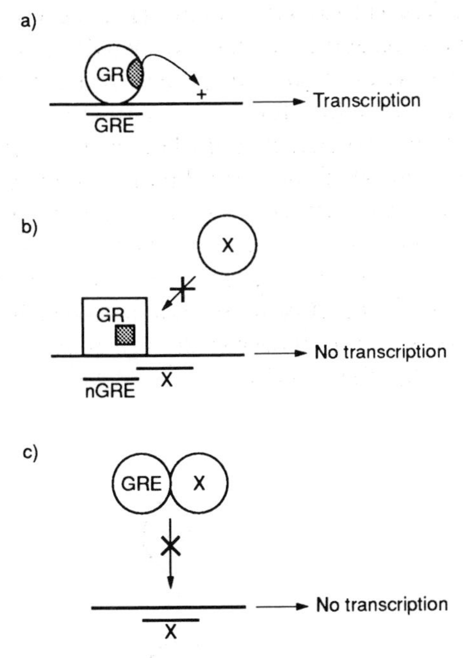

Figure 9.19 Effects of the glucocorticoid receptor on transcription. (a) Binding to a GRE results in an activation of transcription via the activation domain (shaded). (b) Following binding to an nGRE sequence the activation domain of the receptor is not exposed and expression of the gene is repressed by the bound receptor preventing an activator molecule (x) from binding to its appropriate binding site. (c) Inhibition of gene expression can also occur by the receptor interacting with an activator molecule (x) in solution and thereby preventing it binding to its appropriate binding site.

activating protein inhibits gene expression by preventing the DNA binding of another transcription factor with which it associates.

In addition to its gene-activating abilities (Figure 9.19a), the glucocorticoid receptor therefore illustrates two distinct mechanisms of gene repression, involving the inhibition of DNA binding by a positive factor either by the blocking of its binding site (Figure 9.19b) or by formation of a protein–protein complex which prevents DNA binding (Figure 9.19c). It should be noted, however, that all these cases involve the action of an activated receptor which has been freed from hsp90 upon steroid treatment (see Section 4.3.2) and the receptor therefore activates or represses the expression of different genes in a steroid-dependent manner.

As well as cases of this type in which a single factor can act either positively or negatively depending on the gene involved, gene repression involving the inhibition of DNA binding can also be produced by factors which function purely negatively. Thus, as discussed in Section 5.3.2, gene activation by helix–loop–helix proteins such as MyoD can be inhibited by the Id protein, which can dimerize with them by means of its helix–loop–helix domain and then inhibit their DNA binding since it lacks the basic DNA-binding domain.

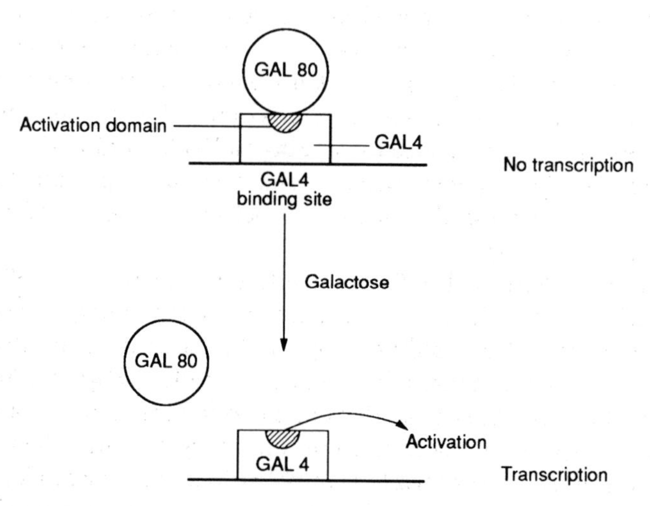

Figure 9.20 Galactose activates gene expression by removing the GAL80 protein from DNA-bound GAL4 protein, unmasking the activation domain on GAL4.

The cases of repression described so far all involve the inhibition of DNA binding, either by blocking the binding site for a factor (Figure 9.16a) or by forming a protein–protein complex (Figure 9.16b). Since DNA binding is a necessary prerequisite for gene activation, this constitutes an effective form of repression. In addition, however, inhibition of transcription can also be achieved by interfering with transcriptional activation by a DNA-bound factor in a phenomenon known as quenching (Figure 9.16c). A simple example of this type is seen in the case of the negatively acting yeast factor GAL80, which inhibits gene activation by the positively acting GAL4 protein. This is achieved by the binding of GAL80 to DNA-bound GAL4, such binding occurring via the 30 amino acids located at the extreme C-terminus of the GAL4 molecule. As these amino acids are located close to the GAL4 activation domain, the binding of GAL80 to GAL4 masks the GAL4 activation domain and hence inhibits the activation of gene expression by GAL4. In response to treatment with galactose, GAL80 dissociates from GAL4, allowing GAL4 to fulfil its function of activating galactose-inducible genes (Figure 9.20). Hence this system provides an elegant means of modulating gene expression in response to galactose, with the activating GAL4 factor being bound to DNA both prior to and after galactose addition but being able to activate gene expression only following the galactose-induced dissociation of the quenching GAL80 factor (Lohr and Hopper, 1985; Johnston *et al.*, 1987).

A related example of quenching, in which the inhibitory factor binds to a DNA sequence adjacent to the quenched factor rather than to the factor itself, is seen in the case of the yeast mating type α-2 protein. Thus, as discussed in Section 5.4.3, the α-2 protein represses the a-specific genes by binding to DNA binding sites in the promoters of these genes which flank the binding sites for the constitutively expressed, positively acting protein PRTF and inhibiting its activity.

In the cases described so far, a negative factor exerts its effect by neutralizing the action of a positively acting factor either by preventing its DNA binding (Figure 9.16a,b) or inhibiting its activation of transcription following such binding (Figure 9.16c). In principle, however, it is possible that some factors inhibit transcription directly, possibly via a discrete inhibitory domain which interacts with the basal transcription complex in a manner analogous to an activation domain but with the opposite effect (Figure 9.16d). Although a number of cases where this may be occurring have been described, in no case has the possibility of the neutralization of a positively acting factor by the inhibitor been eliminated and a direct negative action unequivocally demonstrated (for further discussion see Levine and Manley (1989)).

Nonetheless, even if their action is only mediated by neutralizing a positively acting factor, it is becoming increasingly clear that negatively acting factors play an important role in transcriptional regulation.

9.4 CONCLUSIONS

The fundamental property of transcription factors is their ability to influence the transcription of specific genes either positively or negatively following DNA binding. In this chapter, we have discussed how the activation of transcription is dependent on discrete activation domains whose structure and mechanism of action are gradually being understood, whilst the repression of transcription is generally dependent on the neutralization of a positively acting factor via the inhibition of its DNA binding or activating properties. These properties offer ample scope for gene regulation in different cell types or in different tissues. Thus in addition to the simple activation of gene expression by a positively acting factor present in only one cell type, the effect of a positively acting factor present in several different cell types can be affected by the presence or absence of a negatively acting factor which is active in only one cell type and which inhibits its activity. Similarly, the activation of genes containing a specific binding site can be modulated by the presence in different cell types of distinct and/or overlapping groups of factors, all of which bind to this site but which have activation domains of different strengths or which differ in their requirement for association with additional factors bearing strong activation domains.

Ultimately, however, all such potential mechanisms of gene regulation in response to specific stimuli or in specific cell types are dependent upon mechanisms which control the synthesis or activity of specific transcription factors in different cell types or in response to specific stimuli. These mechanisms are discussed in the next chapter.

REFERENCES

Carey, M., Lin, Y.-S., Green, M. R. and Ptashne, M. (1990). A mechanism for synergistic activation of a mammalian gene by GAL4 derivatives. *Nature* **345**, 361–364.

Courey, A. J. and Tjian, R. (1988). Analysis of Sp1 *in vivo* reveals multiple transcriptional domains including a novel glutamine-rich activation motif. *Cell* **55**, 887–898.

Dahlberg, J. E. and Lund, E. (1988). The genes and transcription of the major small nuclear RNAs. In: *Structure and function of major and minor small nuclear ribonucleoprotein particles* (Birnsteil M., ed.) , pp. 38–70. Heidelberg: Springer-Verlag.

Gill, G. and Ptashne, M. (1987). Mutants of GAL4 protein altered in an activation function. *Cell* **51**, 121–126.

Gill, G. and Ptashne, M. (1988). Negative effect of the transcriptional activator GAL4. *Nature* **334**, 721–724.

Giniger, E. and Ptashne, M. (1987). Transcription in yeast activated by a putative amphipathic alpha helix linked to a DNA binding unit. *Nature* **330**, 670–672.

Goding, C. R. and O'Hare, P. (1989). Herpes simplex virus Vmw65–octamer binding protein interaction: a paradigm for combinatorial control of transcription. *Virology* **173**, 363–367.

Godowski, P. J., Picard, D. and Yamamoto, K. (1988). Signal transduction and transcriptional regulation by glucocorticoid receptor: regulation by glucocorticoid receptor–Lex A fusion proteins. *Science* **241**, 812–816.

Goodbourn, S. (1990). Negative regulation of transcriptional initiation in eukaryotes. *Biochimica et Biophysica Acta* **1032**, 53–77.

Guarente, L. (1988). UASs and enhancers: common mechanism of transcriptional activation in yeast and mammals. *Cell* **52**, 303–305.

Hai, T., Horikoshi, M., Roeder, R. G. and Green, M. R. (1988). Analysis of the role of the transcription factor ATF in the assembly of a functional preinitiation complex. *Cell* **54**, 1043–1051.

Hayashi, S. and Scott, M. R. (1990). What determines the specificity of action of *Drosophila* homeodomain proteins? *Cell* **63**, 883–894.

Hollenberg, S. M. and Evans, R. M. (1988). Multiple and cooperative transactivation domains of the human glucocorticoid receptor. *Cell* **55**, 899–906.

Hope, I. A. and Struhl, K. (1986). Functional dissection of a eukaryotic transcriptional activator, GCN4 of yeast. *Cell* **46**, 885–894.

Horikoshi, M., Carey, M. F., Kakidani, H. and Roeder, R. G. (1988a). Mechanism of action of a yeast activator: direct effect of GAL4 derivatives on mammalian TFIID promoter interactions. *Cell* **54**, 665–669.

Horikoshi, M., Hai, T., Lin, Y.-S., Green, M. R. and Roeder, R. G. (1988b). Transcription factor ATF interacts with the TATA box factor to facilitate establishment of a preinitiation complex. *Cell* **54**, 1033–1042.

Johnston, S. A., Salmeron, J. M. and Pincher, S. E. (1987). Interaction of positive and negative regulatory proteins in the galactose regulon of yeast. *Cell* **50**, 143–146.

Jonat, C., Rahmsdorf, H. J., Park, K.-K, Cato, A. C. B., Gebel, S., Punta, H. and Herrlich, P. (1990). Antitumour promotion and antiinflammation: down modulation of AP-1 (Fos/Jun) activity by glucocorticoid hormone. *Cell* **62**, 1189–1201.

Kakidani, H. and Ptashne, M. (1988). Gal4 activates gene expression in mammalian cells. *Cell* **52**, 161–167.

La Bella, F., Sire, H. L., Roeder, R. G. and Heuntz, N. (1988). Cell-cycle regulation of a human histone H2b gene is mediated by the H2b subtype-specific consensus element. *Genes and Development* **2**, 32–39.

Levine, M. and Manley, J. L. (1989). Transcriptional repression of eukaryotic promoters. *Cell* **59**, 405–408.

Lewin, B. (1990). Commitment and activation at polII promoters: a tale of protein-protein interactions. *Cell* **61**, 1161–1164.

Lin, Y.-S., Carey, M., Ptashne, M. and Green, M. R. (1990). How different eukaryotic activators can co-operate promiscuously. *Nature* **345**, 359–361.

Lohr, D. and Hopper, J. E. (1985). The relationship of regulatory proteins and DNaseI hypersensitive sites in the yeast Gal1–10 genes. *Nucleic Acids Research* **13**, 8409–8423.

Lucibello, F. C., Slater, E. P., Joos, K. U., Beato, M. and Muller, R. (1990). Mutual transrepression of Fos and the glucocorticoid receptor: involvement of a functional domain in Fos which is absent in Fos B. *EMBO Journal* **9**, 2827–2834.

Ma, J. and Ptashne, M. (1987). A new class of yeast transcription activators. *Cell* **51**, 113–119.

Mattaj, I. W., Lienhard, S., Jiricry, J. and De Robertis, E. M. (1985). An enhancer-like sequence within the *Xenopus* U2 gene promoter facilitates the formation of stable transcription complexes. *Nature* **316**, 163–167.

Mermod, N., O'Neil, E. A., Kelley, T. J. and Tjian, R. (1989). The proline-rich transcriptional activator of CTF/NF-1 is distinct from the replication and DNA binding domain. *Cell* **58**, 741–753.

Mitchell, P. J. and Tjian, R. (1989). Transcriptional regulation in mammalian cells by sequence specific DNA binding proteins. *Science* **245**, 371–378.

O'Hare, P. and Goding, C. R. (1988). Herpes simplex virus regulatory elements and the immunoglobulin octamer domain bind a common factor and are both targets for virion transactivation. *Cell* **52**, 435–445.

Preston, C. M., Frame, M. C. and Campbell, M. E. M. (1988). A complex formed between cell components and a herpes simplex virus structural polypeptide binds to a viral immediate early gene regulatory DNA sequence. *Cell* **52**, 425–434.

Ptashne, M. (1988). How eukaryotic transcriptional activators work. *Nature* **335**, 683–689.

Ptashne, M. and Gann. A. A. F. (1990). Activators and targets. *Nature* **346**, 329–331.

Robertson, M. (1988). Homeo boxes, POU proteins and the limits to promiscuity. *Nature* **336**, 522–524.

Sadowski, I., Ma, J., Triezenberg, S. and Ptashne, M. (1988). GAL4–VP16 is an unusually potent transcriptional activator. *Nature* **335**, 563–564.

Scafe, C., Chao, D., Lopes, J., Hirsch, J. P., Henry, S. and Young, R. A. (1990). RNA polymerase II C-terminal repeat influences reponse to transcriptional enhancer signals. *Nature* **347**, 491–494.

Schaffner, W. (1989). How do different transcription factors binding the same DNA sequence sort out their jobs? *Trends in Genetics* **5**, 37–39.

Sharp, P. A. (1991). TFIIB or not TFIIB? *Nature* **351**, 16–18.

Sigler, P .B. (1988). Acid blobs and negative noodles. *Nature* **333**, 210–212.

Sittman, D., Graves, R. A. and Marzluff, N. F. (1983). Histone mRNA concentrations are regulated at the level of transcription and mRNA degradation. *Proceedings of the National Academy of Sciences, USA* **80**, 1849–1853.

Sive, H. L. and Roeder, R. G. (1986). Interaction of a common factor with conserved promoter and enhancer sequences in histone H2b, immunoglobulin and U2 small nuclear (sn) RNA genes. *Proceedings of the National Academy of Sciences, USA* **83**, 6382–6386.

Staudt, L. H., Singh, H., Sen, R., Wirth, T., Sharp, P. A. and Baltimore, D. (1986). A lymphoid-specific protein binding to the octamer motif of immunoglobulin genes. *Nature* **323**, 640–643.

Tanaka, M. and Herr, W. (1990). Differential transcriptional activation by Oct-1 and Oct-2: interdependent activation domains induce Oct-2 phosphorylation. *Cell* **60**, 375–386.

Tanaka, M., Grossniklaus, U., Herr, W. and Hernandez, N. (1988). Activation of the U2 snRNA promoter by the octamer motif defines a new class of RNA polymerase II enhancer elements. *Genes and Development* **2**, 1764–1778.

Tasset, D., Tora, D., Fromenthul, C., Scheer, E. and Chambon, P. (1990). Distinct classes of transcriptional activating domains function by different mechanisms. *Cell* **62**, 1177–1187.

Yang-Yen, H. F., Chambard, J.-C., Sinn, Y.-L., Smeal, T., Schmidt, T. J., Drouin, J. and Karin, M. (1990). Transcriptional interference between c-Jun and glucocorticoid receptor: mutual inhibition of DNA binding due to direct protein-protein interaction. *Cell* **62**, 1205–1215.

CHAPTER TEN

What activates the activators?

10.1 INTRODUCTION

As we have discussed in previous chapters, transcription factors, in addition to their role in constitutive gene expression (Chapter 3), also play a central role in producing the induction of specific genes in response to particular stimuli (Chapter 4) and in the cell type-specific (Chapter 5) or developmentally regulated (Chapter 6) expression of other genes. The ability to bind to DNA (Chapter 8) and influence the rate of transcription either positively or negatively (Chapter 9) are clearly necessary features of transcription factors which regulate gene expression in response to specific stimuli or in specific cell types. Most importantly, however, such factors must also have their activity regulated such that they only become active in the appropriate cell type or in response to the appropriate stimulus, thereby producing the desired pattern of gene expression. The means by which this occurs have already been discussed in the appropriate chapters dealing with specific transcription factors. The aim of this chapter is to categorize the various means by which this regulation of transcription factor activity can be achieved and to illustrate these processes both with examples that have already been discussed and with additional cases involving factors which have not so far been considered.

Two basic mechanisms by which the action of transcription factors can be regulated have been described. These involve either controlling the synthesis of the transcription factor so that it is made only when necessary (Figure 10.1a) or, alternatively, regulating the activity of the factor so that pre-existing protein

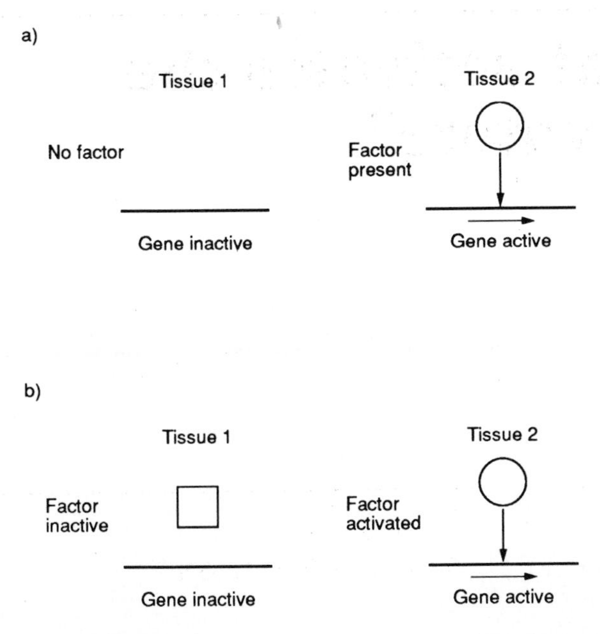

Figure 10.1 Gene activation mediated by the synthesis of transcription factor only in a specific tissue (a) or its activation in a specific tissue (b).

becomes activated when required (Figure 10.1b). These two mechanisms will be discussed in turn.

10.2 REGULATION OF SYNTHESIS

10.2.1 Evidence for the regulated synthesis of transcription factors

Regulating the synthesis of transcription factors such that they are only made when the genes which they regulate are to be activated is an obvious mechanism of ensuring that specific genes become activated only at the appropriate time and place. This mechanism is widely used, therefore, particularly for transcription factors which regulate the expression of cell type-specific or developmentally regulated genes. Thus, as discussed in Section 5.2.2, the octamer-binding protein Oct-2, which stimulates immunoglobulin gene expression in B cells, is detectable in these cells, but is absent from cell types such as HeLa cells which do not express the immunoglobulin genes (Clerc *et al.*, 1988; Scheidereit *et al.*, 1987). Such cell type-specific expression of the Oct-2 protein is paralleled by the presence of the Oct-2 mRNA in B cells and not in HeLa cells (Muller *et al.*, 1988; Staudt *et al.*, 1988). Hence the cell type-specific synthesis of the Oct-2 protein is controlled by regulating the synthesis of the Oct-2 mRNA. Similarly, the MyoD transcription factor and its corresponding mRNA are detectable only in cells of the skeletal muscle lineage in which MyoD plays a critical role in regulating muscle-specific gene expression (Davis *et al.*, 1987; see Section 5.3.2).

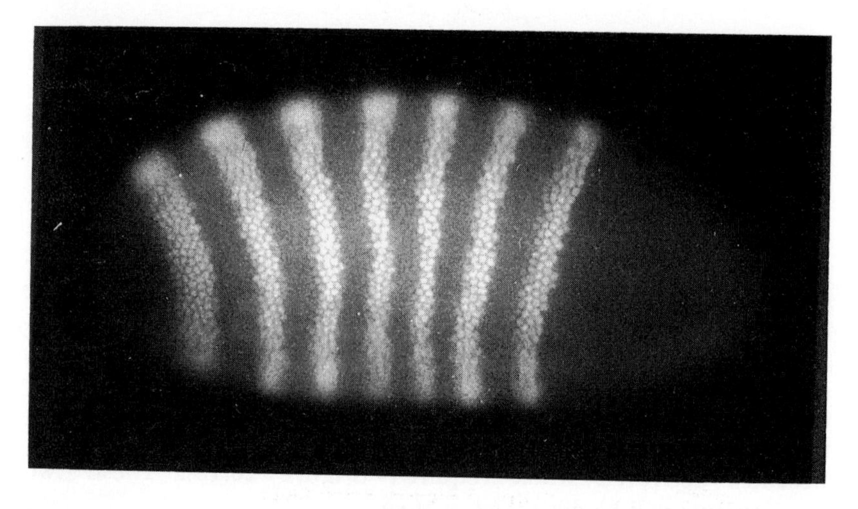

Figure 10.2 Localization of the Ftz protein in the *Drosophila* blastoderm embryo using a fluorescent antibody which reacts specifically with the protein. The anterior end of the embryo is to the left and the dorsal surface to the top of the photograph. Note the precise pattern of seven stripes of Ftz-expressing cells around the embryo.

In addition to its role in controlling cell type-specific gene expression, regulation of transcription factor synthesis is also widely used in the control of developmentally regulated gene expression. Thus numerous studies of the *Drosophila* homeobox transcription factors discussed in Section 6.2, using both immunofluorescence with specific antibodies and *in situ* hybridization, have revealed highly specific expression patterns for individual factors and the mRNAs which encode them, indicating that their role in regulating gene expression in development is dependent, at least in part, on the regulation of their synthesis (review: Ingham, 1988; Figure 10.2). In agreement with this idea, alteration of the levels of the homeobox protein Bicoid within individual cells of the *Drosophila* embryo alters the phenotype of the cell to that characteristic of cells which normally contain the new level of the Bicoid protein (Driever and Nusslein-Volhard, 1988). Hence in a number of cases, where a factor must be

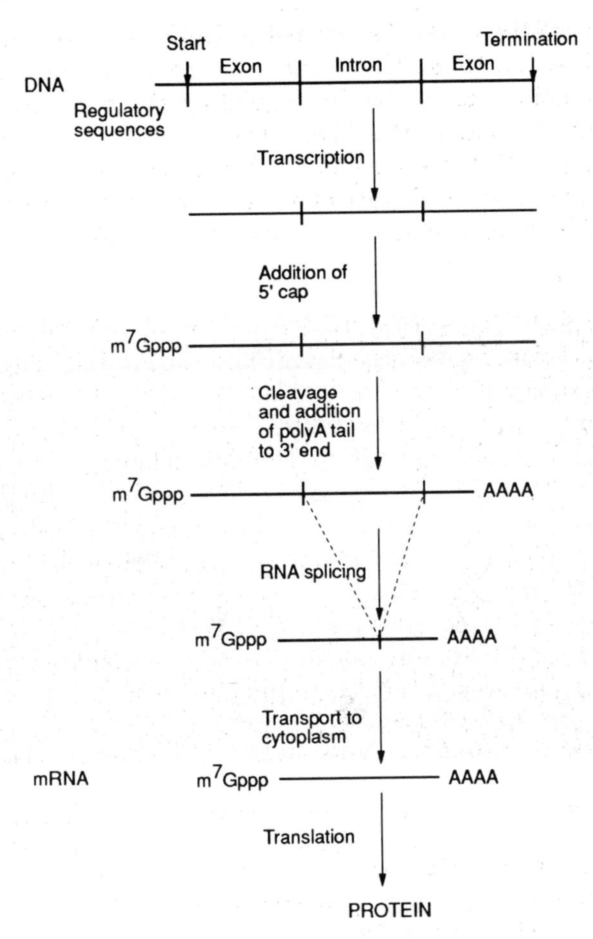

Figure 10.3 Potential regulatory stages in the expression of a gene encoding a transcription factor.

active in a particular cell type or at a specific point in development, this is achieved by the factor being present only in the particular cells where it is required. Clearly, such regulated synthesis of a specific transcription factor could be achieved by any of the methods which are normally used to regulate the production of individual proteins, such as the regulation of gene transcription, RNA splicing or translation of the mRNA (Figure 10.3; for reviews of the levels at which gene regulation can occur, see Darnell (1982) and Latchman (1990a)). Several of these mechanisms of gene regulation are utilized in the case of individual transcription factors and these will be discussed in turn.

10.2.2 Regulation of transcription

As discussed above, a number of cases where the cell type-specific expression of a transcription factor is paralleled by the presence of its corresponding mRNA in the same cell type have now been described. In turn this cell type-specific expression of the transcription factor mRNA is likely to result from the regulated transcription of the gene encoding the transcription factor. Unfortunately, the low abundance of many transcription factors has precluded the direct demonstration of the regulated transcription of the genes which encode them. This has been achieved, however, in the case of the CCAAT box-binding factor C/EBP (Section 3.3.3) which regulates the transcription of several different liver-specific genes such as transthyretin and α_1-anti trypsin. Thus by using nuclear run-on assays to directly measure transcription of the gene encoding C/EBP, Xanthopoulos *et al.* (1989) were able to show that this gene is transcribed at high levels only in the liver, paralleling the presence of C/EBP itself and the mRNA encoding it at high levels only in this tissue (Figure 10.4). Hence the regulated transcription of the C/EBP gene in turn controls the production of the corresponding protein which, in turn, directly controls the liver-specific transcription of other genes such as α_1-anti trypsin and transthyretin.

Interestingly, as well as being used to regulate the relative amounts of a particular factor produced by different tissues, transcriptional control can be used to regulate factor levels within a

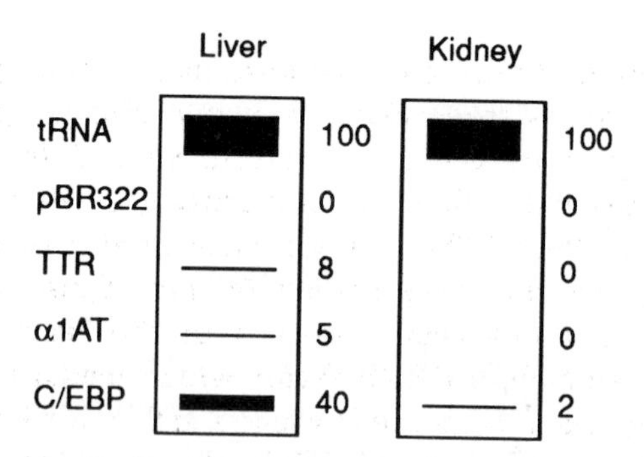

Figure 10.4 Nuclear run-on assay of transcription in the nuclei of kidney and liver. Values indicate the degree of transcription of each gene in the two tissues. Note the enhanced transcription in the liver of the gene encoding the transcription factor C/EBP as well as of the genes encoding the liver-specific proteins transthyretin(TTR) and α_1-antitrypsin (α 1AT). The positive control transfer RNA gene is, as expected, transcribed at equal levels in both tissues, whilst the negative control, pBR322 bacterial plasmid does not detect any transcription.

specific cell type. Thus the levels of the liver-specific transcription factor DBP are highest in rat hepatocytes in the afternoon and evening, with the protein being undetectable in the morning. This fluctuation is produced by regulated transcription of the gene encoding DBP, which is highest in the early evening and undetectable in the morning, whereas the C/EBP gene is transcribed at equal levels at all times. In turn the alterations in DBP level produced in this way produce similar diurnal fluctuations in the transcription of the albumin gene, which is dependent on DBP for its transcription (Wuarin and Schibler, 1990).

Although regulated transcription of the genes encoding the transcription factors themselves is likely, therefore, to constitute the predominant means of regulating their synthesis, it is clear that this process simply sets the problem of gene regulation one stage further back. Thus it will be necessary to have some means of regulating the specific transcription of the gene encoding the transcription factor itself, which in turn may require other transcription factors that are synthesized or are active only in that specific cell type. It is not surprising, therefore, that the synthesis of transcription factors is often modulated by post-transcriptional control mechanisms not requiring additional transcription factors. These mechanisms will now be discussed.

10.2.3 Regulation of RNA splicing

Numerous examples have now been described in eukaryotes where a single RNA species transcribed from a particular gene can be spliced in two or more different ways to yield different mRNAs encoding proteins with different properties (reviews: Leff *et al.*, 1986; Latchman, 1990b). This process is also used in several cases of genes encoding specific transcription factors. Thus, for example, alternative splicing has been observed in the case of the gene encoding the mammalian CREB factor which mediates the induction of specific genes in response to cyclic AMP by binding to specific cyclic AMP response elements (CRE) in the promoters of inducible genes (Montminy and Bilezikjian, 1987). In this case one of the two alternatively spliced mRNAs which are produced contains the full CREB-coding region and can, therefore, encode functional CREB, whilst the other lacks a small region encoding amino acids 88–101 in the intact CREB molecule which play an important role in transcriptional activation by CREB (Yamamoto *et al.*, 1990; Figure 10.5). Hence, although this truncated protein can bind to the CRE, it

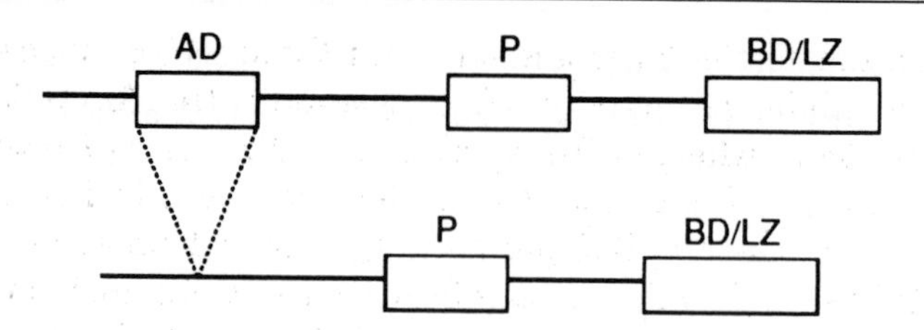

Figure 10.5 Alternative forms of the CREB transcription factor. Note that whilst both forms have the basic domain and leucine zipper (BD/LZ) and the site of cyclic AMP-induced phosphorylation (P; see Section 10.3.4), one form lacks a short region (AD) necessary for activation of transcription by the CREB protein and is therefore unable to activate gene expression following DNA binding.

lacks the activation domain of the molecule and hence cannot activate transcription in response to cyclic AMP.

A similar effect of alternative splicing in producing mRNAs encoding either an intact or a truncated form of a transcription factor is seen in the case of the *era*-1 gene, which encodes a transcription factor that mediates the induction of gene expression in early embryonic cells in response to retinoic acid. In this case, however, whilst one of the two alternative spliced mRNAs encodes the active form of the molecule, the other produces a protein lacking the homeobox region. As the homeobox mediates DNA binding by the intact protein (see Section 6.2.2), this truncated form of the protein is incapable of binding to DNA and activating gene expression (Larosa and Gudas, 1988). A similar use of alternative splicing to create mRNAs encoding proteins with and without the homeobox has been reported for the Hox 2.2 gene (Shen *et al.*, 1991).

Hence in these cases where one of the two proteins encoded by the alternatively spliced mRNAs is inactive, alternative splicing can be used in the same way as the regulation of transcription in order to control the amount of functional protein which is produced.

Interestingly, however, unlike transcriptional regulation, alternative splicing can also be used to regulate the relative production of two distinct functional forms of a transcription factor which have different properties. Thus the E12 and E47 transcription factors which bind to the E2 motif in the immunoglobulin enhancer (see Section 5.2.2) are encoded by the same gene, the two distinct mRNAs encoding E12 and E47 being produced by alternative splicing (Sun and Baltimore, 1991). The mRNAs encoding E12 and E47 are identical except for a single exon which contains the helix–loop–helix and basic domains that mediate dimerization and DNA binding (see Section 8.4) as well as an additional region immediately adjacent to the basic domain (Figure 10.6). Interestingly, in E12 this region adjacent to the basic domain (x in Figure 10.6) appears to contain an inhibitory activity which prevents the formation of E12 homodimers, whereas the equivalent region in E47 does not have this activity. Hence the E47 protein can bind to DNA as a homodimer, whereas the E12 protein cannot form homodimers and can bind to DNA only following the formation of a heterodimer with other helix–loop–helix proteins such as E47 or MyoD. In this case, therefore, alternative splicing does not affect the ability of the protein forms to bind to DNA or activate transcription but results in proteins with different requirements for association with other factors prior to such binding.

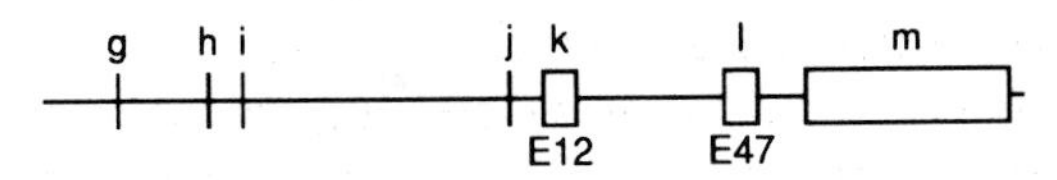

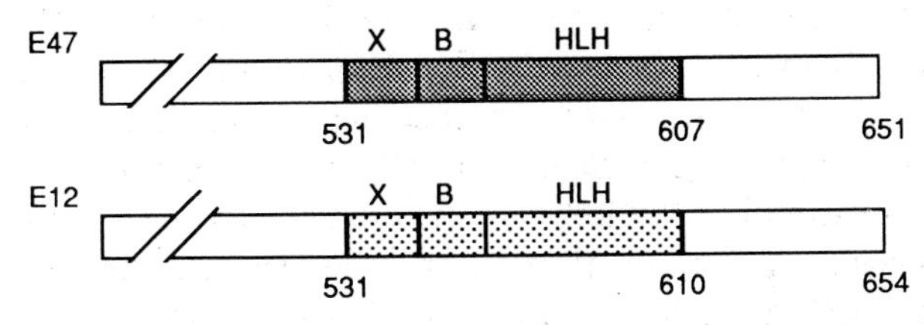

Figure 10.6 (a) Structure of the 3′ region of the gene encoding both the transcription factors E12 and E47. Note that the mRNAs encoding these two proteins are produced by alternative splicing in which the E12 mRNA contains a single E12-specific exon (k), and the E47 mRNA contains a single E47-specific exon (l). All other exons are common to both mRNAs. (b) Structure of the E12 and E47 proteins. Note that the alternative splicing event causes the two proteins to differ in one specific region which contains the basic DNA-binding domain (B), the helix–loop–helix motif (HLH) and a short region (X) which prevents E12 from forming homodimers.

The use of alternative splicing to produce two distinct functional forms of a transcription factor is also seen in the case of the thyroid hormone receptor where, as discussed in Section 4.3.4, alternative splicing produces two forms of the receptor, one of which lacks the ligand-binding domain and, therefore, cannot bind thyroid hormone (Figure 4.19). Although it cannot, therefore, respond to thyroid hormone, this α 2 form of the protein still contains the DNA-binding domain and can, therefore, bind to the specific binding site for the receptor in hormone-responsive genes. By doing so, it acts as a dominant repressor of gene activation mediated by the normal receptor in response to hormone binding. Hence these two alternatively spliced forms of the transcription factor, which are made in different amounts in different tissues, mediate opposing effects on thyroid hormone-dependent gene expression.

In the examples discussed so far, alternative splicing produces forms of the factor with or without a particular property such as the ability to homodimerize or bind hormone. Interestingly, however, in the CREB-related factor CREM, alternative splicing is used to produce two forms of the factor which contain two distinct DNA-binding domains encoded by two different exons in the CREM gene, indicating the potential role of alternative splicing in producing two forms of a single factor with different DNA-binding specificities (Foulkes *et al.*, 1991).

These examples, therefore, illustrate the potential of alternative splicing in generating two forms of a particular transcription factor with different properties which often result in differences in their effects on gene expression.

10.2.4 Regulation of translation

The final stage in the expression of a gene is the translation of its corresponding mRNA into protein. In theory, therefore, the regulation of synthesis of a particular transcription factor could be achieved by producing its mRNA in all cell types but translating it into active protein only in the particular cell type where it was required. However, the observed parallels between the cell type-specific expression of a particular transcription factor and the cell type-specific expression of its corresponding mRNA discussed above (Section 10.2.1) indicate that this cannot be the case for the majority of transcription factors. Nonetheless this mechanism is used to control the synthesis of at least one transcription factor in yeast.

Thus the yeast GCN4 transcription factor controls the activation of several genes in response to amino acid starvation and the factor itself is synthesized in increased amounts following such starvation, allowing it to mediate this effect. This increased synthesis of GCN4 following amino acid starvation is mediated via increased translation of pre-existing GCN4 mRNA (review: Fink, 1986). This translational regulation is dependent upon short sequences within the 5′ untranslated region of the GCN4 mRNA, upstream of the start point for translation of the GCN4 protein.

Most interestingly, such sequences are capable of being translated

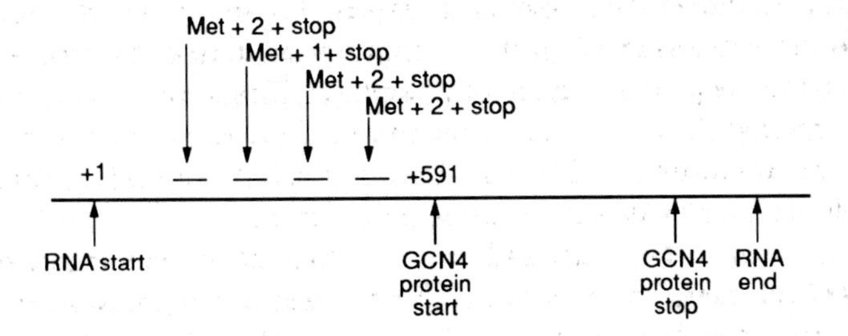

Figure 10.7 Presence of short open reading frames capable of producing small peptides in the 5′ untranslated region of the yeast GCN4 RNA. Translation of the RNA to produce these small proteins suppresses translation of the GCN4 protein. The position of the methionine residue beginning each of the small peptides is indicated together with the number of additional amino acids incorporated before a stop codon is reached.

to produce short peptides of two or three amino acids (Figure 10.7). Under conditions when amino acids are plentiful, these short peptides are synthesized and the ribosome fails to reinitiate at the start point for GCN4 production, resulting in this protein not being synthesized. Following amino acid starvation, however, the production of the small peptides is suppressed and the production of GCN4 is correspondingly enhanced. Hence this mechanism ensures that GCN4 is synthesized only in response to amino acid starvation and then activates the genes encoding the enzymes required for the biosynthetic pathways necessary to make good this deficiency.

10.2.5 Role of regulated synthesis

Regulating the synthesis of a transcription factor constitutes a metabolically inexpensive way of controlling its activity. Thus in situations where the activity of a particular factor is not required, no energy is expended on making it in an inactive form. Such regulation probably takes place predominantly at the level of transcription so that no energy is expended on the production of an RNA, its splicing, transport etc. However, even in cases where regulation occurs at later stages such as splicing or translation, the system is relatively efficient in terms of energy usage, since the step in gene expression which requires the most energy is the final one of translation.

In view of its metabolic efficiency, it is not surprising, therefore, that the regulation of their synthesis is widely used to control the activity of the factors which mediate cell type-specific gene regulation where differences in the activity of a given factor in different cell types are maintained for long periods of time. This mechanism does suffer, however, from the defect that a change in the level of activity of a factor which is controlled purely by a change in its actual amount can take some time to occur. Thus in response to a signal which induces new transcription of the gene encoding a particular factor, it is necessary to go through all the stages illustrated in Figure 10.3, before the production of active factor which is capable of activating the expression of other genes in response to the inducing signal. It is not surprising, therefore, that although some factors such as GCN4 which mediate inducible gene expression are regulated by the regulation of their synthesis, the majority of such factors are regulated by post-translational mechanisms which activate pre-existing transcription factor protein in response to the inducing signal. Thus although mechanisms of this type are metabolically expensive in that they require the synthesis of the factor in situations where it is not required, they have the necessary rapid response time required for the regulation of inducible gene expression. Moreover, unlike transcriptional regulation, they constitute an independent method of gene regulation rather than requiring the activation of other transcription factors in order to activate the transcription of the gene encoding the factor itself.

10.3 REGULATION OF ACTIVITY

10.3.1 Evidence for the regulated activity of transcription factors

In a number of cases, it has been shown that a particular transcription factor pre-exists in an inactive form prior to its activation and the consequent switching on of the genes which depend on it for their activity. Thus, whilst the activation of heat-inducible genes by elevated temperature is dependent on the activity of the heat-shock transcription factor (HSF), this induction can be achieved in the presence of the protein synthesis inhibitor cycloheximide (Zimarino and Wu, 1987). Hence this process cannot be dependent on the synthesis of HSF in response to heat but rather must depend on the heat-induced activation of pre-existing inactive HSF (see Section 4.2.2). Similarly, as discussed in Section 9.3, the yeast GAL4 transcription factor pre-exists in cells prior to galactose treatment, which activates it by causing the dissociation of the inhibitory GAL80 protein.

Although, for the reasons discussed above (Section 10.2.4), the activation of pre-existing transcription factors is predominantly used to modulate transcription factors involved in controlling inducible rather than cell type-specific gene expression, it has also been reported for factors involved in regulating cell type-specific gene expression. Thus, as discussed in Section 5.2.2, the transcription factor NFκB, like Oct-2, plays an important role in the B cell-specific expression of the immunoglobulin κ gene (review: Lenardo and Baltimore, 1989). Unlike Oct-2, however, NFκB is present in an inactive form both in pre-B cells and in a wide variety of other cell types such as T cells and HeLa cells which do not express the immunoglobulin genes. This pre-existing form of NFκB can be activated by treatment with substances such as lipopolysaccharides or phorbol esters. These treatments therefore result in the activation of the immunoglobulin κ gene in pre-B cells which do not normally express it and the expression of other NFκB-dependent genes such as the interleukin-2 α-receptor in T cells. This mechanism in which pre-existing NFκB becomes activated both during B cell differentiation and by agents such as phorbol esters which activate T cells therefore allows NFκB to play a dual role both in B cell-specific gene expression and in the expression of particular genes in response to T cell activation by various agents (Figure 10.8). This effect would otherwise require a complex pattern of regulation in which NFκB was synthesized in response both to B cell maturation and to agents which activate T cells.

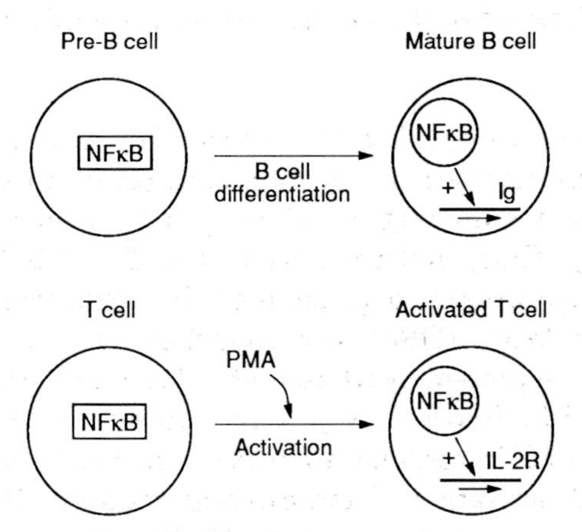

Figure 10.8 Activation of NFκB during B cell differentiation or by agents such as PMA which activate T cells allows it to activate expression of the immunoglobulin κ-chain gene in B cells and the interleukin 2 receptor gene in activated T cells.

Hence modulating the activity of a transcription factor represents a rapid and flexible means of activating a particular factor. Moreover, unlike transcriptional control, such mechanisms allow a direct linkage between the inducing stimulus and the activation of the factor rather than requiring the regulated activity of other transcription factors which in turn activate transcription of the gene encoding the regulated factor. It is not surprising, therefore, that such post-translational mechanisms are widely used. Three basic mechanisms by which such mechanisms can activate factor activity have been described (Figure 10.9) and these will be discussed in turn.

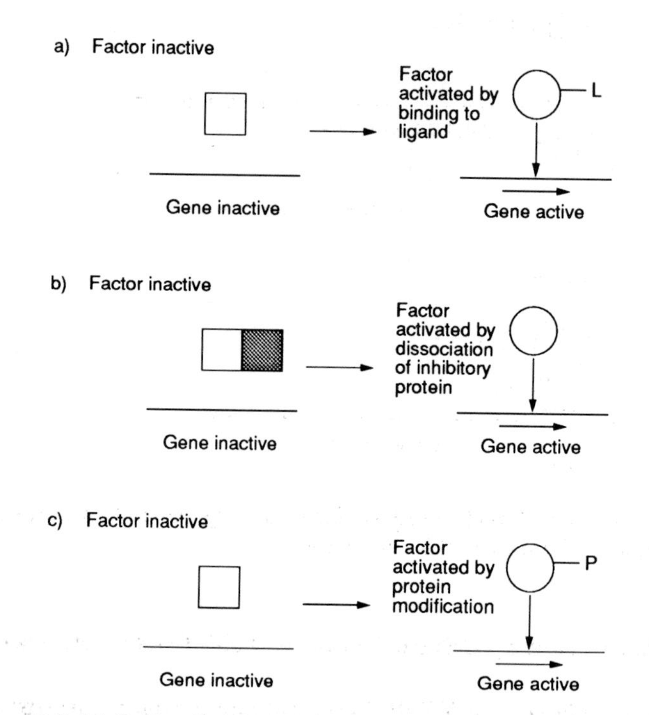

Figure 10.9 Methods of activating a transcription factor in response to an inducing stimulus. This can occur by a ligand-mediated conformational change (a), by removal of an inhibitory protein (b) or by a modification to the protein such as phosphorylation (c).

10.3.2 Activation by protein–ligand binding

As discussed above, one of the principal advantages of regulating the activity of a factor in response to an inducing stimulus is that it allows a direct interaction between the inducing stimulus and the activation of the factor, ensuring a rapid response. The simplest method for this is for an inducing ligand to bind to the transcription factor and alter its structure so that it becomes activated (Figure 10.9a).

An example of this effect is seen in the case of the ACE1 factor which mediates the induction of the yeast metallothionein gene in response to copper. In this case, the transcription factor undergoes a major conformational change in the presence of copper which converts it to an active form which is able to bind to its appropriate binding sites in the metallothionein gene promoter and activate transcription (Figure 10.10; Furst *et al.*, 1988).

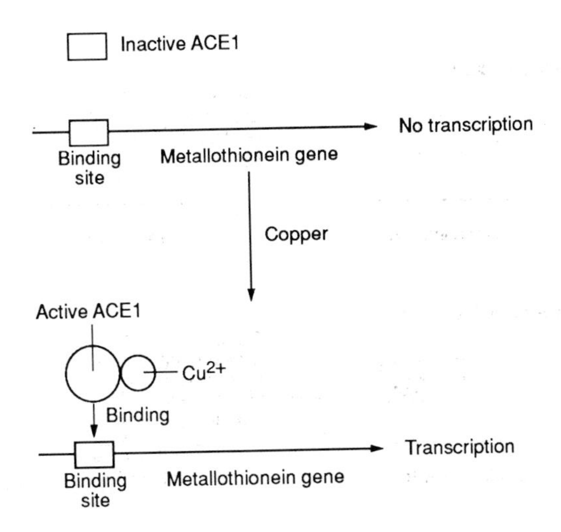

Figure 10.10 Activation of the ACE1 factor in response to copper results in transcription of the metallothionein gene.

10.3.3 Activation by disruption of protein–protein interaction

As discussed above, the NFκB factor only activates transcription in mature B cells or in other cell types following treatment with agents such as lipopolysaccharides or phorbol esters. In agreement with this, no active form of NFκB capable of binding to DNA can be detected in DNA mobility shift assays (see Section 2.2.1) using either cytoplasmic or nuclear extracts prepared from pre-B cells or non-B cell types. Interestingly, however, such activity can be detected in the cytoplasm but not the nucleus of such cells following denaturation and subsequent renaturation of the proteins in the extract. Hence NFκB exists in the cytoplasm of pre-B cells and other cell types in an inactive form which is complexed with another

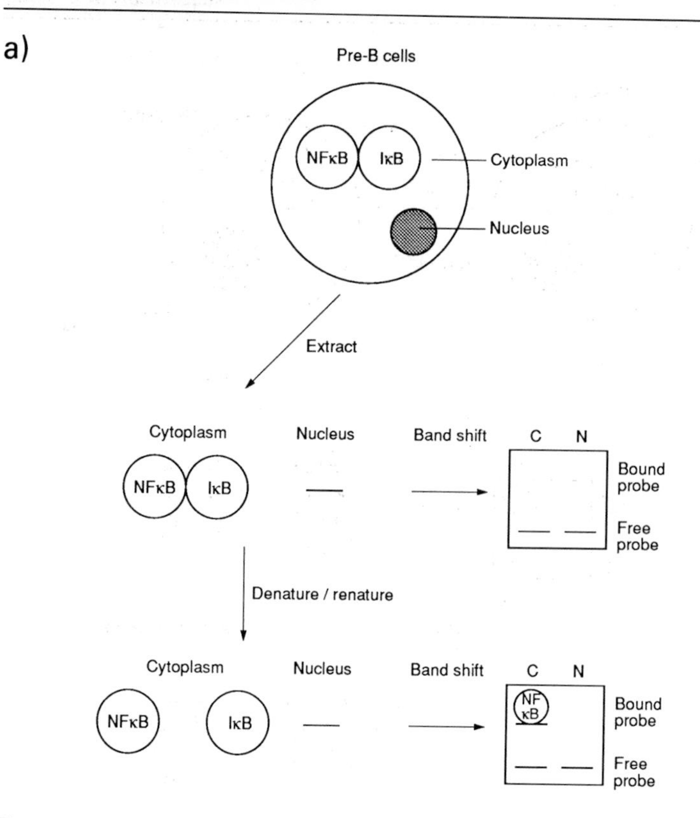

Figure 10.11a Regulation of NFκB. (a) In pre-B cells NFκB is located in the cytoplasm in an inactive form which is complexed to IκB. DNA mobility band shift assays do not therefore detect active NFκB. If a cytoplasmic extract is first denatured and renatured, however, active NFκB will be released from IκB and will be detected in a subsequent band shift assay.

protein known as IκB that inhibits its activity. The release of NFκB from IκB by the denaturation/renaturation treatment, therefore, results in the appearance of active NFκB capable of binding to DNA (Baeurele and Baltimore, 1988a,b; Figure 10.11a).

These findings suggested, therefore, that treatments with substances such as lipopolysaccharides or phorbol esters do not activate NFκB by interacting directly with it in a manner analogous to the activation of the ACE1 factor by copper. Rather they are likely to produce the dissociation of NFκB from IκB resulting in its activation. In agreement with this idea, phorbol ester treatment of cells prior to their fractionation eliminated the latent NFκB activity in the cytoplasm and resulted in the appearance of active NFκB in the nucleus (Figure 10.11b). These substances act, therefore, by releasing NFκB from IκB, allowing it to dimerize and move to the nucleus where it can activate gene expression. Hence this

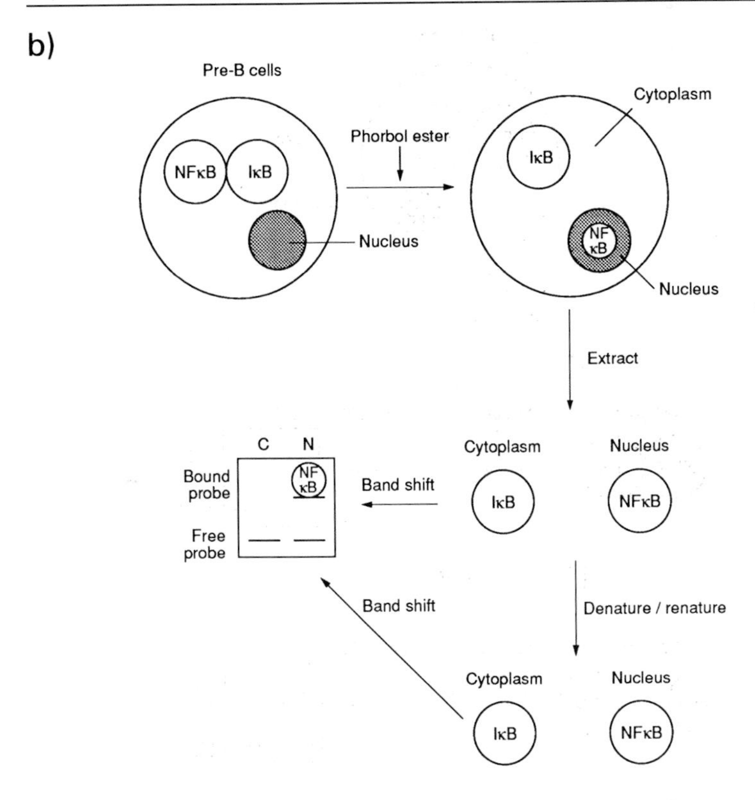

Figure 10.11b In mature B cells, NFκB has been released from IκB and is present in the nucleus in an active DNA-binding form. It can therefore be detected in a DNA mobility shift assay without a denaturation renaturation step, which has no effect on the binding activity.

constitutes an example of the activation of a factor by the dissociation of an inhibitory protein (Figure 10.9b).

As discussed in Section 4.3.2, the activation of members of the steroid–thyroid hormone receptor family in response to treatment with the appropriate hormone also involves the dissociation of the receptors from an inhibitory protein. Thus steroid treatment induces the dissociation of the receptors from the 90-kDa heat-shock protein with which the inactive receptors are associated prior to steroid treatment. This step is essential for DNA binding by the receptors which, although inherently able to bind to DNA, cannot do so until they are released from the complex with hsp90 (Figure 4.10). Most interestingly, the association of hsp90 with the glucocorticoid receptor occurs via the C-terminal region of the receptor, which also contains the steroid-binding domain (Pratt *et al.*, 1988). It has been suggested, therefore, that by associating with the C-terminal region of the receptor, hsp90 masks adjacent domains whose activity is necessary for gene activation by the receptor, for example those involved in receptor dimerization or subsequent DNA binding, thereby preventing DNA binding from occurring. Following steroid treatment, however, the steroid binds to the C-terminus of the receptor, displacing hsp90 and thereby unmasking these domains and allowing DNA binding to occur (Figure 10.12).

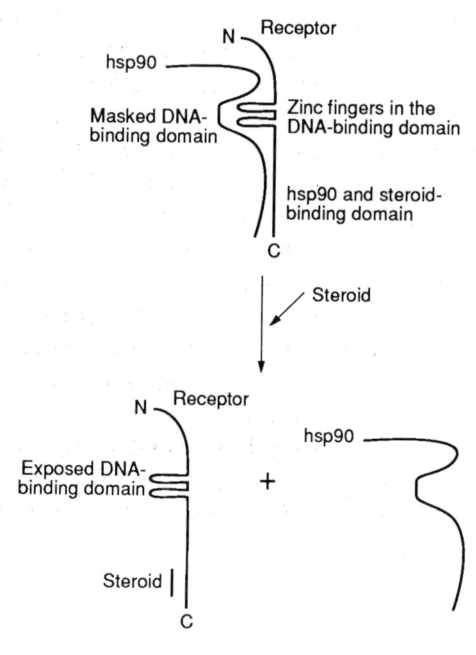

Figure 10.12 Interaction of hsp90 and the glucocorticoid receptor. Hsp90 binds to the receptor via the C-terminal region of the receptor, which also binds steroid and may mask regions of the receptor necessary for dimerization or DNA binding. When steroid is added it binds to the receptor at the C-terminus, displacing hsp90 and exposing the masked regions.

Hence, as with the NFκB factor, activation of the steroid–thyroid receptors involves the dissociation of an inhibitory protein. Paradoxically, however, recent evidence suggests that prior association with hsp90 may also be necessary for the eventual activation of the receptors by hormone. Thus Picard *et al.* (1990) produced yeast cells in which the level of hsp90 has been artificially lowered such that the steroid receptors were not associated with hsp90 prior to steroid treatment. Despite being free of hsp90, however, the receptors in these cells are activated with greatly reduced efficiency upon steroid treatment. This suggests that association with hsp90, as well as preventing activation of the receptors prior to steroid addition, may also alter the structure of the receptor so as to facilitate its activation when steroid is added, possibly by altering the conformation of the receptor so as to increase its ability to bind steroid. Hence in this case the inhibitory protein is acting not only to maintain the transcription factor in an inactive form prior to ligand addition but also to increase the ability of the factor to become activated when ligand is added.

The association and subsequent steroid-induced dissociation of the receptors from hsp90, therefore, appears to play an important role in their activation. As discussed in Section 4.3.2, however, steroid treatment also has an additional effect on the receptors which is essential for transcriptional activation by free receptor. Thus 'domain-swapping' experiments of the type described in Section 2.3.3 have identified transcriptional activation domains in both the glucocorticoid and oestrogen receptors which are only capable of stimulating transcription following hormone addition (Webster *et al.*, 1988). Hence the steroid-induced activation of the receptors appears to involve a conformational change in the receptor which occurs upon steroid binding and which unmasks a steroid-dependent activation domain in the receptor. Activation of the receptors by steroid, therefore, involves both the disruption of protein–protein interaction with hsp90 and a ligand-induced conformational change analogous to that produced in the ACE1 factor by copper. The various effects of steroids on the steroid receptors are summarized in Figure 10.13.

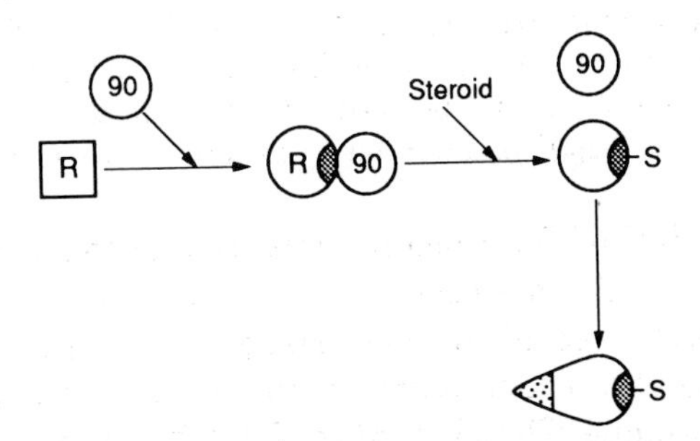

Figure 10.13 Interaction of the receptor (R) with hsp90 and steroid. Note that prior interaction with hsp90 may be necessary to facilitate receptor activation, possibly by enhancing its capacity to bind steroid (shaded region). Subsequent treatment with steroid results in the dissociation of hsp90 and also activates a steroid-dependent activation domain (dotted region).

10.3.4 Activation by protein modification

Many transcription factors are modified extensively following translation by the addition of O–linked monosaccharide residues or by phosphorylation (e.g., Jackson and Tjian, 1988; Gonzalez and Montminy, 1989). Such modifications represent obvious targets for agents that induce gene activation. Thus, such agents could act by altering the activity of a modifying enzyme, such as a kinase. In turn this enzyme would modify the transcription factor, resulting in its activation and providing a simple and direct means of activating a particular factor in response to a specific signal (Figure 10.9c).

One example of this type is provided by the CREB factor which mediates the induction of specific genes in response to cyclic AMP treatment. Thus treatment of cells with cyclic AMP is known to stimulate the activity of the protein kinase A enzyme (Nigg *et al.*, 1985) and in turn this enzyme phosphorylates CREB on a serine residue at position 133 in the molecule (see Figure 10.5; Gonzalez and Montminy, 1989). Interestingly, this serine is located adjacent to the transcriptional activation domain of CREB which is removed by alternative splicing in the truncated form of CREB discussed in Section 10.2.3. Phosphorylation at this adjacent serine stimulates the activity of the activation domain, resulting in the induction of cyclic AMP-responsive genes to which CREB has bound (Yamamoto *et al.*, 1990; Figure 10.14).

Hence in this case the activation of a specific enzyme by the inducing agent directly stimulates the ability of the transcription factor to activate transcription and hence results in the activation of

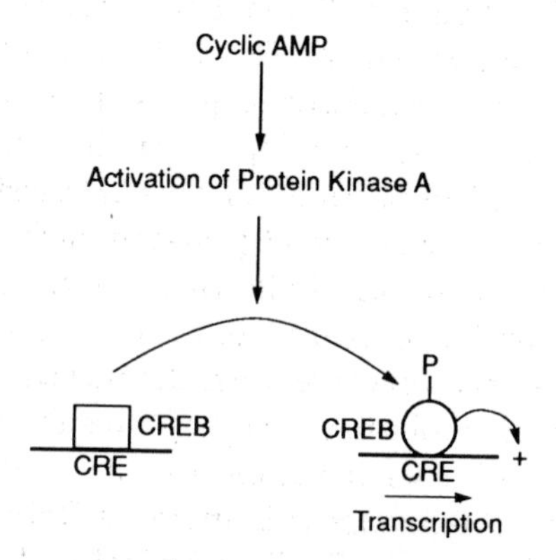

Figure 10.14 Activation of the CREB factor by cyclic AMP-induced phosphorylation. The ability of DNA-bound CREB to activate transcription is produced by the cyclic AMP-dependent activation of protein kinase A which phosphorylates the CREB protein, resulting in its activation.

cyclic AMP-inducible genes. Similarly the phosphorylation of the budding yeast heat-shock factor following exposure of cells to elevated temperature increases the activity of its activation domain, leading to increased transcription of heat-inducible genes (see Section 4.2.2).

In both the cases discussed above, phosphorylation stimulates the activity of the activation domain without affecting other aspects of factor function such as its ability to bind to DNA. In other cases, however, phosphorylation can stimulate other aspects of transcription factor function, producing, for example, increased ability of the factor to bind to DNA such as occurs following phosphorylation of the serum response factor which mediates the induction of several mammalian genes in response to growth factors or serum addition (Manak *et al.*, 1990).

Phosphorylation is also involved in the dissociation of NFκB and its associated inhibitory protein IκB which was discussed above (Section 10.3.3). In this case, however, the target for phosphorylation is the inhibitory protein IκB rather than the potentially active transcription factor itself. Thus following treatment with phorbol esters which activate protein kinase C, IκB becomes phosphorylated. As the phosphorylated form of IκB is unable to bind to NFκB, the NFκB is released from the complex and is free to dimerize, move to the nucleus and activate transcription (Ghosh and Baltimore, 1990). Hence in this case, as before, the inducing agent has a direct effect on the activity of a kinase enzyme but the resulting phosphorylation inactivates an inhibitory factor rather than stimulates an activating factor. This example, therefore, involves a combination of two of the post-translational activation mechanisms we have discussed namely protein modification and dissociation of an inhibitory protein.

In addition to its activation of NFκB, treatment with phorbol esters also results in the increased expression of several cellular genes which contain specific binding sites for the transcription factor AP1. As discussed in Section 7.2, this transcription factor in fact consists of a complex mixture of proteins, including the proto-oncogene products Fos and Jun. Following treatment of cells with phorbol esters, the ability of Jun to bind to AP1 sites in DNA is stimulated and this effect, together with the increased levels of Fos and Jun produced by phorbol ester treatment, results in the increased transcription of phorbol ester-inducible genes. As with the activation of NFκB, phorbol esters appear to increase DNA binding of Jun by activating protein kinase C. Paradoxically, however, it has recently been shown (Boyle *et al.*, 1991) that the increased DNA-binding ability of Jun following phorbol ester treatment is mediated by its dephosphorylation at three specific sites located immediately

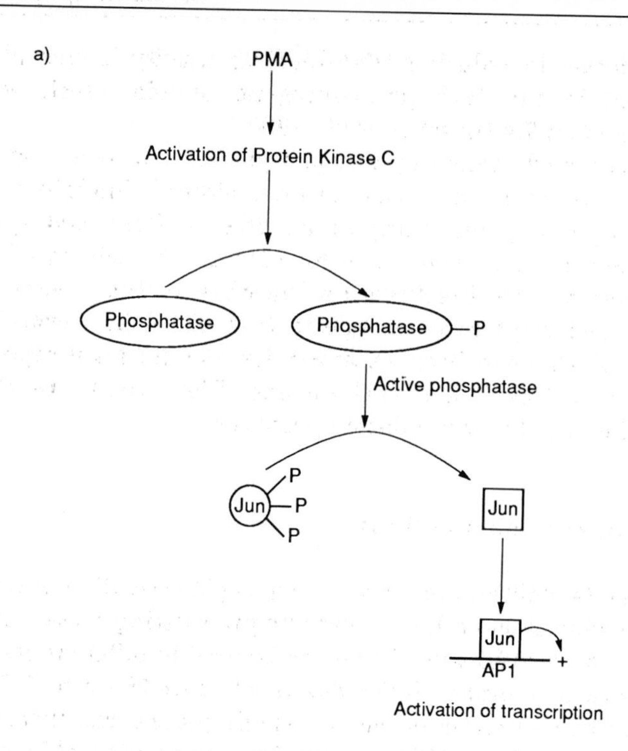

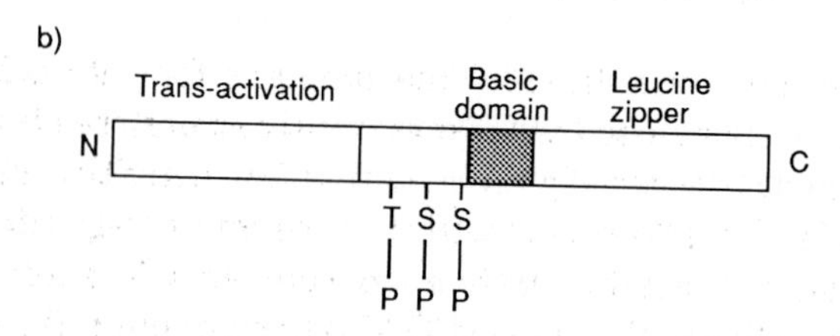

Figure 10.15 (a) Activation of Jun binding to DNA by dephosphorylation. The dephosphorylation of Jun protein following PMA treatment increases its ability to bind to AP1 sites and activate PMA-responsive genes. This is likely to be mediated via the PMA-dependent activation of protein kinase C, which in turn phosphorylates a phosphatase enzyme, allowing it to dephosphorylate Jun. (b) Position in the Jun protein of the two serine (S) and one threonine (T) residues which are dephosphorylated in response to PMA. Note the close proximity to the basic domain (shaded) which mediates DNA binding. The positions of the trans-activation domain and leucine zipper are also indicated.

adjacent to the basic DNA-binding domain, suggesting that protein kinase C may act by stimulating a phosphatase enzyme which in turn dephosphorylates Jun (Figure 10.15).

Such an inhibitory effect of phosphorylation on the activity of a transcription factor is not unique to the Jun protein, a similar effect of phosphorylation in reducing DNA-binding activity having also been observed in the Myb proto-oncogene protein which was discussed in Section 7.4 (Luscher *et al.*, 1990).

Hence protein modification by phosphorylation can have a wide variety of effects on transcription factors, either stimulating or inhibiting their activity and acting via an effect on DNA binding or transcriptional activation domains or by affecting the activity of an inhibitory protein. The directness and rapidity of this means of transcription factor activation suggests that the other forms of modification which have been observed for specific transcription factors such as glycosylation (Jackson and Tjian, 1988) are also likely to be the targets for regulatory processes.

10.3.5 Role of regulated activity

In addition to its ability to produce a very rapid activation of gene expression, modification of the activity of a pre-existing protein also allows specific targets for modification to be used in different cases. Thus in the case of phosphorylation discussed above (Section 10.3.4) we have seen how in different cases a single process can increase either the DNA-binding ability or the trans-activation ability of specific factors.

Clearly, therefore, post-translational mechanisms for activating pre-existing protein could be used to independently stimulate either the DNA-binding or the transcriptional activation activities of a single factor in different situations within a complex regulatory pathway. Indeed, such a combination of mechanisms is actually used to regulate the activity of the yeast GAL4 transcription factor. Thus, as discussed in Section 9.3, the activation of transcription of galactose-inducible genes by GAL4 is mediated by the galactose-induced dissociation of the inhibitory GAL80 protein which exposes the activation domain of DNA-bound GAL4. Interestingly, however, this effect only occurs when the cells are grown in the presence of glycerol as the main carbon source. By contrast however, in the presence of glucose, GAL4 does not bind to DNA and the addition of galactose has no effect (Giniger *et al.*, 1985). Hence by having a system in which glucose modulates the DNA binding of the factor and galactose modulates the activation of bound factor, it is possible for glucose to inhibit the stimulatory effect of galactose. This ensures that the enzymes required for galactose metabolism are only induced in the presence of glycerol and not in the presence of the preferred nutrient, glucose (Figure 10.16).

Such a system in which two different activities of a single factor

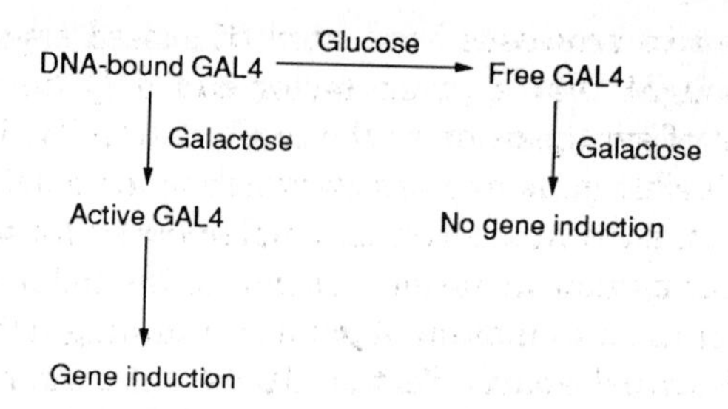

Figure 10.16 Effects of glucose and galactose on GAL4 activity. Note that whilst galactose stimulates the ability of DNA-bound GAL4 to activate transcription, this effect does not occur in the presence of glucose, which results in the release of GAL4 from DNA.

are independently modulated could clearly not be achieved by stimulating the *de novo* synthesis of the factor, which would simply result in more of it being present. Hence, in addition to its rapidity, the activation of pre-existing factor has the advantage of flexibility in potentially being able to generate different forms of the factor with different activities. It should be noted, however, that this effect can also be achieved by alternative splicing of the RNA encoding the factor (Section 10.2.3) which can, for example, generate forms of the protein with and without the DNA-binding domain, as in the case of the *Era*-1 factor, with and without the activation domain, as in the case of CREB, with and without the ability to form a homodimer, as in the case of E12 and E47, or with and without the ligand-binding domain, as in the case of the thyroid hormone receptor.

10.4 CONCLUSIONS

In this chapter we have discussed how the regulation of gene expression by transcription factors is achieved by the regulated synthesis or by the regulated activity of these factors. Although there are exceptions, the regulation of synthesis of a particular factor is used primarily in cases of factors which mediate tissue-specific or developmentally regulated gene expression, where a factor is only required in a small proportion of cell types and is never required in most cell types. In contrast, however, the rapid induction of transcription in response to inducers of gene expression is primarily achieved by the activation of pre-existing inactive forms of transcription factors that are present in most cell types, since this process, although more metabolically expensive, provides the required rapidity in response.

Although these two processes have been discussed separately, it should not be thought that a given factor can only be regulated either at the level of synthesis or at the level of activity. In fact, in many cases of inducible gene expression which involve activation of pre-existing factors, such activation is supplemented by the slower process of synthesizing new factor in response to the inducing agent. Thus in the case of the stimulation of genes containing AP1 sites by phorbol esters discussed above (Section 10.3.4), the phorbol ester-induced increase in the DNA binding of pre-existing Jun protein is supplemented by increased synthesis of both Fos and Jun following phorbol ester treatment, and such newly synthesized Fos and Jun will clearly eventually become a major part of the increased AP1 activity observed following phorbol ester treatment (see Section 7.2). Similarly the activation of NFκB by dissociation from IκB following treatment with substances such as phorbol esters which activate T cells (see Sections 10.3.3 and 10.3.4), has recently been shown to be supplemented by increased synthesis of NFκB and its corresponding mRNA following T cell activation. Hence in these cases the rapid effects of post-translational processes in activating gene expression are supplemented by *de novo* synthesis of the factor, which, although slower, will enhance and maintain the effect.

This combination of regulated synthesis and regulated activity is also seen in the case of factors which mediate tissue-specific gene expression and which are synthesized in only a few cell types. Thus in the case of the MyoD factor which regulates muscle-specific genes, the factor and its corresponding mRNA are synthesized only in cells of the muscle lineage (see Section 5.3.2). The activation of MyoD-dependent genes which occurs when myoblast cells within this lineage differentiate into myotubes is not, however, mediated by new synthesis of MyoD, which is present at equal levels in both cell types. Rather, it occurs due to the decline in the level of an inhibitory protein Id, resulting in the post-transcriptional activation of pre-existing MyoD and the transcription of MyoD-dependent genes. Hence in this case regulation of synthesis is used to avoid the wasteful production of MyoD in cells of non-muscle lineage, whilst the activation of pre-existing MyoD ensures a rapid response to agents which induce differentiation within cells of the muscle lineage. Hence in a number of cases a combination of both regulated synthesis and regulated activity allows the precise requirements of a particular response to be fulfilled rapidly but with minimum unnecessary wastage of energy.

In summary, therefore, the different properties of regulated synthesis and regulated activity allow these two processes, both independently and in combination to efficiently regulate the complex processes of inducible, tissue-specific and developmentally regulated gene expression.

REFERENCES

Baeurele, P. A. and Baltimore, D. A. (1988a). Activation of DNA-binding activity in an apparently cytoplasmic precursor of the NF-kappa B transcription factor. *Cell* **53**, 211–217.

Baeurele, P. A. and Baltimore, D. (1988b). I kappa B: a specific inhibitor of the NF kappa B transcription factor. *Science* **242**, 540–546.

Boyle, W. J., Smeal, T., Defize, L. H. K., Angel, P., Woodgett, J. R., Karin, M. and Hunter, T. (1991). Activation of protein kinase C decreases phosphorylation of c Jun at sites that negatively regulate its DNA binding activity. *Cell* **64**, 573–584.

Clerc, R. G., Corcoran, L. M., LeBowitz, J. H., Baltimore, D. and Sharp, P. A. (1988). The B-cell-specific Oct-2 protein contains POU box and homeo box-type domains. *Genes and Development* **2**, 1570–1581.

Darnell, J. E. (1982). Variety in the level of gene control in eukaryotic cells. *Nature* **297**, 365–371.

Davis, H. L., Weintraub, H. and Lassar, A. B. (1987). Expression of a single transfected cDNA converts fibroblasts to myoblasts. *Cell* **51**, 987–1000.

Driever, W. and Nusslein-Volhard, C. (1988). The bicoid protein determines position in the *Drosophila* embryo in a concentration-dependent manner. *Cell* **54**, 95–104.

Fink, G. R. (1986). Translational control of transcription in eukaryotes. *Cell* **45**, 155–156.

Foulkes, N. S., Borrelli, E. and Sassone-Corsi, P. (1991). CREM gene: Use of alternative DNA binding domains generates multiple antagonists of cAMP-induced transcription. *Cell* **64**, 739–749.

Furst, P., Hu, S., Hackett, R. and Hamer, D. (1988). Copper activates methallothionein gene transcription by altering the conformation of a specific DNA binding protein. *Cell* **55**, 705–717.

Ghosh, H. and Baltimore, D. (1990). Activation of NF kappa B by phosphorylation of its inhibitor I kappa B. *Nature* **344**, 677–682.

Giniger, E., Varnam, S. M. and Ptashne, M. (1985). Specific DNA binding of GAL4, a positive regulatory protein of yeast. *Cell* **40**, 767–774.

Gonzalez, G. A. and Montminy, M. R. (1989). Cyclic AMP stimulates somatostatin gene transcription by phosphorylation of CREB at serine 133. *Cell* **59**, 675–680.

Ingham, P. W. (1988). The molecular genetics of embryonic pattern formation in *Drosophila*. *Nature* **335**, 25–34.

Jackson, S. P. and Tjian, R. (1988). O-glycosylation of eukaryotic transcription factors: implications for mechanisms of transcriptional regulation. *Cell* **55**, 125–133.

Larosa, G. J. and Gudas, L. J. (1988). Early retinoic acid-induced F9 teratocarcinoma stem cell gene ERA-1: alternative splicing creates transcripts for a homeobox-containing protein and one lacking the homeobox. *Molecular and Cellular Biology* **8**, 3906–3917.

Latchman, D. S. (1990a). *Gene Regulation: A Eukaryotic Perspective.* London: Unwin Hyman.

Latchman, D. S. (1990b). Cell-type specific splicing factors and the regulation of alternative RNA splicing. *The New Biologist* **2**, 297–303.

Leff, S. E., Rosenfeld, M. G. and Evans, R. M. (1986). Complex transcriptional units: diversity in gene expression by alternative RNA processing. *Annual Review of Biochemistry* **55**, 1091–1117.

Lenardo, M. J. and Baltimore, D. (1989). NF-kappa B: a pleiotropic mediator of inducible and tissue-specific gene control. *Cell* **58**, 227–229.

Luscher, B., Christenson, E., Litchfield, D. W., Knebs, E. G. and Eiserman, R. N. (1990). *Myb* DNA binding inhibited by phosphorylation at a site deleted during oncogenic activation. *Nature* **344**, 517–522.

Manak, J. R., de Bisschop, N., Kris, R. M. and Prywes, R. (1990). Casein kinase II enhances the DNA binding ability of serum response factor. *Genes and Development* **4**, 955–967.

Montminy, M. R. and Bilezikjian, L. M. (1987). Binding of a nuclear protein to the cyclic AMP response element of the somatostatin gene. *Nature* **328**, 175–178.

Muller, M. M., Rupert, S., Schaffner, W. and Matthias, P. (1988). A cloned octamer transcription factor stimulates transcription from lymphoid-specific promoters in non-B cells. *Nature* **336**, 544–551.

Nigg, E. A., Hilz, H., Eppenberger, H. M. and Dutly, F. (1985). Rapid and reversible translocation of the catalytic subunit of cAMP dependent protein kinase type II from the Golgi complex to the nucleus. *EMBO Journal* **4**, 2801–2806.

Picard, D., Khursheed, B., Garabedian, M. J., Fortin, M. G., Lindquist, S. and Yamamoto, K. R. (1990). Reduced levels of hsp90 compromise steroid receptor action *in vivo*. *Nature* **348**, 166–168.

Pratt, W. B., Jolly, D. J., Pratt, D. V., Hollenberg, S. M., Giguere, V., Cadepond, F. M., Schweizer-Groyer, G., Catelli, M-G., Evans, R. M. and Bauileu, E.-E. (1988). A region in the steroid binding domain determines formation of the non-DNA binding 9S glucocorticoid receptor complex. *Journal of Biological Chemistry* **263**, 267–273.

Scheidereit, C., Heguy, A. and Roeder, R. G. (1987). Identification and purification of a human lymphoid-specific octamer-binding protein (OTF-2) that activates transcription of an immunoglobulin promoter *in vitro*. *Cell* **51**, 783–793.

Shen, W.-F., Detmer, K., Simonitch-Easton, T., Lawrence, H. J. and Largman, C. (1991). Alternative splicing of the Hox 2.2 homeobox gene in human hematopoetic cells and murine embryonic and adult tissues. *Nucleic Acids Research* **19**, 539–545.

Staudt, L. M., Clerc, R. G., Singh, H., Le Bowitz, J. H., Sharp, P. A. and Baltimore, D. (1988). Cloning of a lymphoid-specific cDNA encoding a protein binding the regulatory octamer DNA motif. *Science* **241**, 577–580.

Sun, X.-H. and Baltimore, D. (1991). An inhibitory domain of E12 transcription factor prevents DNA binding in E12 homodimers but not in E12 heterodimers. *Cell* **64**, 459–470.

Webster, N. J. G., Green, S., Jun, J. R. and Chambon, P. (1988). The hormone-binding domains of the estrogen and glucocorticoid receptors contain an inducible transcription activation function. *Cell* **54**, 199–207.

Wuarin, J. and Schibler, U. (1990). Expression of the liver-enriched transcriptional activator protein DBP follows a stringent circadian rhythm. *Cell* **63**, 1257–1269.

Xanthopoulos, K. G., Mirkovitch, J., Decker, T., Kuo, C. G. and Darnell, J. E. Jr (1989). Cell-specific transcriptional control of the mouse DNA binding protein mC/EBP. *Proceedings of the National Academy of Sciences, USA* **86**, 4117–4121.

Yamamoto, K. R., Gonzalez, G. A., Menzel, P., Rivier, J. and Montminy, M. R. (1990). Characterization of a bipartite activation domain in transcription factor CREB. *Cell* **60**, 611–617.

Zimarino, V. and Wu, C. (1987). Induction of sequence-specific binding of *Drosophila* heat-shock activator protein without protein synthesis. *Nature* **327**, 727–730.

CHAPTER ELEVEN

Conclusions and future prospects

In the few years since the first isolation of the genes encoding specific transcription factors, enormous progress has been made in understanding the nature and role of these factors. Thus the roles of specific factors in processes such as constitutive (Chapter 3), inducible (Chapter 4), tissue-specific (Chapter 5) and developmentally regulated (Chapter 6) gene expression have been defined, as have their involvement in diseases such as cancer (Chapter 7). Moreover, by studying these factors in detail, it has proved possible to analyse how they fulfil their function in these processes by binding to specific sites in the DNA of regulated genes (Chapter 8) and activating or repressing transcription (Chapter 9), as well as the regulatory processes which result in their doing so only at the appropriate time and place (Chapter 10). Moreover, the regions of individual factors which mediate these effects and the critical amino acids within them which are of importance have been identified in a number of cases.

What is now required is to link this information with detailed structural information on the nature of the specific motifs within individual factors. This will allow a precise definition of the manner in which, for example, a specific amino acid sequence mediates binding to a specific DNA sequence and why particular alterations in this amino acid sequence either alter the DNA sequence which is bound or abolish binding altogether. Clearly many of these questions will involve not only the solving of the structure of one region of a given protein but also an understanding of how it interacts with a specific region of another protein. Thus, for example, the analysis of how particular activation domains mediate transcriptional activation will involve an understanding of both their structure and that of the appropriate region of the other proteins whose activity they stimulate.

Such studies of the interaction between different transcription factors are clearly more complex than the study of individual factors and much less information is therefore available. Ultimately, however, studies on the interactions between different factors offer the key to the eventual understanding of transcription factor function. Thus, in addition to mediating the effects of individual factors in, for example, transcriptional activation, such interactions can also be used to alter the target genes which are bound by a particular factor, as in the case of the interaction between the yeast a1 and α2 proteins (see Section 5.4.3), or to alter the range of genes which are activated by a specific factor, as in the case of the mammalian Oct-1 protein and the viral VP16 protein (see Section 9.2.1). Hence by altering the specificity of particular factors, interactions of this type are likely to play a crucial role in the complex regulatory networks which allow a relatively small number of transcription factors to control highly complex processes such as development (see Chapter 6).

Ultimately, therefore, the understanding of transcription factor function will require a knowledge of the interactions between different factors which is as good as that now available for individual factors as well as detailed structural studies of the domains in individual factors and the manner in which they interact with other domains in the same factor or in other factors. Clearly much work remains to be done before this is achieved. The rapid progress in the last few years suggests, however, that an eventual understanding in molecular terms of the manner in which transcription factors control highly complex processes such as *Drosophila* and even mammalian development can be achieved.

Index

A residues 25
a gene products 108
a1 gene 111
α gene products 108
α1 gene 111
α1 protein 266
a2 gene 115
α2 gene 111, 114
α2 protein 266
Abd-B 141
Acanthamoeba 32, 46, 48
ACE1 251, 252, 255
Acidic domains 215–17
Activating factor, interaction with other transcription factors 224–7
Activation domains 209–18, 220–4
Activation of transcription 209–27
Activity, regulation of 248–60
α_1-Antitrypsin (α_1-AT) 242
Antennapedia 125, 126, 128, 129, 140–2, 145
Antibodies 89
AP1 11, 12, 156–61, 229, 230, 257
ATF 225–6
ATGCAAAT 7, 8, 18, 91
Avian erythroblastosis virus (AEV) 162
5-Azacytidine 96, 97

B cell-specific expression of immunoglobulin genes 89–90
B cells 8, 10
Bacteriophage protein 33
Bicoid protein (Bcd) 138–9
Bithorax 142

C. elegans 140
Cancer 153–76
CCAAC 54
CCAAT box 4, 52, 53, 55, 56
CCAAT box-binding factor (C/EBP) 34, 242
CCAAT box-binding factor (CBF) 56
CCAAT box-binding proteins 52–6
CCAAT box transcription factor (CTF) 53
CCAAT displacement protein (CDP) 56
CCAAT/enhancer binding protein (C/EBP) 53, 55, 56, 242, 243
Cell type-specific gene expression 87–9
Cell type-specific transcription 87–123
Cellular oncogenes 153–4
C3H 10T1/2 cells 95
Chromatin 77
Cloned genes 34–8
Complementary DNA (cDNA) 31–5, 38
Constitutive transcription 41–61
Constitutive transcription factors 51–2
CP1 54
CP1A 54
CP1B 54
CP2 54
CP2A 54
CP2B 54
CREB 243, 244, 256, 260
CREM 246
CTF/NF1 53, 55, 76, 219
Cyclic AMP 1, 7, 243, 244, 256, 257
Cyclic AMP response element (CRE) 78, 243
Cycloheximide 66, 248

DBP 243
Deformed 140
Deoxyribonuclease I. *See* DNase I
Developmentally regulated gene expression 124–6
2,4-Dinitrophenol (DNP) 65
DNA
  competitor 19, 20
  labelled 20
  non-sequence-specific 28
  non-specific carrier 29
  total bacterial 28
  unlabelled 19, 20
DNA binding 17, 25, 32, 36–8, 70–7, 83, 101, 146, 165, 177–211, 220, 221, 229, 231–6, 251, 253, 254, 257–60
DNA carrier 27
DNA codons 31
DNA–DNA binding 33
DNA mobility shift assay 16–20, 91
DNA–protein binding 33
DNA–protein interactions 23, 25, 26
  methods for studying 16–27
DNA sequence 1–15, 27, 34, 37, 62, 71, 109, 132, 181, 187, 193, 265

DNase I 66, 75, 76
DNase I footprinting assay 20–3, 72
Domain-swapping experiments 38, 210, 255
Drosophila 5, 63–5, 68, 75, 83, 103, 109, 124–40, 142, 144, 145, 220, 240
Drosophila bithorax 141
Drosophila Krox-20 protein 186
Drosophila Kruppel protein 182, 184

E12, 103, 245, 260
E47 103, 245, 260
EF-1 87, 88
Eng 137
Engrailed 140
Enhancers 8–10
Era–1 260
ErbA 82
ERE 80
Eve protein 130–1
Exonuclease III 64, 65

Fos 156–61
Fushi-tarazu gene (Ftz) 126, 130, 137

G proteins 154
G residues 25, 26
GAL4 211, 215–18, 220, 221, 224, 225, 232, 233, 248, 259
GAL80 232, 233, 248
GCN4 37, 215, 246–8
Gene cloning 31–4
Gene promoter 2–3
GGGCGG 52
Glucocorticoid-dependent genes 230
Glucocorticoid-inducible genes 78
Glucocorticoid-mediated inhibition 229
Glucocorticoid receptor 74, 79, 192, 229–31
Glucocorticoid response 7, 78
Glutamine-rich domains 218–19
GRM 112

H2B 56
Haemophilia B 55
HAP2 54
HAP3 54
Heat-inducible transcription 63–9
Heat-shock factor (HSF) 63–9, 75–7, 248
 activation by heat 66–9
Heat-shock promotor element (HSE) 63, 65, 66, 69
Heat-shock response 6
HeLa cells 29, 54, 94, 249
Helix–loop–helix motif 101, 199, 201
Helix–loop–helix proteins 103, 105, 245
Helix–turn–helix motif 178–82, 198
Hepatitis B 56
HLA class II genes 153
HML 107
HMR 107
HO gene 107–8, 110
Homeobox 126, 178–2
Homeobox–containing genes 140–6
Homeobox–containing POU proteins 146–8
Homeobox proteins as transcription factors 128–32
Homeobox transcription factors 133–9
Hormone binding 70
Hormone receptors, activation 71–5
Hox 2 cluster 142, 144, 145
hsp70 3–7, 10, 64
hsp82 65
hsp90 73, 74, 254, 255
Human immunodeficiency virus (HIV) 22, 23
Hunchback 139
4-Hydroxytamoxifen 74

IkB 94
Immunoglobulin genes, B cell-specific expression of 89–90
Immunoglobulin promoter and enhancer elements 91–4
Inducible gene expression 62–86
Interleukin-2 α-receptor 94, 249

Jun 156–61, 258

Knirps 139
Kruppel 139

Lambda β-galactosidase 33
Leucine zipper 194–201

Mating type switching, control of 107–8
Metal response elements (MRE) 7
Metallothionein 3, 4, 7, 11, 52, 251
Methylation interference assay 23–7
MMTV-LTR 72
mRNA 31, 240–4, 246
Multi-cysteine zinc finger 187–94
Muscle-lineage genes (MLG) 102, 104

Muscle terminal differentiation markers (MTDM) 104
Myb 165–72, 259
MyoA 97, 98
MyoD 97, 98, 100, 103–5, 118, 232, 240, 245, 261
  identification 94–101
  regulation of 101–4
MyoH 97, 98

NF1 29, 30
NFkB 93, 94, 104, 118, 249, 251, 252, 254, 257

Oct-1 18–20, 91, 92, 146, 148, 203, 204, 211–15, 266
Oct-2 7, 18–20, 91–3, 104, 118, 146, 148, 211–15
Oestrogen response element (ERE) 79–80
Oligonucleotide probes 31, 32
10T1/2 cells 95–7

P box 112–13
P′ box 112–13
Pit-1 7, 8
Pituitary cells 8
PMA 258
POU proteins 146–8
Pre-B cells 249
Proline-rich domains 219
Protein-ligand binding 250–1
Protein modification 256–9
Protein-protein complex, binding to DNA 81
Protein-protein interaction 73, 251–5
Protein purification 27–31
Proximal sequence element (PSE) 213
PRTF (pheromone receptor transcription factor) 112

Regulation
  of activity 248–60
  of RNA splicing 243–6
  of synthesis 239–48
  of transcription 4–8, 242–3
  of translation 246–7
Repression of transcription 227–33
RNA polymerase 11, 41–3, 177, 222, 225
  activation of 223–4
RNA polymerase I 42–6
RNA polymerase II 3, 41, 42, 49–51, 223
RNA polymerase III 42, 46–9, 182
RNA splicing 1, 2
  regulation of 243–6
RNA transcript 1

Schizosaccharomyces pombe 68
Serum response factor 31, 32
SIN3 108
SL1 46
Sp1 11, 27–9, 32, 52, 55
Squelching 221–2
Stable transcriptional complex 43–51
Steroid hormone receptors 73
Steroid hormones 78
Steroid-inducible genes, activation of 77
Steroid-inducible transcription 69–81
Steroid receptor-steroid hormone complex 76
Steroid receptors 69–71
  activation of 75
Steroid-thyroid hormone receptor family 187
Steroid-thyroid hormone receptor superfamily 71
Steroid-thyroid hormone receptors 188, 192
SWI4 108, 110
SWI5 108, 186
SWI6 108, 110
Synthesis, regulation of 239–48

T cells 94, 249
TAATGARAT 211
TATA box 3, 4, 10, 49–51, 56, 65, 66, 91, 213, 225
TCAATTAAAT 135
TCAATTAAATGA 130
TFIIA 50, 51
TFIIB 51, 227
TFIID 49–51, 56, 66, 68, 76, 77, 213, 222, 224–7
TFIIE 51, 225
TFIIF 51
TFIIIA 47–9, 182, 185, 205
TFIIIB 47–9
TFIIIC 47, 48
TGTGGA/TA/TA/TG 53
Thyroid hormone response element (TRE) 79
TIF-1 44–6
Tissue-specific CCAAT box-binding factors 55

Tissue-specific gene regulation 55
Transcription
  activation of 75–8, 209–27
  importance of 1–2
  regulation of 4–8, 242–3
  repression of 227–33
  sequences involved in basic process 3
Transcriptional activation 70
Translation, regulation of 246–7
Transthyretin (TTR) 242
Two-cysteine two-histidine finger 182–7

Ultrabithorax 126
Ultrabithorax (Ubx) protein 128–9, 131–4
Upstream binding factor (UBF) 45–6
Upstream promoter elements 4, 11, 51–6

Viral ErbA protein 163
Viral v-*erbA* gene 162
VP16 211, 213, 215, 226, 227, 266

Xenopus 10, 49

Yeast mating type 105–6, 116–17

Zinc finger motif 182–94